Digital Manufacturing
The Industrialization of "Art to Part" 3D Additive Printing

Digital Manufacturing
The Industrialization of "Art to Part" 3D Additive Printing

Edited by

Chandrakant D. Patel
HP Chief Engineer and Senior Fellow, HP Inc., Palo Alto, CA, United States

Chun-Hsien Chen
HP-NTU Digital Manufacturing Corporate Lab, School of Mechanical and Aerospace Engineering, Nanyang Technological University, Singapore

Elsevier
Radarweg 29, PO Box 211, 1000 AE Amsterdam, Netherlands
The Boulevard, Langford Lane, Kidlington, Oxford OX5 1GB, United Kingdom
50 Hampshire Street, 5th Floor, Cambridge, MA 02139, United States

Copyright © 2022 Elsevier Inc. All rights reserved.

No part of this publication may be reproduced or transmitted in any form or by any means, electronic or mechanical, including photocopying, recording, or any information storage and retrieval system, without permission in writing from the publisher. Details on how to seek permission, further information about the Publisher's permissions policies and our arrangements with organizations such as the Copyright Clearance Center and the Copyright Licensing Agency, can be found at our website: www.elsevier.com/permissions.

This book and the individual contributions contained in it are protected under copyright by the Publisher (other than as may be noted herein).

MATLAB® is a trademark of The MathWorks, Inc. and is used with permission. The MathWorks does not warrant the accuracy of the text or exercises in this book. This book's use or discussion of MATLAB® software or related products does not constitute endorsement or sponsorship by The MathWorks of a particular pedagogical approach or particular use of the MATLAB® software.

Notices
Knowledge and best practice in this field are constantly changing. As new research and experience broaden our understanding, changes in research methods, professional practices, or medical treatment may become necessary.

Practitioners and researchers must always rely on their own experience and knowledge in evaluating and using any information, methods, compounds, or experiments described herein. In using such information or methods they should be mindful of their own safety and the safety of others, including parties for whom they have a professional responsibility.

To the fullest extent of the law, neither the Publisher nor the authors, contributors, or editors, assume any liability for any injury and/or damage to persons or property as a matter of products liability, negligence or otherwise, or from any use or operation of any methods, products, instructions, or ideas contained in the material herein.

ISBN: 978-0-323-95062-6

For Information on all Elsevier publications
visit our website at https://www.elsevier.com/books-and-journals

Publisher: Matthew Deans
Acquisitions Editor: Brian Guerin
Editorial Project Manager: Rafael Guilherme Trombaco
Production Project Manager: Prasanna Kalyanaraman
Cover Designer: Miles Hitchen

Typeset by MPS Limited, Chennai, India

Contents

List of contributors	xiii
Preface	xv
Acknowledgments	xvii

1. A historical perspective on industrial production and outlook
Chor Hiong Tee

1

Abbreviations		1
1.1	**Introduction**	2
1.2	**Preindustrialization**	3
	1.2.1 Craft production	3
	1.2.2 Agricultural revolution	3
1.3	**First Industrial Revolution**	4
	1.3.1 Mechanization	4
	1.3.2 Laissez-faire capitalism	4
	1.3.3 Social and environmental impact	5
1.4	**Second Industrial Revolution**	6
	1.4.1 Division of labor	6
	1.4.2 Mass production	7
	1.4.3 Batch production	8
1.5	**Third Industrial Revolution**	9
	1.5.1 Automation	11
	1.5.2 Numerical control	12
	1.5.3 Industrial robots	14
	1.5.4 Early computers	15
	1.5.5 Group technology	18
	1.5.6 Modern computers	23
	1.5.7 Computer system architecture	23
	1.5.8 Computer-aided applications	26
	1.5.9 Computer-integrated manufacturing	33
	1.5.10 Product development process	36
	1.5.11 Additive manufacturing	41
	1.5.12 Sustainability in manufacturing	42
1.6	**Forth Industrial Revolution**	46
	1.6.1 Industrie 4.0	47
	1.6.2 Cyber-physical system	49
	1.6.3 Factory of the future	50

v

vi Contents

1.7 Summary	52
References	53

2. Digital product design and engineering analysis techniques 57

Tianyu Zhou, Weidan Xiong, Yuki Obata, Carlos Lange and Yongsheng Ma

Abbreviations	57
2.1 Introduction	58
2.2 Product design process	58
2.3 3D digital form creation	59
2.3.1 3D digital forms	61
2.3.2 Form modeling	64
2.3.3 Case study	66
2.4 Intent-based systemic design	69
2.4.1 Functional feature approach	70
2.4.2 Feature-based computer-aided design modeling	73
2.4.3 Two typical decision-making types: retrieval and inspirational	80
2.5 Engineering analysis	82
2.5.1 Computational fluid dynamics simulation	83
2.5.2 Case study	86
2.6 Current challenges and future work	89
2.6.1 Current challenges	91
2.6.2 Expected future work	92
2.7 Summary	94
References	94

3. Design methodologies for conventional and additive manufacturing 97

Xue Ting Song, Jo-Yu Kuo and Chun-Hsien Chen

List of abbreviations	97
3.1 Introduction	97
3.1.1 Design for Assembly	98
3.1.2 Design for Manufacturing	98
3.2 Design methodologies for conventional manufacturing	100
3.2.1 Design for Manufacturing and Assembly guidelines	100
3.2.2 Design for Manufacturing and Assembly procedures	101
3.2.3 Applications of Design for Manufacturing and Assembly	108
3.2.4 Limitations of Design for Manufacturing and Assembly	110
3.3 A paradigm shift	112
3.3.1 Design for X	112
3.3.2 Design for Additive Manufacturing	113
3.3.3 Trend of hybrid manufacturing production	116

Contents **vii**

3.4 Design methodologies for additive manufacturing 117
    3.4.1 Notable Design for Additive Manufacturing research works 117
    3.4.2 Design stages of a general Design for Additive Manufacturing framework 119
    3.4.3 Challenges of Design for Additive Manufacturing 136
3.5 Summary 140
References 140

4. Additive manufacturing for digital transformation 145

Yu Ying Clarrisa Choong

List of abbreviation 145
4.1 Introduction to additive manufacturing 146
    4.1.1 Definitions and terminologies 146
    4.1.2 Overview of the additive manufacturing market 147
    4.1.3 Industry drivers for additive manufacturing adoption 150
4.2 Additive manufacturing process chain 152
    4.2.1 Level of additive manufacturing implementation 152
    4.2.2 Design, optimization, and simulation 154
    4.2.3 Material selection 155
    4.2.4 Manufacturing 155
    4.2.5 Postprocessing 155
    4.2.6 Process monitoring and validation 157
4.3 Additive manufacturing technologies and processes 158
    4.3.1 Vat photopolymerization 158
    4.3.2 Material extrusion 164
    4.3.3 Material jetting 165
    4.3.4 Sheet lamination 168
    4.3.5 Powder bed fusion 170
    4.3.6 Binder jetting 173
    4.3.7 Directed energy deposition 174
4.4 Case studies of additive manufacturing during the COVID-19 pandemic 176
    4.4.1 Providing rapid emergency responses 177
    4.4.2 Mass customizations 178
    4.4.3 Agile operations and accelerated productions 178
    4.4.4 Preserving sustainability and continuity 178
4.5 Summary 180
References 180

5. Simulation and optimization for additive manufacturing 183

How Wei Benjamin Teo, Kim Quy Le, Kok Hong Gregory Chua and Hejun Du

Abbreviations 183
Symbols 183

viii Contents

5.1	**Introduction**	185
	5.1.1 Macroscale modeling	186
	5.1.2 Mesoscale modeling	187
	5.1.3 Microscale modeling	187
	5.1.4 Parameters optimization	188
	5.1.5 Objectives	188
5.2	**A review of models employing in additive manufacturing simulations**	188
	5.2.1 Powder interaction	189
	5.2.2 Heat transfer and melt pool dynamics	191
	5.2.3 Light source simulation	195
	5.2.4 Crystallization/microstructure simulation	198
	5.2.5 Summary	206
5.3	**Topology optimization**	207
	5.3.1 Structural optimization	208
	5.3.2 Types of topology optimization methodologies	210
	5.3.3 Topology optimization workflow for additive manufacturing	211
	5.3.4 Available commercial software for topology optimization	214
5.4	**Summary**	216
	References	216

6. Polymer materials for additive manufacturing 221
Jia An

	List of abbreviation	221
6.1	**Introduction**	221
	6.1.1 Molecular material—related classifications	222
	6.1.2 Molecular structure—related classifications	223
	6.1.3 Polymer classification for additive manufacturing	223
6.2	**Thermosets**	224
	6.2.1 Curing	224
	6.2.2 Curing characteristics	225
	6.2.3 Dynamic covalent bonds	229
6.3	**Thermoplastics**	229
	6.3.1 Polymer melt	229
	6.3.2 Rheological properties	231
	6.3.3 Thermal properties	232
6.4	**Printability in 3D printing**	234
	6.4.1 Layering	234
	6.4.2 Energy and material bonding	236
6.5	**Characteristics of 3D printed parts**	240
	6.5.1 Porosity	241
	6.5.2 Anisotropy	241
	6.5.3 Heterogeneity	242

Contents **ix**

6.6	Summary	243
6.7	Further recommendation	243
	References	243

7. Metal additive manufacturing 247
Chao Cai and Kun Zhou

	Abbreviations	247
7.1	Introduction	248
7.2	Classification of metal additive manufacturing technology	249
	7.2.1 Powder bed fusion	249
	7.2.2 Direct energy deposition	252
	7.2.3 Binder jetting	254
	7.2.4 Sheet lamination	255
7.3	Preparation and characterization techniques for metal additive manufacturing feedstock	258
	7.3.1 Powder preparation techniques	258
	7.3.2 Powder characterization techniques	260
7.4	Mechanical properties standard testing for metallic additive manufacturing components	263
	7.4.1 Tension	263
	7.4.2 Compression	263
	7.4.3 Hardness	266
	7.4.4 Fatigue performance	268
7.5	Defects in metallic additive manufacturing components	268
	7.5.1 Defect categories	270
	7.5.2 Defects detection techniques	277
7.6	Postprocessing	281
	7.6.1 Removal of adhesive powders, support structures, and substrate plates	281
	7.6.2 Heat treatment	283
	7.6.3 Surface finishing	283
7.7	Applications	284
	7.7.1 Aerospace	285
	7.7.2 Automotive industry	288
	7.7.3 Healthcare	290
7.8	Conclusion and perspectives	293
	References	294

8. The emerging frontiers in materials for functional three-dimensional printing 299
Jia Min Lee, Swee Leong Sing, Guo Dong Goh,
Guo Liang Goh, Wei Long Ng and Wai Yee Yeong

	List of abbreviations	299
8.1	Introduction	300
8.2	Composites materials for aerospace industry	302

x Contents

	8.2.1	Overview of the composite industry	303
	8.2.2	Composites for three-dimensional printing	304
	8.2.3	Challenges and potentials in composites materials for aerospace industry	307
8.3	**Biomaterials for bioprinting**		**310**
	8.3.1	Overview of bioprinting	310
	8.3.2	Bioinks for bioprinting	311
	8.3.3	Challenges and potential in bioprinting of biomaterials	313
8.4	**Ceramics for biomedical implants**		**315**
	8.4.1	Overview of three-dimensional printed ceramic implants	315
	8.4.2	Ceramic materials by three-dimensional printing for biomedical implants	316
	8.4.3	Challenges and potential in ceramics for three-dimensional printing	319
8.5	**Conductive materials for electronic printing**		**322**
	8.5.1	Overview of three-dimensional printed electronics	322
	8.5.2	Materials for three-dimensional printing of electronics	325
	8.5.3	Challenges and potential in three-dimensional printing electronics	328
8.6	**Summary and moving forward**		**331**
References			**332**

9. Three-dimensional (3D) printing for building and construction 345

Mingyang Li, Xu Zhang, Yi Wei Daniel Tay,
Guan Heng Andrew Ting, Bing Lu and Ming Jen Tan

List of abbreviations			**345**
9.1	**Introduction**		**345**
	9.1.1	Digital transformation and automation in building and construction	345
	9.1.2	Short history of construction three-dimensional printing	347
	9.1.3	Technology trends and needs—why 3D printing?	348
9.2	**Current concrete printing technologies**		**350**
	9.2.1	Gantry-based systems	350
	9.2.2	Arm-based systems	352
	9.2.3	Multirobot printing systems	354
	9.2.4	Printing process control	355
9.3	**Fresh and harden properties of three-dimensional printable concrete**		**355**
	9.3.1	Different materials used and their effect on three-dimensional printing technology	356
	9.3.2	Fresh properties of three-dimensional printable concrete materials	360
	9.3.3	Harden properties of three-dimensional printable materials	363

Contents **xi**

	9.3.4 Three-dimensional concrete printing parameters	366
9.4	**Three-dimensional concrete printed applications and case study**	**368**
	9.4.1 Applications of three-dimensional printing in building and construction	370
	9.4.2 3D concrete printing technology developed by NTU Singapore	372
9.5	**Sustainable raw materials in concrete printing**	**373**
	9.5.1 Sustainable materials for cement replacement	375
	9.5.2 Sustainable materials for natural sand replacement	376
	9.5.3 Sustainable materials in spray-based three-dimensional printing	378
9.6	**Summary**	**381**
	References	**381**

10. Process monitoring and inspection 387

Tuan Tran and Xuan Zhang

	Abbreviations	**387**
10.1	**Introduction**	**388**
10.2	**Signals, sensors, and techniques for process monitoring**	**388**
	10.2.1 Optical signals	389
	10.2.2 Thermal signals	394
	10.2.3 X-ray signals	397
	10.2.4 Acoustic signals	399
	10.2.5 Other signals	402
10.3	**Applications in additive manufacturing processes**	**404**
	10.3.1 PBF processes	404
	10.3.2 DED processes	414
	10.3.3 Material extrusion processes	415
	10.3.4 Other additive manufacturing processes	420
10.4	**Quality and feedback control**	**422**
	10.4.1 Process parameters	422
	10.4.2 Signal processing and feedback control	422
	10.4.3 Applications of machine learning in additive manufacturing process and process monitoring	428
10.5	**Standards and toolkits**	**431**
	10.5.1 Standards	431
	10.5.2 Toolkits	435
10.6	**Insights and future outlook**	**439**
10.7	**Summary**	**439**
	References	**440**

Index	443

List of contributors

Jia An Singapore Centre for 3D Printing, School of Mechanical and Aerospace Engineering, Nanyang Technological University, Singapore

Chao Cai State Key Laboratory of Materials Processing and Die and Mould Technology, Huazhong University of Science and Technology, Wuhan, China; HP-NTU Digital Manufacturing Corporate Lab, School of Mechanical and Aerospace Engineering, Nanyang Technological University, Singapore

Chun-Hsien Chen HP-NTU Digital Manufacturing Corporate Lab, School of Mechanical and Aerospace Engineering, Nanyang Technological University, Singapore

Yu Ying Clarrisa Choong LRQA Limited, Singapore

Kok Hong Gregory Chua Singapore Centre for 3D Printing, Nanyang Technological University, Singapore

Hejun Du HP-NTU Digital Manufacturing Corporate Lab, Nanyang Technological University, Singapore; Singapore Centre for 3D Printing, Nanyang Technological University, Singapore; School of Mechanical and Aerospace Engineering, Nanyang Technological University, Singapore

Guo Dong Goh Singapore Centre for 3D Printing, School of Mechanical and Aerospace Engineering, Nanyang Technological University, Singapore

Guo Liang Goh Singapore Centre for 3D Printing, School of Mechanical and Aerospace Engineering, Nanyang Technological University, Singapore

Jo-Yu Kuo Department of Industrial Design, National Taipei University of Technology, Taipei, Taiwan

Carlos Lange Department of Mechanical Engineering, University of Alberta, Edmonton, AB, Canada

Kim Quy Le HP-NTU Digital Manufacturing Corporate Lab, Nanyang Technological University, Singapore

Jia Min Lee HP-NTU Digital Manufacturing Corporate Lab, School of Mechanical and Aerospace Engineering, Nanyang Technological University, Singapore

Mingyang Li Singapore Centre for 3D Printing, School of Mechanical and Aerospace Engineering, Nanyang Technological University, Singapore

Bing Lu Singapore Centre for 3D Printing, School of Mechanical and Aerospace Engineering, Nanyang Technological University, Singapore

Yongsheng Ma Department of Mechanical and Energy Engineering, Southern University of Science and Technology, Shenzhen, China

xiv List of contributors

Wei Long Ng HP-NTU Digital Manufacturing Corporate Lab, School of Mechanical and Aerospace Engineering, Nanyang Technological University, Singapore

Yuki Obata Division of Mechanical and Aerospace Engineering, Faculty of Engineering, Hokkaido University, Sapporo, Hokkaido, Japan

Swee Leong Sing Singapore Centre for 3D Printing, School of Mechanical and Aerospace Engineering, Nanyang Technological University, Singapore; Department of Mechanical Engineering, National University of Singapore, Singapore

Xue Ting Song HP-NTU Digital Manufacturing Corporate Lab, School of Mechanical and Aerospace Engineering, Nanyang Technological University, Singapore

Ming Jen Tan Singapore Centre for 3D Printing, School of Mechanical and Aerospace Engineering, Nanyang Technological University, Singapore; HP-NTU Digital Manufacturing Corporate Lab, School of Mechanical and Aerospace Engineering, Nanyang Technological University, Singapore

Yi Wei Daniel Tay Singapore Centre for 3D Printing, School of Mechanical and Aerospace Engineering, Nanyang Technological University, Singapore

Chor Hiong Tee HP-NTU Digital Manufacturing Corporate Laboratory, School of Mechanical and Aerospace Engineering, Nanyang Technological University, Singapore; Rehabilitation Research Institute of Singapore, School of Mechanical and Aerospace Engineering, Nanyang Technological University, Singapore

How Wei Benjamin Teo HP-NTU Digital Manufacturing Corporate Lab, Nanyang Technological University, Singapore

Guan Heng Andrew Ting Singapore Centre for 3D Printing, School of Mechanical and Aerospace Engineering, Nanyang Technological University, Singapore

Tuan Tran HP-NTU Digital Manufacturing Corporate Lab, School of Mechanical and Aerospace Engineering, Nanyang Technological University, Singapore, Singapore

Weidan Xiong Guangdong Laboratory of Artificial Intelligence and Digital Economy (SZ), Shenzhen University, Shenzhen, China

Wai Yee Yeong HP-NTU Digital Manufacturing Corporate Lab, School of Mechanical and Aerospace Engineering, Nanyang Technological University, Singapore

Xu Zhang Singapore Centre for 3D Printing, School of Mechanical and Aerospace Engineering, Nanyang Technological University, Singapore

Xuan Zhang Singapore Centre for 3D Printing, School of Mechanical and Aerospace Engineering, Nanyang Technological University, Singapore, Singapore

Kun Zhou HP-NTU Digital Manufacturing Corporate Lab, School of Mechanical and Aerospace Engineering, Nanyang Technological University, Singapore

Tianyu Zhou Department of Mechanical Engineering, University of Alberta, Edmonton, AB, Canada

Preface

The 21st century cyber-physical age is driving a seamless integration of digital and physical systems. In that context the current factories are undergoing digital transformation to become digital factories. These digital factories can be best described as an integration of operational technologies—the manufacturing machines and processes—with information technologies. Indeed, the rise of digital factories is synonymous with the fourth industrial revolution. However, the game changer in the fourth industrial revolution is the integration of 3D additive printing as operational technology into the factories to fabricate parts. Now, the digital factory can be called a 3D digital factory made up of advanced operational technologies such as 3D additive printing devices, robotics, and information technologies.

The 3D digital factories of tomorrow bear a great promise in creating a novel 21st century production—consumption model that services the needs of the customers based on their individualized needs. Indeed, this on-demand ability of 3D digital factories not only provides customized outcomes for the customers but also reduces waste, a key advantage in light of the social, economic, and ecological trends. Furthermore, the ability of 3D additive manufacturing technologies to create physical contours and shapes that were hitherto not possible opens up immense possibilities in system optimization to drive performance and energy efficiency. As an example, a designer can consider shape optimization at design and analysis time for a pump impeller to maximize the "air to water" pump efficiency—a ratio of pump work (product of pressure drop and volume flow in Joules per second) and power into the pump (Joules per second). Once the impeller is optimized in digital design, the designer can send the digital file to a 3D additive printer in a digital factory. The part turned around in a matter of hours can now be tried out before committing to large numbers. Indeed, the designer can "print" multiple impellers in one batch, a feat hitherto not possible.

We contend that to imbue a holistic knowledge of 3D additive printing, one must take an end-to-end "art to part" perspective that covers design, device, and the 3D digital factory. This book is the first volume of the Digital Manufacturing handbook series. The series will consist of two separate volumes intended to cover the continuum of design, device, and the 3D digital factory. The objective is to provide the essential knowledge necessary for the 21st century cyber-physical engineer to be well versed in the 3D "art

xvi Preface

to part" digital manufacturing pipeline. Volume 1 is focused on design and the range of 3D additive manufacturing technologies. It covers the fundamentals and mechanistic insights of key technologies and material science, before ending with computational simulation, characterization techniques, monitoring, and inspection processes. Volume 2 focuses on the different devices and agents at the factory level including autonomous robots, artificial intelligence, machine learning, and cyber-physical systems. It also presents new business models working toward a sustainable net zero operations and economy. The chapters are mostly industry-focused with relevant real-world case studies to educate students and train industry professionals and leaders.

- Chapters 1−4 are for everyone who could be a practitioner of engineering, finance, marketing, or a business leader planning to build 3D digital factories. It starts from a historical perspective in digital manufacturing (Chapter 1) to the development of various digital design practices and engineering analysis techniques (Chapter 2). The subsequent chapters delve into the evolution of design methodologies in manufacturing (Chapter 3) with the advent of key additive manufacturing technologies (Chapter 4).
- Chapters 5−10 are for those who want to go deeper into specific areas, such as simulation and design optimization (Chapter 5), polymer materials (Chapter 6), metals (Chapter 7), biomedical and conductive materials (Chapter 8), concrete printing (Chapter 9), and process monitoring (Chapter 10). These chapters will require undergraduate level knowledge in any field of engineering or science.

Dissection of the 3D print engine reveals that it is a stack of science, engineering fundamentals, and technologies. Therefore our objective is to provide a state-of-the-art handbook which incorporates all aspects of 3D printing with immense depth in science and engineering fundamentals, and breadth in a range of technologies. We hope these books will serve as a one-stop resource for those with interest in 3D additive printing to practitioners of any art.

Acknowledgments

This book is supported under the RIE2020 Industry Alignment Fund—Industry Collaboration Projects (IAF-ICP) Funding Initiative, as well as cash and in-kind contribution from the industry partner, HP Inc., through the HP-NTU Digital Manufacturing Corporate Lab.

The editors would like to express their deepest gratitude to Khoo Li Pheng for his vision of this project. They thank all the authors who have contributed to the development of this book. This project would not have been possible without their partners from the Singapore Centre of 3D-Printing (SC3DP), Nanyang Technological University, Singapore.

Meanwhile, the editors are grateful to the directors of HP-NTU Digital Manufacturing Corporate Lab, Michael J. Regan and Tan Ming Jen, for their strong supports. They extend a great many thanks to the editorial team members: Kuo Jo-Yu, Song Xue Ting, Frankie Tee Chor Hiong, and Clarrisa Yu Ying Choong.

Lastly, the editors are thankful to the colleagues at Elsevier, particularly, Brian Guerin (Acquisition Editor), Rafael Guilherme Trombaco (Editorial Project Manager), and Prasanna Kalyanaraman (Production Project Manager) for their warm support and painstaking efforts, which have ensured the smooth publication of this book.

Chapter 1

A historical perspective on industrial production and outlook

Chor Hiong Tee[1,2]

[1]HP-NTU Digital Manufacturing Corporate Laboratory, School of Mechanical and Aerospace Engineering, Nanyang Technological University, Singapore, [2]Rehabilitation Research Institute of Singapore, School of Mechanical and Aerospace Engineering, Nanyang Technological University, Singapore

Abbreviations

2D	2-dimensional
3D	3-dimensional
AEM	Assemblability Evaluation Method
AM	additive manufacturing
APT	Automatically Programed Tool
CAD	computer-aided design
CAE	computer-aided engineering
CAM	computer-aided manufacturing
CAD/CAM	computer-aided design and manufacturing
CAPP	computer-aided process planning
CIM	computer-integrated manufacturing
CNC	computer numerical control
CPS	cyber-physical system
CPU	central processing unit
DCS	distributed computer system
DfMA	Design for Manufacture and Assembly
DNC	Direct numerical control
ENIAC	Electronic Numerical Integrator and Computer
ERP	enterprise resource planning
FMS	flexible manufacturing system
GT	group technology
GUI	graphic user interface
I4.0	Industrie 4.0
IC	integrated circuit

Digital Manufacturing. DOI: https://doi.org/10.1016/B978-0-323-95062-6.00009-7
© 2022 Elsevier Inc. All rights reserved.

2 Digital Manufacturing

ICT	Information and Communication Technology
I/O	input/output
IoT	Internet of Things
IR4	Fourth Industrial Revolution
MIT	Massachusetts Institute of Technology
NC	numerical control
OS	operating system
RAM	random-access memory
ROM	read-only memory
SLA	service-level agreement

1.1 Introduction

Imagine a day in the future; you are seated in a quiet cafe sipping a cup of aromatic Darjeeling tea and biting into a soft, creamy slice of butter cake. You delicately savor in the familiar texture and sweetness, filling your senses with a delightful reminiscence that brought a grin to your face. Curiously, this existential connection between memories and the physical presence is conceived by a 3-dimensional (3D) printer. Printed layers of molecules bound together to form chemical chains and structures, triggering different receptors in your nervous system to kindle back that intimate moment in time. The concept of printing custom-designed edible "food for the soul" does sound absurd today, but can it be our reality in the future?

Physiological needs, such as food, water, sleep, homeostasis, sex, and air, are the most fundamental biological requirements for physical survival, according to American clinical psychologist Abraham Maslow [1]. Sadly, our current state of production practices and limited natural resources are inadequate to sustain and provision everyone on planet Earth any longer. The United Nations Food and Agriculture Organization estimated that the world population would surpass 9.1 billion by 2050 [2]. What if we are able to design and 3D print "sustenance for our body," right down to the exact dosage of micronutrients without exhausting the Earth's natural resources. Can that be our reality one day too? Indeed, can such need-based provisioning be extended to the creation of physical goods? What if we are able to design and 3D print a physical object using the right materials, in appropriate proportions given the ultimate functional performance, to achieve a given objective when the need arises?

Throughout history, we have rationalized technology as an application of science [3]. Our ancestors designed mechanical tools and invented technology to help ease the burden of physical labor and give them access to raw materials and energy resources to make objects they desired. Today, technology has empowered us with the capacity to push past boundaries and construct a world beyond our imagination. Indeed, the modern technology-driven society we have created has brought about great improvements in quality of life.

However, the incessant population growth, coupled with increased per capita consumption and dwindling natural resources will pose great challenges to the quality of our current livelihood and the lives of future generations. Externalities, such as environmental pollution, are becoming a burden to society. We cannot expect to meet the future needs of society simply by extending existing infrastructures and conducting business as usual with a reactive stance toward resource and environmental concerns. In the future the role of technology will be to reinvent the consumption—production model and manufacturing processes to mitigate the impact on the environment and slow down the consumption of the finite pool of global natural resources [4].

Is technology innovation a boon or a bane to humanity? To find the answer, we need to understand the different relative forms of technology, how every discovery and invention frame and transform the course of our survival, livelihood, and habits over the centuries and, finally, where the art of innovations will bring us in the future.

1.2 Preindustrialization

Back in the Stone Age, our ancient forefathers used rocks to create simple hand tools and weapons for hunting. Centuries later, they acquired the skills of farming and learned to build hand-operated contraptions to harness natural energy and aid crop production. Farm animals were put to use in the field to ease intensive labor work. Agriculture was a technology that soon became their primary form of economic livelihood. Closed communities were forged and the society at large became known as the agrarian society [5].

1.2.1 Craft production

Europe was widely ruled under classical-styled feudalism between the 10th and 13th centuries. To receive protection, most farmers would work and stay on land owned by venal lords and noblemen, hence leaving many of them with little savings. To supplement their needs, small groups of entrepreneurs responded by making farming tools and household items to trade at the local market. Products were often handcrafted and tailored to the customers' needs that led to the conception of a new trade called the cottage industry. Many of them eventually went on to become highly skilled craftsmen and artisans who were able to command high prices for their work as supplies were usually smaller than demand. Handcrafted goods were, however, inconsistent in quality, had low reliability, and required manual assembly.

1.2.2 Agricultural revolution

Farmers continued to devise better tools and explored more effective farming techniques to yield healthier crops, better harvests, and more food supplies.

4 Digital Manufacturing

Europe saw a sudden spike in its population as the infant mortality rate dropped because the people were eating healthier and contracting fewer diseases. Most of them eventually became independent farmers and small landowners.

Great Britain implemented the Enclosure Acts in the early 17th Century, comprising a series of United Kingdom Acts of Parliament. It granted legal property rights to private possession of the land that was previously considered public [6]. As a result, wealthy lords and land proprietors were able to purchase and claim private ownership of common fields and large complex farms. Small-scale farmers were consequently forced to give up their land and traveled into urban towns and cities to look for work. Many of them ended up toiling away as laborers in industrial establishments as the country experienced its First Industrial Revolution.

1.3 First Industrial Revolution

An industrial revolution is generally defined as the increase of industrial production that was brought about and characterized by the use of machines and new energy sources. Great Britain became the birthplace of Western industrialization in the 1700s, mainly because it had a politically stable society, an ample supply of slave labor and a vast deposit of natural resources such as coal and iron ore that were cheap to mine [7].

As Great Britain rose to become the world's dominant colonial power, colonies around the globe under the British Empire rule provided additional sources of raw materials to these factories and operated as marketplaces to trade manufactured goods [7]. Machines and mechanization helped improve transportation and facilitated deeper mining for more natural resources. Industrialization soon replaced agriculture as the foundation of society's economic structure.

1.3.1 Mechanization

Burning coal provided the necessary heat energy to convert a tank of water into steam. The steam was converted to mechanical energy and channeled out to power machines in industrial mills and factories. Therefore steam became the new form of energy source that substituted conventional water, wind, and animal-power operations. Steam-powered machinery such as the spinning jenny, a cotton spinning machine that revolutionized the textile industry by increasing human productivity and efficiency, helped boost cost-effective manufacturing [7]. Machine-made goods soon replaced handcrafted products as they were cheaper to produce and more consistent.

1.3.2 Laissez-faire capitalism

In the late 1700s, Scottish economist Adam Smith published his seminal book *The Wealth of Nation*. He postulated a theory that every individual in the society, when acted actively in looking out for themselves and their self-interests, would end up helping the community at large [8]. They had created wealth for themselves and for the country, the citizens of which worked hard to better themselves and their finances. In addition, he stressed that it was important to embrace the idea of a free market that was subsequently construed as an economic system based on supply and demand with little or no government control. The government should only play the role of the "invisible hand" to three key functions: protecting national borders, enforcing civil law, and engaging in public works like education. A group of French economists known as the Physiocrats shared a similar sentiment and later coined the term "laissez-faire" that translated as "leave alone."

Industrialists and wealthy capitalists, popularly known as the bourgeoisie at the time, perceived the idea of a free market as an opportunity to gain more profit and collectively pumped their riches into constructing many privately owned factories, mines, and mills. They inevitably gained control over what their establishments produced, how the goods were made, and who they traded. Consequently shifting the economic decision-making and means of production away from government interventions and into their hands. Being in power, wealthy individuals enjoyed privileges in tariffs and subsidies, shaped trade regulations, and imposed dominance on the poor and powerless.

1.3.3 Social and environmental impact

Many towns and cities saw an influx of job seekers as more and more farmers abandoned their land in the countryside. The majority of them ended up toiling away in industrial establishments owned by greedy capitalists who were largely motivated by production, selfish profits, and individualistic principles [7]. Sadly, as the British government contributed very little in governing the production British's natural resources and the livelihood of its people at that time, many of the laborers and even children ended up being gravely exploited by their employers. Death caused by industrial accidents happened very often as most of them were not trained to operate steam-powered machineries safely and warned of the inherent dangers. Moreover, living conditions and sanitation were severely inadequate and poorly provisioned. The mortality rate escalated with the widespread of unknown diseases. Burning of coal in the steam-powered factories choked the environment and dangerously affected the air quality, while the sky was often engulfed in thick clouds of black smog emitted from their chimneys.

The distribution of wealth, at the time, was very much dictated by the anarchic rules of the market and the economy. The trend resulted in a continual widening of income gaps between the proprietors (the capitalists) and the laborers (the proletariat). By the early 1800s, socialism eventually became formalized in Europe as a response to the economic system and its apparent

6 Digital Manufacturing

shortcomings. It aspired to resolve socioeconomic discrepancies and create a more equitable distribution of income among the people.

Karl Marx, a German philosopher, economist, and sociologist, wrote *Das Kapital* to critically expose the capitalist system and the ways workers were exploited by the capitalist mode of production [9]. He advocated the idea of a socialist society where a community of free individuals would work with the means of production in common and that all individuals' labor power would be consciously applied as the combined labor power of the community. As a result, the productive forces would increase with the all-round development of the individual and all the springs of cooperative wealth would flow more abundantly [9]. Interestingly, Marx's work was a contrast to the classical political-economic writings of Smith and his contemporaries.

Regardless, the British population and economy continued to grow and flourish with the continual rise of industrialization. Meanwhile, the rest of Europe and the United States followed suit with a strong focus on pure science under their game plan and pathed a variety of technological inventions soon after [7]. Such innovations would ultimately expand existing industries and create new ones, such as steel, oil and gas, and the power grids.

1.4 Second Industrial Revolution

Advances in material science engineering and discoveries of new energy sources led Germany and the United States to become global leaders of innovative industrialization in the early 1800s [7]. Steel, being lighter and cheaper to produce, replaced iron as an alternative material in the construction and manufacturing industries. Combustion engines running on natural oil and gas substituted steam engines that drove the growth of the automobile and aviation industries. Electricity brought new possibilities in the development of telecommunications, while power distribution centers managed energy supplies to homes and factories.

Applied science opened many opportunities to all, especially for the manufacturing industry. By the mid-1800s, established factories in the United States were facing supply issues as demands for product quality and quantity grew. Motivated by the prospects and benefits of rationale thinking, business proprietors adopted the scientific approach in early manufacturing and operations management in hope to boost their factories' efficiency and productivity.

1.4.1 Division of labor

Before the advent of the Second Industrial Revolution, business proprietors were generally oblivious and unconcerned about the way goods were manufactured on the factory shop floor [10]. They entrusted all technical tasks to the senior workers, who took on the responsibility of formulating the ideal

A historical perspective on industrial production and outlook Chapter | 1 **7**

process planning based solely on their skillsets and work experiences. The plans were often tedious and convoluted. Such word of mouth closed work processes soon affected not only the shop floor operations but also the product quality and quantity. Junior workers were unable to reference important information as there were no properly documented standard guidelines and benchmarks. Business proprietors realized that this could be an anthropogenic problem and eventually implemented a more rationalized approach to resolve the situation.

In the early 1900s, American author Frederick W. Taylor published a monograph called *The Principle of Scientific Management*, where he laid out his views on how an organization could be coordinated through scientific management practices to benefit everyone [11]. Intrigued by Taylor's approach, business proprietors adopted Taylor's set of principles into the planning and operations on the factory shop floors. Many began replacing manual construction of individual products built by highly skilled hands with mechanized production processes. Product parts and human labor became fragmented and simplified, followed by a careful systemization of work sequences. This selective breakdown of tasks and sequences was eventually known as the division of labor [10].

Ultimately, unskilled workers could use a set of standardized machine tools, jigs, and gauges to repeatedly manufacture identical parts by simply following a series of instructions. The precisely machined pieces would provide a smoother and quicker assembly afterward. Precision manufacturing in the automotive industry allowed operators to produce an interchangeable part that replaces a specific faulty component of a car or an engine [12], hence saving them time and costs of replacing it with an entirely new one instead. The concept of interchangeable parts manufacturing was later coined as the American system of manufacturing [13]. It ensured not only better product quality control but also increased the volume and speed of production, reducing labor and manufacturing costs.

1.4.2 Mass production

Indeed, the concept of division of labor, precision manufacturing, interchangeable parts, and other production methods based on scientific studies and reasoning helped manufacturing establishments increase their profit and productivity. Many eventually started to streamline their manufacturing processes and implemented the idea of the manual assembly line. As a result, factories could churn out large quantities of standard products in a relatively short amount of time. This manufacturing process was widely known as mass production [13].

A manual assembly line would consist of a series of sequentially arranged machines. A base product part would travel from one machine to the next along a mechanical conveyor. Operators assigned at each station

8 Digital Manufacturing

would perform specific tasks to build on the workpiece progressively until the job was completed at the end of the line. Although the manufacturing process of the assembly line had led to higher productivity through creating a sense of flow and continuity, it did present some significant flaws and disadvantages [14]. Over relying on one particularly planned system meant that any complications that cropped up within the system could bring the entire assembly line to a halt. In addition, having to focus on making one particular part design per line would eventually create a lot of waste and leave little room for flexibility. Modifying dedicated machines along the assembly line to cater to design changes would be time-consuming and costly. Many operators also reported that they suffered repetitive stress injuries due to the rote and repetitious nature of the work. As a result, the quality and consistency of products from factories became inferior and erratic [14].

The study of scientific management philosophy had nonetheless produced many early mathematical formulations of models for manufacturing operations and production management, such as the economic order quantity in batch production and statistical quality control [15]. The latter would later impact the field of quality control and laid the foundation for the acceptance of probability and statistical concepts in forecasting, inventory control, and many other fields. It had contributed significantly to the development of supply chain management and the design of ecosystems in modern factories decades later.

1.4.3 Batch production

Manufacturing facilities in the United States became a lot coordinated and sophisticated in the mid-1900s as factories implemented new approaches to optimize their production process [10]. They reorganized the factory shop floor configuration and added newly developed machinery into the assembly line. For mass production, special-purpose equipment was custom-built to produce a specific part's design or a single product in vast quantities. In contrast, general-purpose equipment was designed to be more flexible and adaptable to handle a variety of jobs in one production run, often in lesser amounts. The latter would become an outcome of the development in flexible manufacturing [10].

While factories that produced a low volume of custom-designed pieces were called "job shops," medium-volume production of discrete items in a number of standard design variations was known as batch production. In batch production, products were made in sequential batches by running one batch of a single design variant through the assembly line before switching the tooling or machine over to produce the next batch of a different design and so on.

Although batch production offered flexibility in producing a variety of designs, the overall activity was later found to be somewhat wasteful and inefficient [10]. The main reason was because the assembly line would often be packed with randomly sequenced batches of workpieces, each of them was dissimilar and unrelated in design. Swapping the machine tooling and recalibrating the general-purpose equipment for each design change would incur unnecessary processing time to the overall production. Finally, each workpiece had to be manually loaded and positioned before machining and then offloaded before being routed to other machines. Job sequencing and process planning were highly dependent on the skills and work experience of the operators, and any delays would result in a buildup of backlog with equipment idling along in the assembly line.

Business proprietors realized that these nonvalue-adding activities were detrimental to their factories' overall product output [10]. Many of them struggled with high inventory carrying costs that were accumulated from small to medium-volume production. Hence, they decided to engage the help of scientists and engineers to develop innovative manufacturing strategies that would ultimately improve and change the organizational design of factories and their manufacturing processes. Workstations were introduced and the shop floor reconfigured according to a new manufacturing approach of grouping similar part designs known as group technology (GT). Concurrent development in the area of automation, computers, and manufacturing philosophies also contributed significantly to improving the overall operating efficiencies, product quality, and productivity of the factory.

1.5 Third Industrial Revolution

Progressive countries around the world experienced an economic boom following the end of the Second War. The population in the United States expanded in the mid-1900s as Americans felt confident and optimistic about their livelihood again [16]. Rates of unemployment and inflation fell while wages in every industry climbed. The country went through upward socio-economic mobility in the golden age of capitalism as more and more people rose to join the middle class with plenty of money to spend [16]. They embraced the modern lifestyle and developed exquisite tastes in the variety of products they consumed. As a result, the consumer market saw a steady surge in demand for wider product selections and personalization. Business proprietors noticed the prospect for profits and decided to reengineer their factories' existing production systems. They adopted new manufacturing philosophies and elements of automation and flexibility into the production line.

The rise of automation, computers, computer-aided systems, and Information and Communication Technology (ICT) significantly impacted the electronics and manufacturing industries toward the late 1900s [17]. Factories relying on conventional analog signals in their daily operations eventually switched to digital data and began digitizing their processes. Digitalization, an enabling approach of leveraging digital technologies and

10 Digital Manufacturing

digital data, greatly improved their communication and productivity. Digital transformation, on the other hand, gave them a competitive edge to evolve and react to the rising dynamic market trends.

Development in advanced ICT introduced the Internet and cloud computing. Both vastly revolutionized the way companies and individuals create, store, and share information locally and remotely. It also pushed and flattened the hierarchical structure of organizations and helped spread the power of decision-making to everyone. As useful knowledge and valuable resources became readily accessible over the web to individuals, the manufacturing industry will eventually become decentralized and democratized in the future.

Product customization disrupted the way consumer goods were traditionally manufactured and assembled to keep up with changing market demands. The product architecture of designed goods took on the modular and adaptable approach to give consumers the freedom to "mix-and-match" and create the best configuration of their preference [18]. Companies started to emphasize the design outcomes based on individual needs and fancies, and ultimately reshaped the way products were designed, manufactured, marketed, and delivered.

The emergence of 3D printing technology made prototyping physical models with complex designs quicker and safer by depositing materials that swiftly "grow" or build objects up [17]. Independent designers and makers could design, print, and assemble highly customized products at the convenience of their homes or offices using consumer 3D printers. They could make use of the ICT to collaborate with other creators, vendors, and factories globally or vice versa and market, mass produce, distribute, and sell their designs, hence, coining the term social cloud manufacturing or social platform manufacturing [19].

As the global population continued to grow exponentially beyond the end of the 20th century, demands for raw materials and consumer goods intensified while the highly profitable manufacturing industry thrived. The strain of manufacturing bearing on nature's resources to keep pace with economic growth eventually triggered concerns over environmental sustainability, greenhouse gas emissions, and energy consumption [4]. Factories often operated on a "take, make, dispose" linear economic model, also known as the cradle-to-grave approach, had resulted in reckless massive extraction of raw materials from the ground and wasteful energy consumption over time [20]. The activity had steadily depleted the earth's limited natural resources and choked the environment with toxic landfills and harmful pollutions, causing catastrophic changes to our climate.

Concerns over the sustainability of our livelihood on the planet and the prospect of future generations prompted many global leaders, government bodies, academic institutions, and influential corporations to converge at several international environmental summits. They debated, deliberated, and proposed several policies and action plans to address challenges in renewable energy sources, sustainable design and manufacturing practices, efficient waste management, and education.

1.5.1 Automation

Automation can be simply defined as a relatively self-operating technology and loosely traced back to its elemental forms during the First Industrial Revolution [10]. However, the true origin of automation in manufacturing is still debatable. An automated system will comprise three basic elements, power to drive a process, a program of instructions to direct a process, and a control system to carry out the instructions.

Take the spinning jenny used to spin cotton into threads for textile production, for instance. Steam would drive the spinning process itself and operate the machinery according to a set of actions carried out by the worker. The spinning wheels were crude forms of control systems that performed a set of routines or program of instructions framed in the worker's mind and perfected from years of experience. However, the technology depended heavily on human actions and did not accurately establish what true automation aimed to achieve. It was not until the mid-1900s that development in the concept of programmable automation brought innovations and opportunities to manufacturing systems and production lines.

A simple automated manufacturing system is designed to carry out operational tasks such as processing, assembly, inspection, and material handling along the production line. Jobs are often performed with reduced levels of human participation [10]. However, some systems will execute several of these jobs simultaneously depending on their processing sequence. Highly automated systems, on the other hand, will replace human labor with robots, thus improving the efficiency and accuracy of the production while reducing the cost of labor and unnecessary wastage.

Various types of automated systems are configured to serve specific manufacturing needs. In fixed automation the processing sequence is often straightforward and fixed according to a particular machine configuration. Automated assembly machines and machining transfer lines are examples of machines that utilize fixed automation. Special-purpose equipment is usually very expensive as they are custom engineered for high production rate. Hence, the setup is suited for mass production.

In programmable automation the processing sequence is more complicated and strictly controlled by a set of instructions called a code or program that is fed into the system. It will read and interpret the code before instructing the machines to perform the directed tasks. New codes can be written and fed into the system to "reprogram" the machines whenever there is a need to carry out different tasks. Computer numerical controlled machines and robots with programmable logic controllers are examples of machines

12 Digital Manufacturing

that can execute programmable automation. General-purpose equipment is typically used as it provides enough flexibility to deal with variations and changes in product configurations when required. Thus this setup is more suited for batch production and flexible manufacturing.

1.5.2 Numerical control

A typical factory in the early 1900s would include specialized machines deployed along mechanical conveyor lines that were managed and operated by human operators. However, as manufacturing processes evolved and design of workpieces became increasingly complex and intricate over the years, human errors, and operational inefficiencies started to crop up and eventually brought productivity down.

To resolve issues of poor product quality and rising manufacturing costs while cutting down on unproductive time wastage, business proprietors adopted programmable automation into their factory operations. It would control machines with minimal human intervention and replace the need for coordinated control of skilled machine operators, a skill set that took years to master. The technology eventually led to the development of numerical control (NC) machines. NC is a term used to define a form of programmable automation in which a program is coded with a set of alphanumeric data and applied to control the mechanical actions of machine tools or other equipment, often with predictable accuracy [10].

As the world fought its second war in the early 1900s, there was an increased demand to develop and produce sophisticated aircrafts for both military and civilian use. The US Air Force discovered that to make airplanes fly higher and faster, vital aircraft parts needed to be light and structurally sound with perfectly machined surface contours. John Parsons, a contractor who worked for the US Air Force, designed a machine by connecting a jig bore to a simple controller. He fed the controller with punched cards that contained numerical position data. They were instructions for the jig bore to control the movement of the cutting tools. The machine managed to carve out complex helicopter blades successfully.

The punched cards he used were squares of stiff paper carefully perforated to compose a string of codes or a set of data that would be interpreted as instructions by a controller for various operational sequences. The cards were later replaced with punched paper tapes and magnetic tapes to enable a faster transfer and higher storage of numerical data. Interestingly, the use of punched cards as data storage and processing was, in fact, a borrowed concept from early computer hardware that was in development around the same time. The evolution of computer hardware and computing had strong influence in shaping the NC technology and its applications, even today.

Parsons named his system the "Card-a-matic" milling machine due to the way he used the punched cards as a means to store numerical data for

A historical perspective on industrial production and outlook Chapter | 1   13

processing. Although the machine was a success, it was cumbersome to operate. The US Air Force later commissioned the Massachusetts Institute of Technology (MIT) to expand on Parsons' concept and develop a milling machine prototype that could be programmed and was capable of processing tasks automatically. Researchers in MIT ended up retrofitting an existing three-axis vertical Hydro-Tel milling machine with a hardwired electromechanical controller.

The entire setup was massive, with the controller occupying a floor space much bigger than the milling machine itself. The controller housed both analog and digital components, consisting of 292 vacuum tubes, 175 electromechanical relays, and numerous moving parts. The use of vacuum tubes and relay circuits were appropriated technologies from the ongoing development of early computers. They contained electrodes for controlling electron flow and were used as switches or amplifiers to carry out quick calculations and send signals.

The machine was lauded as the first working prototype of a continuous-path NC milling machine and was even presented to the public. During its operation a particular program was punched into the tape and fed to the controller. The system would move its axes into positions according to the instructions coded on the tape. Simultaneously, a clock modulated the traveling speed of the axes at any given time by sending out electric impulses at various intervals to the motor controllers. The machine was able to make intricate cuts with extremely high accuracy that no other machines, nor humans, could have achieved with nearly as much ease.

Despite the impressive demonstration, the manufacturing community had doubts about the system and did not show any favorable interests in adopting it. Many felt that the entire setup was too costly and the operation too complicated. Business proprietors felt skeptical about the machine's reliability as the system setup was very complex. They were also worried about their staffs' qualifications and capacity to program, debug, operate, and maintain the machine. Sadly, the poorly received system eventually led to a formal conclusion of the project between the US Air Force and MIT.

Nevertheless, the development of automation and NC technologies continued to evolve over the years as manufacturing engineers and computer scientists collaborated and made considerable progress. Advances in computer technology contributed significantly to the development of NC machines [21]. Rather than having physical punch cards or tapes made manually, software programs were developed using primitive coding language that automatically generated coded instructions after the human operators entered the desired parameters into a computer. It greatly reduced the chance for potential human errors and shortened the time taken to manufacture parts. The coding process eventually led to the development of a standardized language called the Automatically Programed Tool (APT) language. The APT program would generate instructions for NC machine tools to create complex

14 Digital Manufacturing

parts by coordinating the cutting tools to move in space. Pioneered by American computer scientist Douglas Ross at MIT, he developed the APT project to extend the potential of the NC technology and build the future of computer programming concepts, computer graphics, and designs.

Unsurprisingly, the development of industrial robots followed quite closely with NC as both operated along similar technologies. Although they were specifically designed to carry out different tasks separately, both relied on the concept of coordinated control of multiple axes (or joints for robots) for precision movement and relied on dedicated programmable controllers to carry out automated tasks.

1.5.3   Industrial robots

A typical industrial robot is a general-purpose, programmable machine that possesses specific anthropomorphic characteristics such as a mechanical arm [10]. It has the ability to emulate human operators and carry out various industrial tasks, particularly in harsh and hazardous conditions. These tasks may range from pick and place, welding, and pointing to the placement of objects. The machine cell layout on factory shop floors is often designed around the capabilities of the robot's speed and reach and has become an essential element for most manufacturing processes.

The concept of a robot can be traced all the way back to ancient Egypt, but its application in factories did not begin until the mid-1900s [10]. Two influential creators made their original contributions to robotics technology around the same time. One of them was a British inventor named Cyril Kenward, who came up with a manipulator that could move in an x—y—z axis system. The other was an American inventor named George Devol, who built a hydraulically driven robot prototype and called it the "programed article transfer" to handle objects. Together with his colleague Joseph Engelberger, Devol later borrowed similar working principles from NC and built on his prototype to develop their first industrial robot.

The robot, named Unimate by the inventors, had a programmable hydraulic manipulator arm that could perform risky repetitive tasks and was mostly deployed in the automotive industry. The robot lifted and stacked hot pieces of metal from a die-casting machine, transported die castings from an assembly line, and welded them onto car chassis. These tasks were usually deemed dangerous for human operators as they might risk losing a limb during an accident or be poisoned from exposure to the toxic fumes [10].

Although the Unimate robot made quite an impression on potential investors and business proprietors during public demonstrations, the reception of the robot in the manufacturing setting suffered a similar fate as the "Card-a-matic" milling machine. The operational cost for the robot was worked out to be about twice the average hourly wage of a worker during that time [10]. Besides, the entire robot setup was bulky, costly to manage, and had limited industrial applications. It was only after the development and incorporation of computer technologies and programming software years later that industrial robots and NC machines became widely adopted in manufacturing operations and factories around the world.

1.5.4 Early computers

The chronological development of early computers is generally unclear but was debated to have begun in the mid-1900s following the invention of the first-generation electronic computer at the University of Pennsylvania [24]. Widely known as the Electronic Numerical Integrator and Computer (ENIAC), the machine was initially designed and built by American physicist John Mauchly and electrical engineer J. Presper Eckert for military applications during the Second World War. The ENIAC would compute complicated values for artillery range tables. The machine contained roughly 18,000 vacuum tubes, 1500 relay modules, and other electromechanical components. It weighed a massive 30 tons and filled a room of 20 by 40 ft. With all its processing units hung on a frame, the machine was also popularly known as the mainframe computer.

The ENIAC was built with a card reader and hole puncher to facilitate interactions with human operators. The reader would decode instructions on punched cards made by the operator and at the same time record and store information by punching holes on the paper cards for future reference. Operators of the machine were often expected to pick up the complicated machine's language code before they could program its mechanisms to function accordingly. The learning curve was, however, extremely steep for an untrained nonspecialist. To make matters worse, the machine was often unstable, highly unreliable and would generate a massive amount of heat during its operation.

The invention of the transistor in the mid-1900s made a huge impact and revolutionized the entire electronic industry [3]. It replaced the vacuum tubes in the original electronic computers as they were smaller, more reliable, and economical. The transistor was made from a special class of materials with electrical properties sandwiched between a conductor and an insulator. Compared to vacuum tubes, this "semiconducting" material required less power to function and produced relatively lesser heat during operation. Shortly, a category of smaller and more affordable machines called transistor computers emerged [24].

The second-generation transistor computers went on to replace the bulky controllers used in running the original NC machines and gave the limited NC technology more control and autonomy. They were also used as signal controllers in the development of the first industrial robot built around the same time. Transistor computers, being more cost-effective and efficient,

16 Digital Manufacturing

were also widely applied in the running business operations and administration in the financial industry.

Although transistors had effectively reduced the risk of accumulated heat damaging the computer, certain underlying thermal issues still existed that could not be easily resolved. The internal circuit board of a computer was typically made of various electronic components such as transistors, resistors, and capacitors, all intricately hardwired and soldered together. They were highly heat-sensitive and would malfunction after prolonged operations, thus causing the entire electronic circuit to shut down without warning. Furthermore, the circuit board would often be clustered with other components, wires, and connections, making troubleshooting for errors and detecting any faulty components extra tedious and time-consuming.

The invention of the integrated circuit (IC) chips a couple of years later resolved all that and more [3]. It cleverly incorporated all the essential electronic components into an internal circuit board and modeled them out in a single flat block of semiconducting material. The result was a more compact circuit chip that would significantly reduce the number of wires on the circuit board and the possibility of a faulty connection caused by overheating. In addition, the IC chips consumed less power and generated less heat compared to its predecessor. The third generation of computers built with IC chips, also known as minicomputers, emerged soon after [24]. They were smaller, more stable, and had ample memory storage to retain information for computing processes, programs, and data.

1.5.4.1 Programmable logic controller

The advent of minicomputers provided factories with the capabilities to automate and control their existing manufacturing processes and machines. Conventional machine controllers were often made up of relays, coils, counters, timers, and other similar components, making them cumbersome to operate and maintain. Programmable logic controllers, developed in the late 1900s, would replace these conventional hardwired controllers in logic control systems [21]. They were ruggedized minicomputers that could be easily programed and reprogramed with operational instructions. These commands would be stored in their rewriteable memory space, retrieved, and executed on demand for various logic, sequencing, timing, counting, and arithmetic functions through their input or output modules. Compared to conventional logic controllers, the programmable logic controllers had a greater variety of functional controls and accessibilities to the various processes and machines. They were also more reliable, easier to maintain, and took up less floor space in the factories.

1.5.4.2 Computer numerical control

The popularity of NC technology received quite a boost in the late 1900s when minicomputers were implemented into earlier NC machines [10].

Known as the computer numerical control (CNC) machines, the minicomputers would replace the bulky controllers and serve as the central control unit for standard operations and memory storage to drive the movement of machine tools with speed and precision. With ample processing power and memory space, different commands and programs could be stored, retrieved, and processed automatically in the minicomputer before sending them out to the machine tools for machining. In addition, operators could leave the machines to run independently after the initial setup and move on to work on other jobs. Hence, reducing the number of machine operators needed to accomplish the same amount of work done manually otherwise [10].

Compared to earlier NC machines running on punched cards, CNC machines were easier to operate and more adaptable to flexible manufacturing. Hence, CNC machines became widely implemented in many workstations that formed the highly automated GT machine cells. By the late 1900s the development of IC microchips, microprocessors, and microcomputers soon replaced minicomputers as a cheaper, smaller, and more powerful alternative. CNC machines became a lot more accessible and affordable for everyone.

1.5.4.3 Direct numerical control

Before computers were introduced into analog NC machines, factories would struggle to fabricate complex designs that demanded the machines to process a large amount of control data [10]. Machine operators often reported that the punched cards and tapes approach of handling heavy milling instructions in individual machines was time-consuming and inadequate. To address these issues, factory engineers later proposed the idea of linking several NC machines assigned with similar jobs directly to a dedicated central process controller. Milling instructions could then be transmitted to the machines simultaneously in real time, instead of the machinists loading them individually. In addition, the central process controller could be stationed and managed by a principal operator in an office, while the rest of the machinists on the shop floor could focus their time and attention on running various machines and other operational tasks.

The engineers developed a primitive "network" system called the direct numerical control (DNC) by using a mainframe computer as the process controller to control several connected NC machines remotely [10]. The system allowed more extensive programs to be fed into the NC machines continuously and made machining critical, complex designs and made off-site control of the machines possible. Although the prospect of the DNC system seemed promising, it was poorly received by many factory and business proprietors. The system essentially required intensive processing power that only the massive mainframe computers could achieve. However, mainframe computers at the time were not only costly but also highly unstable and

18 Digital Manufacturing

unreliable. An error in the system would immediately stall the fleet of connected NC machines and put the entire production line to a halt.

1.5.4.4 Distributed computer system

The DNC system eventually became obsolete when CNC technology and more reliable computers became more accessible [10]. It was succeeded by a more advanced network system called distributed NC or distributed computer system (DCS).

A basic DCS setup would comprise a central computer connected to an assembly of CNC machines in a closed network environment. Unlike the DNC system, the central computer could selectively load heavy milling instructions or programs into a single or group of CNC machines' local memory. With the aid of the built-in computers, the CNC machines would begin the milling process automatically and independently, leaving the central computer free to process other operational tasks and schedules. In addition, as all the CNC machines were connected within a closed network environment, any machine malfunction could be quickly resolved by reprogramming other idling machines to take over the job, thus sustaining the continuity of the production line with minor disruptions to delivery schedules.

The success of the DCS pushed and supported many potential developments in the manufacturing industry, particularly the integration of computer systems in other operations and processes. It also played a significant role in linking and establishing the communications between workstations and other automated components within an automated, flexible manufacturing system (FMS).

1.5.5 Group technology

During the Second World War, factories manufacturing military equipment would often label and arrange their assembly lines according to similar designs and machine sequences when making specific groups of components [10]. By planning ahead and grouping-related parts together, the number of equipment changeovers needed to produce different designs on a single workstation was greatly reduced. Hence, the entire manufacturing process experienced fewer interruptions.

A Russian scientist named Sergei P. Mitrofanov noticed that the total amount of time spent on setting up machines between batches was significantly reduced by almost 80% when compared to the same set of tasks done on an assembly line sequenced otherwise with different and unrelated designs [25]. He concluded his observations and published a book entitled *Scientific Principles of Group Technology* in the mid-1900s. In his book, Mitrofanov introduced a manufacturing philosophy known as group technology, where he described the merits of identifying and grouping similar or near-identical parts into a "part family" and using this as a trait to establish a common procedure for designing and manufacturing them. A part family could be a collection of related parts that were comparable in geometric shapes or sizes, or both, or processes, such as milling and drilling operations. The benefits of this method included better process flow, increased productivity, and greater cost savings [25].

In practice, different designed parts could be manufactured faster with minimal need for elaborate equipment changeover and machine downtime between design changes. In addition, with proper planning and measures beforehand, careless duplications of existing designs could also be avoided [15]. Hence, the drafting process became less tedious and more effective. Part designs were subsequently assigned with numbers, classifications, and codes after they were sorted and grouped to allow for quick referencing and retrieval in the future. With previous project work already stored in the factory's inventory records, new part designs that had similar attributes to an existing old drawing can be readily retrieved and referenced. Minor modifications or alterations could then be easily made on the final design without the likelihood of miscommunications between operators and designers. Hence, it would reduce the designer's time and effort to rework the finished design and ultimately save money for the company.

1.5.5.1 Cellular manufacturing

A factory setup running on GT may comprise several workstations clustered together into a single cell unit called a machine cell. Every machine cell is designed to make a specific part, product family, or a limited group of families. Hence, each cell is like a mini-factory making up a larger factory that covers all production requirements. Compared to the traditional assembly line, this approach to cell unit organization is more efficient and widely known as cellular manufacturing [22].

Production running on cellular manufacturing was typically sequential along a dedicated production line. Manually operated single machine cells were simple to manage but liable to unexpected equipment breakdown, hence jamming the production line and caused unexpected delays. A multiple machine cell, also known as a flexible manufacturing cell, comprised two or more processing workstations designed to produce a larger variety of family parts simultaneously. Hence, the equipment had the flexibility to change over and take up any tasks in the case of a breakdown. Error recovery time was, however, limited to the number of functioning equipment available.

Operators assigned to work within each machine cell would often learn and operate on a narrow set of functional tasks to produce designs characterized by the part family. Over time, they became very familiar and proficient at the task and made better quality parts at a faster rate. As a result, labor productivity improved as parts took lesser time to pass through the

20 Digital Manufacturing

production cycle. Besides meeting delivery deadlines, operators working under a particular machine cell would also take account for the quality of the parts they produced. Thus process planning became a lot more manageable in cellular manufacturing.

With GT, related parts were recognized, itemized, and grouped together before sending to different machine cells accordingly. Material handling became less tedious as workpieces would only move between the machine cells and not across the entire shop floor. Hence, they were less likely to be misplaced or missing. In addition, to facilitate efficient processing and machining of a parts' family for flexible manufacturing, group tooling was put together to cut down on the number of individual tools required for machining and the time required for tool change between parts. Thus reducing the total machine setup time and manufacturing lead time substantially.

1.5.5.2 Automated flexible manufacturing system

The idea of an automated, FMS was first conceived by a British engineer named David Williamson back in the mid-1900s [10]. He proposed to develop an automated milling system where a group of NC machines would operate within a simple "network" environment to churn out various part designs continuously for 24 hours, with more than half of the day running independently without human operators. The system would comprise a central mainframe computer connected to several NC machines fitted with tool magazines that could hold various milling tools for different machining operations. The tool magazines provided a variable component for the NC machines to produce different part designs whenever desired. Although the system was flexible enough to create a variety of designs effectively, it could not be claimed to be fully automated as it required human participation to intermittently oversee specific operational tasks.

An automated FMS can be perceived as an extension of earlier definitions of cellular manufacturing, but with elements of automation and flexibility incorporated. While both approaches share similar applications in GT, an automated FMS will typically house more workstations with an elaborated array of supporting hardware that requires specific skillsets to operate and manage [10]. Hence, it has the versatility to produce different part designs efficiently while quickly adapting to new ones whenever desired. The system can also cope with any unexpected equipment malfunctions or breakdowns automatically and respond promptly to changes in the production forecast without compromising delivery schedules.

The configuration of an automated FMS is highly dependent on the intended processing sequence and special handling requirements of the various part designs during production [22]. Therefore the design and type of workstations, choice of supporting hardware, and skill proficiency of human operators are important elements to consider in determining the overall productivity and efficiency of the factory's output.

1.5.5.2.1 Workstation

The workstations in most automated FMS are often made up of automated general-purpose equipment such as CNC machines. They are interconnected to a DCS that uses a central computer to control and coordinate various processes in the system. The DCS will download discrete operational instructions to different workstations. The machines will start milling various part designs simultaneously as each of them can operate independently under the command of a unique NC program temporarily stored on an onboard computer. The workstations, being in a closed-loop DCS environment, can be easily reprogramed to take over the tasks of other workstations that have stopped functioning. Hence, an automated FMS with a surplus of workstations can buffer any production delays and avoid compromising operators' working hours and delivery schedules [22].

1.5.5.2.2 Supporting hardware

The supporting hardware in an automated FMS typically comprises an automated material handling and storage system, the DCS, and other nonmachining hardware components. Transfer modules, usually part of the material handling system, are responsible for moving and holding workpieces between workstations and across the factory shop floor. Fig. 1.1 illustrates a shop floor configuration of automated workstations with different transfer modules such as automated roller conveyor, automated guided vehicles, and industrial robotic arm in an automated FMS [22].

The size and geometry types of the part design to be produced often determine the selection of transfer modules deployed in an FMS configuration. In addition, elements such as the economics and compatibility of the produced parts with other hardware components that make up the system are also considered [22]. In general, automated roller conveyors and automated guided vehicles are used for transporting regular or nonrotational parts, as they can be firmly secured on specially designed pallet fixtures. On the other hand, if weight is not a limiting issue, robotic arms are preferred to handle irregular or rotational parts. These are harder to stabilize and require special attention.

The DCS is a communication network system comprising a central computer connected to various hardware components, such as individual workstations, the automated material handling and storage devices, functional local sensors, and other operational stations [22]. The central computer is responsible for interfacing and coordinating activities between machines and hardware components. It can ensure that different part designs are processed, produced, and transferred accordingly to schedule so that the overall

22 Digital Manufacturing

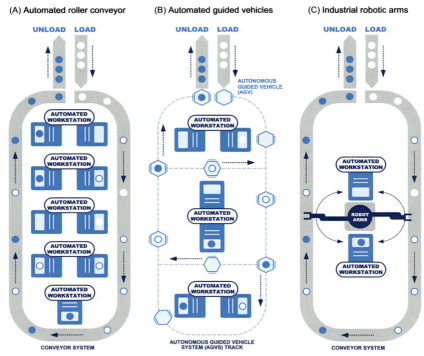

FIGURE 1.1 Shop floor configuration of automated workstations with different transfer modules in an automated FMS [22]. *FMS*, Flexible manufacturing system.

operation will run smoothly without any compromise. If one of the machines or hardware components would malfunction suddenly during production, the central computer will respond quickly with a recovery plan to minimize any wastage and delays. It does this by interacting with the built-in microprocessor or control units of other functioning equipment to coordinate and take over the task of the faulty machine.

Other nonmachining hardware components incorporated along with various processes in an automated FMS may depend on the system's functional needs and capabilities. These may include special workstations built for quality inspection, testing, heat treatment, workpiece post-processing, and other special operations [22].

1.5.5.2.3 Human operators

Although the automated FMS is designed to be an autonomous system, it still requires some form human interactions to function efficiently due to certain operational limitations of its hardware and software [22]. Machine operators are required to provide close supervision to operational activities, conduct periodic system maintenance, and implement sporadic repairs to faulty equipment. They will ensure the system is well calibrated and functioning efficiently to prevent any unnecessary backlog and delays from occurring. In addition, software engineers are needed to perform part programming, computing, and other operational demands of various computers functioning within the system.

1.5.6 Modern computers

Modern society transformed remarkably following the appearance of computers powered by integrated microchips. Electronic and computer technology innovations enabled engineers to develop smaller and more efficient ICs by cramming millions of microelectronic components onto a tiny chip. The first microprocessor chipset was created in the late 1900s by integrating a series of tiny chips to function like the computer's processing module. As a result, the size of computers shrunk significantly while their computational power improved with higher processing efficiency and memory capacity [24].

The success of the microprocessor had revolutionized the world of computing as it made modern computers smaller, faster, and more reliable. Microcomputers, as these were popularly called partly due to their petite size and bearing a micro chipset as its processor, became the fourth generation of computers that were quickly adopted in the manufacturing industry. With better performance and a developed systematic approach to specialized programming languages, the task of coding and debugging the operating system (OS) and its application programs became a lot simpler and friendlier to the operators. In addition, commercial investors saw the high growth potential of the microprocessor and funded heavily in the research work to push its development. As a result, microcomputers became even more compact and affordable, that individual consumers and hobbyists could purchase one for their homes or office. Hence, the term "personal computer" was coined, and it flooded the consumer market.

From CNC to DCS and automated FMS, the pervasive use of computers and computer systems has indeed become a point of departure for the future of integrated manufacturing. Computer systems will gradually take up more critical roles in the organization and implementation of automated operations in the manufacturing industry. They will assist in the control, monitor, and execution of various manufacturing functions, such as product design and development, manufacturing process and planning, and even carry out different business activities related to manufacturing.

1.5.7 Computer system architecture

A simple computer system architecture is basically made up of both hardware and software components. The hardware comprises all physical

24 Digital Manufacturing

elements in the computer and various devices connected to it. The software, on the other hand, refers to the computer programs or sets of instructions that direct the hardware to execute a particular task [23].

1.5.7.1 Hardware

The central processing unit (CPU) is often called the brain of the computer because this is where all computational and processing activities are executed. It is also the central module where communications from other peripherals or equipment converge. These peripherals can be input or output devices. Fig. 1.2 summarizes the relationship between the CPU and various peripheral hardware devices. Input devices such as the keyboard, cameras, scanners, sensors, and other equipment generate and send digital information to the CPU. Conversely, output devices such as display monitors, 2-dimensional (2D) printers, CNC machines, robotic arms, and others apparatus receive digital information from the CPU and execute their assigned tasks.

The CPU has built-in internal memory storage that enables the computer to retain information for basic system operations. Data can be processed and saved in the form of random-access memory (RAM), where information is deleted when the computer is switched off, and read-only memory (ROM), where information is retained permanently. However, both forms of internal data storage can be heavily bound by cost issues. Hence, peripheral devices such as external hard disks and flash drives are cheaper alternatives that can provide additional storage space for reading and writing large data sets or programs.

In a CNC machining setup, the CPU will fetch a set of system commands from its internal ROM to instruct the input/output (I/O) module to retrieve files containing a large amount of NC data from its attached external device's storage memory. The NC data are subsequently loaded into its RAM for processing before sending them to the output devices via the I/O module. Information in the RAM is periodically purged when it is no longer needed to make room for more tasks. On the other hand, the ROM that stores the OS and application programs are vital for booting up the computer and ensuring normal functions. It is never cleared unnecessarily.

Interfacing is a common term used in computing to describe the interaction between hardware components via the I/O modules in an integrated system. There are two broad categories of interfaces, serial and parallel. For serial interfaces, data are communicated one bit at a time in a sequence, while the information in parallel interfaces is consolidated as packages in bytes and transmitted simultaneously. Serial interfaces are simple to set up as they require fewer connections between hardware, but the communication speed is slow. Examples of serial interfacing are devices such as the mouse and keyboard. They are connected serially to the CPU.

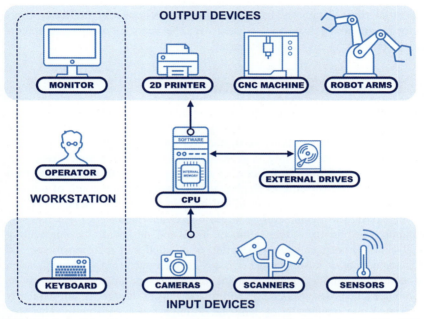

FIGURE 1.2 Relationship between CPU and various peripheral hardware devices [23]. *CPU*, Central processing unit.

In parallel interfaces, the communication between hardware is faster as more information is allowed to be exchanged between each transmission cycle. However, the setup is often not as straightforward and may require many data lines. An example of parallel interfacing is the DCS in an automated FMS where different hardware components are connected within a network. The central computer is linked via parallel interfaces to various equipment and processes, as well as to other satellite computers working on different operations. Interfaces between software programs and process controls need to be relevant and reliable to ensure that information flow between hardware is fast and stable.

1.5.7.2 Software

The OS and various application programs make up the software component of a computer system. The OS is a set of information that commands the CPU to process essential system operations. It also carries out specific tasks with various application programs to execute specific tasks and functions in connected hardware. Application programs are written in a variety of coding formats known as computer languages. MIT developed the APT to aid NC programming, while Unimation came up with the variable assembly language as a proprietary robot-oriented language to control their proprietary

26 Digital Manufacturing

industrial robot, Unimate [10]. These are two examples of the many computer programming languages created over the years for various computer-aided control systems in manufacturing.

1.5.8 Computer-aided applications

The commercial success and market value of a manufactured product often depend on how well it is designed [10]. Therefore the design process of that product can never be wholly separated from its manufacturing as both are functionally, technologically, and economically interdependent. Computers can play essential roles in facilitating and integrating both functions as they are powerful and versatile to receive and manage a large amount of infomation and process them from start to finish effectively and efficiently [17].

In a design to manufacturing workflow, design engineers often reply on computers to aid the visualization of ideas and design drafts, engineering analysis of concept models, production of detailed part representations, and documentation of information necessary to manufacturing. The application of computers and computer systems in the design process is known as computer-aided design (CAD). Information from design is subsequently translated into manufacturing language through process planning. This step is often challenging but a critical bridge between both functions. A simple product may present many different plans that can cause undue confusion. Hence, intelligent or expert systems are used in the planning process to automate and manage multiple tasks. The application of computers and computer systems in process planning is known as computer-aided process planning (CAPP). In manufacturing, computers are built on other developed technologies such as NC, robotics, material handling, and storage systems to assist, control, and automate their physical processes. The application of computers and computer systems in manufacturing is referred to as computer-aided manufacturing (CAM).

Besides aiding the engineering functions, computers and computer systems can play essential roles in supporting various business and management needs [17]. These may include operations in sales and marketing, sales forecasting, order entry, cost accounting, customer billing, and other business functions. Computers have contributed to the relief of many labor-intensive administration and financial tasks that require tedious calculations, processing, and reporting. They are also useful in facilitating communication between internal departments or with external customers and vendors.

1.5.8.1 Computer-aided design

In the agrarian society, when a customer needed an object to be made, he would often verbally describe a generic shape and size of what he needed to the maker. The maker would, in return, craft it according to what he perceived in his mind. However, if the customer needed something more unique with specific instructions, the maker might just make a simple drawing on the sandy ground of the workshop to visualize and verify his verbal message [10].

As industrialization took over, objects were no longer handcrafted but carefully designed and manufactured by a team of specialists and machines in a fast-paced and productive factory environment. Design engineers would use a formalized design process and CAD system to generate an accurately represented engineering model of the object that included all specifications required for manufacturing. The details were later communicated and rendered between groups of personnel within the design-manufacturing workflow.

1.5.8.1.1 Computer-aided design system

CAD is defined as the creation, modification, analysis, and documentation of an engineering drawing or design on a computer to facilitate the design process [10]. Technical drawings used to be hand drawn on pieces of paper using pens, markers, or pencils by a drafting specialist to detail the orthographic or auxiliary views of a workpiece [26]. Fig. 1.3 shows a drafting table displaying a 2D technical drawing of a 3D workpiece in different orthographic projections.

The manual drafting process was often tedious, time-consuming, and highly prone to human errors. The experience was greatly improved when computers and computer systems became available and were integrated into the process. 2D drawings were artfully rendered in an interactive computer graphics setup called the CAD system. The system served as a graphic user

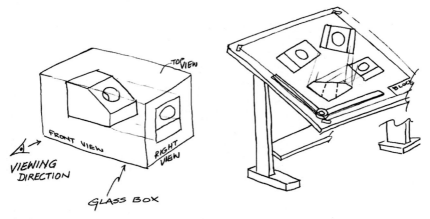

FIGURE 1.3 A drafting table showing a 2D technical drawing of a 3D workpiece [26]. *2D*, 2-Dimensional; *3D*, 3-dimensional. *Courtesy Chandrakant D. Patel, 2019.*

28 Digital Manufacturing

interface (GUI) between the design engineer and the CPU. It contained all the necessary CAD applications for various design and analytical functions. It provided design engineers a virtual canvas to sketch, modify, duplicate, analyze, simulate conditions, and render engineering concepts quickly by creating 3D geometric representations with specific software programs [26]. Fig. 1.4 shows the GUI of a CAD system displaying the modeling process of a workpiece from a simple 2D profile to a 3D geometric representation by using special program operations such as Extrusion and Boolean Difference.

The CAD system would effectively increase the drafting efficiency and productivity of design engineers by facilitating them through the design process. The process also aided in creating a database for various part families and design specifications as information of the geometric models, design rules, and product features could be readily documented and stored in the system's memory. These records would be conveniently retrieved and drafted out automatically for future presentation and manufacturing whenever required.

Besides supporting the design process, the CAD system could allow physical information of a particular section within the product part to be extracted and analyzed virtually using various analytical engineering programs. The investigative process was known as computer-aided engineering

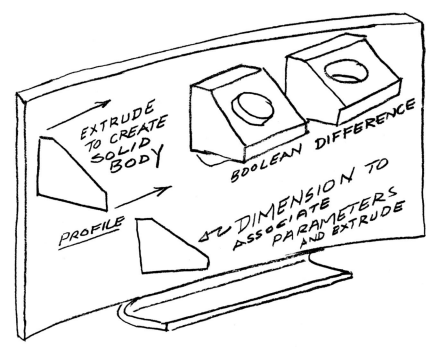

FIGURE 1.4 GUI of a CAD system. *CAD*, Computer-aided design; *GUI*, graphic user interface. *Courtesy Chandrakant D. Patel, 2021.*

(CAE). Engineering operations such as finite-element analysis, heat-transfer analysis, stress analysis, fluid dynamic analysis, and dynamic simulation of mechanisms could be repeatedly performed with the intention to tweak the overall form, reduce total mass, and optimize the performance before sending the final design for manufacturing [26]. In addition, test parameters and analytical results would be documented and stored in the system's database for future work and referenced by other engineers.

Before the introduction of computers, calculating analytical functions of engineering designs was onerous, time-consuming, and usually reserved for a group of trained senior specialists. As computers and various engineering programs became more affordable and readily accessible, more junior engineers started to pick up the new skills of working with the CAD system. They eventually felt more confident in the job and were more willing to execute further rigorous analysis on their designs.

1.5.8.1.2 Design process

The design process is one of the well-established problem-solving approaches that many design engineers drew on to tackle and manage large projects with overwhelming concerns [10]. The approach is similar to the scientific method, where the task in question is broken down and addressed in a series of manageable steps so that ideas and solutions can be organized and refined, respectively. However, unlike the scientific method of observing and conducting experiments, the process requires the design engineer to design, fabricate, and test out the proposed idea. It is an iterative activity that checks the suitability of the design concept again and again [10].

Step 1 in the design process for a product typically starts with recognizing a consumer needs in the market or a specific design flaw in an existing product that requires rectifying. In step 2, the inquiry and problem are addressed to define a set of specifications. In step 3, different design ideas and engineering elements are consolidated to synthesize a crude representation of the product. In step 4, the proposed design is analyzed mathematically and esthetically for any weaknesses. Both steps are closely related and highly interactive. They are repeated many times until the final design is mechanically accurate and visually acceptable. The number of iterations is, however, often subjected to the design engineer's creativity, sense of esthetics, experience, and accessibility to the various design tools and resources. Therefore the outcome can be highly subjective and inconsistent.

In step 5, the final design is carefully evaluated by comparing it with the specifications set earlier in step 2. A physical mock-up is constructed to serve as an acid test for the design concept. It gives design engineers and target users a physical form to see, touch, and play with so that they can evaluate it intimately. Any issues arising from the evaluation process will be reported back and the recommended design changes made in step 3. Steps 3,

4, and 5 are repeated until all major concerns are resolved. The entire reiterative process is often time-consuming and laborious as physical mock-ups are often made by hand that can be subjective and inconsistent.

The final design of the product is presented to the marketing and management teams in step 6. It often involves tedious documentation and filing of various marker renderings, material specifications, assembly lists, and other necessary paperwork. Fig.1.5(A) illustrates the six conventional steps taken sequentially in a design process described by Shigley [10].

As the use of computers and programmable software grew in popularity among design engineers, it has subsequently facilitated and reduced several laborious operations in the design process. Fig.1.5(B) illustrates the development of various CAD/CAE processes in the CAD system and their contributions to the conventional design process.

In step 3', a generative CAD-based design exploration method can create an assortment of 3D geometric models defined with specific constraints and limits [27]. The design engineer then selects the most desirable design and sieves it through several constraint settings to refine different regions on the model. In step 4', computational software will simulate and analyze the model to generate different child variants that the design engineer can again select and refine. Both steps are repeated in the CAD system until the selected model is mechanically optimized and esthetically desirable.

Once decided, the final design is reviewed and evaluated in step 5'. Detailed 3D renderings of the model are created in a virtual environment and presented to target users to gather feedback. Virtual prototyping is a quick

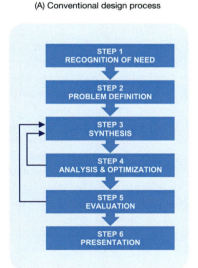
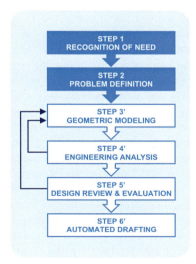

FIGURE 1.5 Conventional versus computer-aided design process [10].

A historical perspective on industrial production and outlook Chapter | 1   **31**

and cost-effective to explore consumer trends and design esthetics. Any feedback and changes to the design can be swiftly communicated and executed back in step 3'. The CAD system database can store all design-related details, records, and documents and are easily retrieved in the future for revisions and presentation in step 6'.

Indeed, the approach to a design challenge is an engineering feat and a creative endeavor. Computers can aid the tedious repetition of predicted operations, but human experiences, perceptions, and imaginations are still essential in realizing an innovative design. Therefore design engineers need to acquire both the knowledge and emotional skills to exercise various steps in the design process effectively and creatively. Fig. 1.6 traces the creative design process of an easy stool design from concept development sketch (A) to virtual prototyping (B) to physical mock-up in 1:1 scale (C) and final product presentation (D).

1.5.8.2 Computer-aided process planning

After a part or product is designed, it is sent for manufacturing . The manufacturing process may involve using different types of machines and preparing resources like materials and labor. Hence, the sequence of activities, individual operations, and related information must be carefully mapped out and recorded. A route sheet is typically used to document these plans and information. The planning of various tasks within the design-manufacturing workflow is known as process planning [23].

Process planning is not an easy task. Manufacturing a relatively simple part or product can present several possible plans and selecting the most

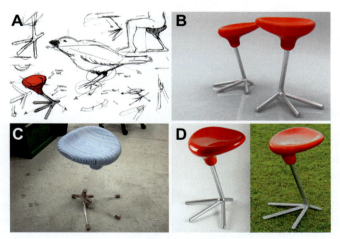

FIGURE 1.6   A creative design process of an easy stool design inspired by nature. *Courtesy the author, 2008.*

32 Digital Manufacturing

viable and cost-effective one can be challenging and time-consuming. Route sheets were traditionally prepared manually but were usually lengthy and sometimes confusing. Following the advent of computers and computers systems, attempts were made to computerize the process planning tasks. It resulted in the development of two CAPP systems, the retrieval and generative types.

The retrieval type, also known as variant type, relies on the concept of GT [23]. It starts by sorting out existing standard process plans of parts according to their coding and classification in a family and storing the information in a central database. Relevant plans are subsequently retrieved and appropriately modified whenever a different part within a family needs to be processed. The retrieval type of CAPP is suited for cellular manufacturing as each cell can be designed and laid out to produce parts that are grouped within a family.

In contrast, the generative type of CAPP will develop individual plans from scratch rather than retrieving and referencing existing ones [23]. The operator starts by entering information, such as part geometry, materials, and other related factors, into the CAPP system whenever a different part needs to be processed. An intelligent or expert system will automatically generate a unique plan for the part based on the input information using algorithms by which process sequences are synthesized. The generative type of CAPP is suited for integrated manufacturing systems as data from various functional units can be exchanged and processed automatically, without the need for any human interventions.

1.5.8.3 Computer-aided manufacturing

When minicomputers were widely incorporated into NC technology, CNC machines quickly became a staple in many established factories as a cost-effective solution for various complex machining tasks and processes. Hence, the term computer-aided machining was better known by many as the initial application of computers in manufacturing [23].

Manufacturing engineers and machine operators would pick up the skills of NC part programming and become responsible for creating part programs or sets of instructions for controlling the slide and tool movements of NC machines and other auxiliary functions. However, they often encountered problems computing lengthy and tedious calculations in the process due to the limited processing power of the minicomputers. It got incredibly challenging when working on geometries that were elaborated and highly complicated. Fortunately, the issue was resolved when the microchip, microprocessor, and microcomputers became available. The inventions not only accelerated the calculation process to make programming easier and faster but also replaced key functions of the machine's control units.

It did not take long before the use of computers grew beyond other non-engineering functions and took up roles in administrations, planning and logistics within the manufacturing process. The term computer-aided machining was no longer adequate to fully describe the extensive application of computers and the functions they carry out in the workflow. Hence, a better-fitting term called CAM was coined.

CAM could be broadly divided into two categories, manufacturing control and manufacturing planning [10]. In manufacturing control, computer systems were developed to manage and control the physical activities on the shop floor. The roles included management functions such as process monitoring and control, quality control, shop floor control, and inventory control. In manufacturing planning, computers were not connected directly to the manufacturing process. Instead, they played supporting roles to provide "offline" information for various computer-aided design and manufacturing (CAD/CAM) applications that ensured effective planning and management of production activities on the shop floor. The roles included operational functions such as process planning (see CAPP), cost estimating, production and inventory planning, and computer-aided line balancing.

1.5.9 Computer-integrated manufacturing

The emergence of modern computers and computer systems in the late 1900s had greatly transformed many traditional factories through reshaping their internal organizational structure, operational processes, and external relationships with the society [10]. Machine operators leveraged computers to direct the movement of automated machines such as CNC machines and industrial robots. Computers and computer systems assisted manufacturing engineers in the planning and operational activities of automated manufacturing systems such as the FMS by interfacing, controlling, and managing a web of connected equipment. They also aided design engineers in the design process by creating a conducive environment for creative work and facilitated effective communication with other coworkers.

In a distributed computer network, ICT played critical roles in monitoring and controlling the storage and distribution of information, software programs, and data across various workstations, platforms, and domains. The operational processes of machines and product design data could be conveniently stored, retrieved, and shared among users within the computer system network connected to a central database when required. In addition, ICT provided support and assistance for various information-related administrative and business functions, such as documenting customers' orders, production scheduling, data collection, record keeping, product performance tracing, inventory checks, and shipment tracking.

Indeed, computers, computer systems, and ICT would rationalize, coordinate, and integrate all functions in the design-manufacturing-business workflow while managing data records and information throughout every aspect of the organization [28]. The term computer-integrated manufacturing (CIM)

34 Digital Manufacturing

was eventually coined to effectively consolidate and encapsulate this idea of a single-integrated manufacturing system.

American engineer Joseph Harrington first proposed the concept of CIM in the mid-1900s. He conducted a study on the feasibility of connecting various automated workstations or islands of automation through a network of computers and subsequently published a book called *Computer Integrated Manufacturing* [29]. Sadly, the manufacturing industry did not perceive his grand idea as revolutionary. Many business proprietors had already implemented a similar approach to integrated manufacturing in their establishments, but on a simpler and smaller scale as computers were not readily available. However, when computers eventually become more advanced and widely accessible, they saw the potential benefits of CIM and started adopting it in its operations [23].

The CIM philosophy, when fully implemented, helped factories yield better productivity with shorter product development and production lead time. Organized automation and computer systems significantly improved the design process and overall product quality while reducing direct labor costs and stock inventories. Hence, factories were able to react quickly to the volatile consumer market and keep production costs low.

The Computer and Automated Systems Association of the Society of Manufacturing Engineer or CASA/SME developed and published the first graphical representation of the CIM strategy in the late 1900s with the aim to help the business proprietors grasp its concept and scope. The framework underwent several changes and expansions over the years to include many other functions. Fig. 1.7 shows a better developed CIM architecture that incorporated all aspects of the manufacturing companies and their administration, including human resource and finance management [23].

The success of CIM is attributed to the apparent use of computers, computer systems, and ICT. In addition, manufacturing philosophies such as GT, specialized business process and management software, and intelligent or expert systems were developed and adopted into various workflow also contributed substantially to the support and effective execution of several functions within the system.

GT plays a major role in the CIM philosophy [17]. Product characteristics are converted into descriptive information and stored as common data in the information resource management and communication database (Fig. 1.7) for sharing among everyone involved in the production process. Design engineers from the product/process unit can easily retrieve any parts for design referencing or editing by using the CAD system, while operation planners from the manufacturing, planning, and control unit can quickly look at various manufacturing processes and resources to forecast future needs via the CAPP system. In addition, machine operators from the factory automation unit can access specific machine tool selections and NC programs when they need to recall previous machining protocols through the CAM system. Even

A historical perspective on industrial production and outlook Chapter | 1 35

FIGURE 1.7 A CIM framework developed by CASA/SME [23]. *CASA/SME*, Computer and Automated Systems Association of the Society of Manufacturing Engineer; *CIM*, Computer-integrated manufacturing.

the business proprietor can easily retrieve and reference the entire shop floor layout from the central database for future improvement and reconfiguration.

In a highly functional and fast-moving CIM environment, personnel across different units and divisions depended on the enterprise resource planning (ERP) process to help them make multiple decisions and manage complex tasks at various levels at any given time. The ERP process uses a specialized business process and management software with multiple computer hardware components to organize and integrate all essential business processes, manufacturing resources, and communication services into a unified entity known as the ERP system [30]. The system can render assistance in design and manufacturing operations, such as GT and the CAD/CAM system. It can also assist in business intelligence, assets management, and e-commerce services, including customer services, sales and marketing,

36 Digital Manufacturing

procurement, production planning, distribution, accounting, human resource, and corporate performance and governance.

Lastly, intelligent computer systems such as expert systems can enhance various automated processes, make intelligent decisions, and solve complex challenges within the CIM environment [23]. An expert system is typically developed from a branch of computer-based artificial intelligence that can enable associated personnel to emulate the same level of services and problem-solving capabilities like a real expert will offer. It utilizes a knowledge base and an inference engine interactively to develop solutions or responses unique to a problem or challenge. Hence, the personnel can work with sensors and machines that are interfaced directly with several discretely designed expert systems to perform specific functions and operations across different domains. Certain expert systems are created to help factory managers make logical decisions at different management levels, while others can aid machine operators in predicting equipment maintenance through sensor feedback. Some expert systems can support design engineers in developing product concepts from various consumers' input. Others assist manufacturing engineers in generating the most effective production process plans based on past production data.

1.5.10 Product development process

The product development process can be perceived as an elaborated extension of the design process, where multiple teams of dedicated people and elaborate tasks are involved [31]. The tasks can range from identifying a market need, formulation of requirements and consumers' feedback, generation of ideas, design conceptualization, manufacturing, product evaluation, sales and marketing, etc. The earliest approaches to the concept of product development were probably conceived during the Second Industrial Revolution where there was a relentless push for the commercialization of various inventions. It was only around the mid-1900s that the need for a well-grounded process became necessary and subsequently formalized [10].

During the golden age of capitalism in the mid-1900s, many rose to join the middle class with excess money to splurge. Factories started to overproduce a variety of consumer goods blindly just to keep up with the demands but without a clear business strategy. Products reaching the market were often modifications of existing designs but not radical inventions. As a result, their production lines were overburdened with inconsistencies and unnecessary wastage while the shop floor was choked with repetition and nonproduct value-adding activities. Business proprietors became worried and decided to overhaul the inefficient manufacturing systems and redress the production strategies in their establishments in an attempt to resolve the situation and increase productivity. A systematic product development approach was established and eventually led to its formalized process.

A historical perspective on industrial production and outlook **Chapter | 1** 37

Similar to the division of labor approach, complex processes were restructured and sorted according to their functions and department, such as marketing, research, design, manufacturing, and sales. Hence, laborious work and complex operations could be consolidated, streamlined, and better managed with well-defined job scopes and responsibilities. Specialists from one department would focus on a specific task and pass the finished work to another department to follow through. By doing so, they could study, gather, and analyze information according to the specific needs of consumers and develop mass customized products that were more customer-centric and targeted for better brand positioning. With better brand management, flexible manufacturing, and mass customization, business proprietors were able to lower the overall production costs and keep manufacturing lean with minimal wastage.

1.5.10.1 Sequential engineering

A sequential engineering approach to developing products comprises a series of activities where tasks and information would flow sequentially from one department to the next in a unidirectional, daisy chain fashion [22]. Fig. 1.8 shows the sequential engineering approach to the product development process where specialists from one department would complete a specific task or assignment before handing it "over-the-wall" to the subsequent department to follow through.

As business proprietors continued to aggressively push more product varieties out in the global market to give themselves a unique and competitive edge, the time-to-market of new products became shorter and more intense [17]. Factories soon realized the struggle to keep up with demands for shorter development time and lower production costs. The sequential engineering approach to product development quickly became irrelevant as the process was time-consuming and the output was prone to design deficiencies and inadequacies.

Although the sequential workflow might seem orderly and systematic, specialists often found it difficult to interact and exchange sufficient information effectively "over-the-wall" with specialists in other departments.

FIGURE 1.8 A sequential engineering approach to product development [22].

38 Digital Manufacturing

Many of them would struggle to resolve ambiguities and confusion caused by miscommunication, delays, and duplications of work. Due to lack of mutual interactions, flawed and incomplete tasks were often thrown back "over-the-wall" to the previous department for amendment, hence creating unnecessary backlogs, lapses, and delays in the production schedules and delivery dates [22].

Department managers and specialists gathered their complains and shortcomings, and decided to explore a new approach of working in parallel with one another through concurrent interactions, rather than waiting for one task to be completed before proceeding to the next. As a result, they were able to simultaneously share knowledge and insights, exchange relevant information, and contribute to the decision-making process [22]. Any confusions and disagreements in the design-manufacturing workflow would be resolved cooperatively and concurrently before they start to surface. The integrated approach eventually led to a new manufacturing philosophy known as concurrent engineering in the late 1900s.

1.5.10.2 Concurrent engineering

In a concurrent engineering approach to the design-manufacturing workflow, specialists from the manufacturing department would advise the design department on how to design the product and its components better to facilitate their manufacturability and assembly in the later stages of the production. In return, they could gain insights into the product's design details from the design engineers to initiate a more effective manufacturing plan. On the other hand, specialists from the design department would gather detailed information on the capability of manufacturing resources from the manufacturing department and subsequently focus on creating designs based on the manufacturability viewpoint. They could also customize their final design according to the production and assembly facilities available.

Concurrent interactions between specialists in the design and manufacturing departments were indeed instrumental in refining the product development process. Furthermore, integrating other associated functions such as sales and marketing, quality engineering, and vendors supplying raw materials and critical components in the initial design stage of the development process was advantageous as they could contribute valuable inputs and advice. It was estimated that the fundamental decisions made during the design stage of the product development process would improve the overall productivity and reduce more than half of the product cost [32]. Fig. 1.9 illustrates the concurrent engineering approach of various associated functions in the product design stage can help improve productivity and shorten the product launch time compared to the sequential engineering approach [22].

Advances in computers, computer systems, and ICT in the late 1900s aided the development of the CAD/CAM systems and other communication

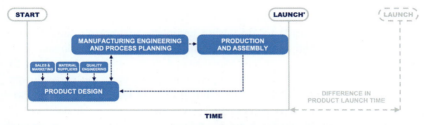

FIGURE 1.9 A concurrent engineering approach to product development [22].

platforms by enabling effective workflow and communication links between managers and specialists in different departments. Design engineers were able to access and acquire detailed information on the manufacturing capabilities of the factory's existing resources from the CAD/CAM system's network database and make quick decisions on the manufacturability of their designs in the initial design stage. Decisions made might include the choice of materials, part geometry, tolerances, surface finish and more. With this information on hand, they were able to avoid designing features that were unnecessarily expensive to manufacture and make the optimum choices of materials to minimize production costs. In addition, the CAD/CAM system gave design engineers the space to reconfigure product parts into subassemblies and explore various assembly methods. The design-manufacturing interaction with the CAD/CAM system eventually facilitated the successful development of multiple design schemes and methods. Some of them were universally applied to nearly any design task, while others were only applicable in specific processes.

When the trend of automation hit full trottle in many factories during the late 1900s, two American engineers named Geoffrey Boothroyd and Peter Dewhurst decided that they should help product design engineers quantify their designs for the ease of automatic feeding and insertion for the assembly of parts. Hence, they formulated a methodology known as Design for Manufacture and Assembly (DfMA) by relying on the use of certain design principles and guidelines to consider the manufacturability and assemblability of a product and its components [28].

The DfMA method enabled design engineers to identify and reduce the number of assembly operations by decreasing unnecessary part components in the initial design stage of the development process. It also guided them in designing parts that would make assembly operations easier to perform. Boothroyd and Dewhurst went on to publish a handbook called *Product Design for Manufacture and Assembly* based on these two basic design principles and subsequently developed an improved computerized version a few years later. The DfMA program was later integrated into the CAD/CAM system and implemented across a broad range of manufacturing companies.

40 Digital Manufacturing

Coincidentally, Hitachi Ltd. also developed the Hitachi Assemblability Evaluation Method (AEM) to aid design engineers to improve their design quality and simplify the automatic insertion action of part components [28]. Conceived by Japanese engineers Miyakawa, Ohashi, and Iwata, the AEM was devised to assess the assembly operations of parts at the early design stage of the development process. It assigned penalty points for each assembly step when two separate components were mated. The evaluation began with 100 points and would be gradually reduced according to the number of separate motions and motion axes needed to achieve a complete union. More points were deducted from a twist-and-turn operation than a straight push. Finally, assemblies that ended with less than 90 points would require the part to be redesigned. The AEM was eventually integrated into the CAD/CAM system following its development, and variations of the methodology were introduced years later.

1.5.10.3 Collaborative engineering

By the late 1900s, market globalization, shorter delivery times, fast-evolving customer demands, and complex supply chains had already posed numerous challenges and intensified the competitiveness in many international companies. Systems, resources, consumers, and services were no longer restricted within certain geographical confines of physical and time zones as advanced computers, computer systems, ICT became widely available and adopted globally. These technological innovations had helped create and connect conducive network-oriented design spaces for sharing of collaborative work among multiple creative individuals worldwide [33].

With increasing demands for sophisticated and consumer-centered products and services, conventional manufacturing philosophies applied in developing mass customized products could no longer effectively support the dynamics of collaborative teamwork spanning across distributed spaces, levels, and disciplines [34]. Soon, it became clear that there should be a need to address this synergy in the product development process formally. A new manufacturing philosophy known as collaborative engineering was subsequently conceived. It expanded the foundations presented in concurrent engineering and established new perspectives and goals that extended across diverse specialists and managers working in multiple engineering domains.

The collaborative engineering approach to developing products typically involves applying collaborative activities within design, manufacturing, and operations. Collaboration is a process where multiple individuals work together interdependently to achieve a higher goal than is possible for any person to accomplish alone [35]. While concurrent engineering encourages teammates to share resources and outcomes through coordination and cooperation, collaborative engineering entails going beyond the mentioned activities. It requires everyone to achieve a common goal together [36].

It emphasizes the need for shared responsibilities, coaction in attaining results, and search for consensus.

Collaborative engineering can shorten the product development time, improve product innovation, enhance the integration and utilization of enabling technologies, increase product and process quality, add total value, and lower development and production costs [36]. With better teamwork and technology integration, local and remote team members can create new ideas together and make faster turnaround decisions collectively because they can communicate and understand each other more effectively. Their morale and attitude toward responsibilities also improved as everyone can gain mutual respect and appreciation among one another. Everyone will produce better task-work results and ultimately boost the overall competitiveness of manufacturing companies.

The subsequent development of socio-technical systems and advanced ICT in the early 2000s has encouraged collaborative activities to be widely implemented across various engineering and management domains, such as product design, manufacturing, operations, enterprise-level collaboration, and supply chain management [37]. The application of collaborative engineering in product design and development process is known as computer-supported collaboration design. It describes the process of designing a product through close participation and collaboration among a remote web of multidisciplinary product developers. They may include small design studios, independent designers, and makers with specific domain knowledge and experience. With the aid of the Internet, cloud computing, and 3D printing technology, they can develop the entire product life cycle by themselves and team up with larger manufacturing companies for off-site mass production, consequently, coining a new manufacturing philosophy known as social cloud manufacturing [19].

1.5.11 Additive manufacturing

The application of 3D printing technology in manufacturing is referred to as additive manufacturing (AM). AM technologies underwent intense development in the early 2000s that led to the characterization and classification of various AM processes based on the types of special material used and the proprietary techniques developed to build up the printed part.

According to its working principle, AM offers design freedoms and the capabilities to fabricate geometric contours that were hitherto not possible in a cost-effective manner [26]. It opens up new opportunities, such as examining complex geometric structures that can mimic the properties of biological materials [38]. Through the study of biomimicry or biomimetic abstractions, AM can create innovative physical parts and products with the desired bioinspired functionalities and properties that emulated nature's genius [39]. Furthermore, with advancement in technology and breakthrough in material

42 Digital Manufacturing

research, AM-produced parts can sometimes outshine their equivalents that are otherwise made using traditional production processes [38], therefore making AM a highly feasible and desirable technology for critical applications in the aerospace, automotive, and health-care industries. Such applications can range from individualized articles to customized parts for space missions, tailored drugs to printed finger food, and specialized medical implants to living human tissue [40].

Besides offering possible functional benefits over traditional manufacturing, AM technologies have also been recognized to provide several potential sustainability advantages [41]. Being an additive process, AM can efficiently generate less or zero waste during manufacturing compared to conventional subtractive techniques. Part geometries created for AM are often optimized and lightweight, reducing the amount of material used for manufacturing and the energy consumed while using the product. AM can offer on-demand capabilities, where spare parts are printed as and when required, hence reducing unnecessary inventory wastage. Finally, with the rise of social cloud manufacturing, transportation in the supply chain is cut down as manufacturing services are decentralized with accessible AM facilities worldwide.

Overall, traditional manufacturing will be disrupted and conventional business models will eventually become irrelevant when emerging technologies such as 3D printing, artificial intelligence, and advanced robotics become readily available in the near future [42]. 3D printing digital factories will appear and function differently from traditional factories. Design engineers will acquire a different set of design rules and methodologies to work effectively in an AM environment [43]. These guidelines are unique to AM and essential to ensure the manufacturability of parts and products with intricate geometries. Noiseless 3D printers will line neatly in rows to replace loud CNC milling machines. They will work collaboratively with each other or with other devices to build complex parts that are impossible to carve by any conventional means. Manual handling and assembly will no longer be necessary as different materials can be readily printed together in a single operation due to their unique building process. Autonomous robots will post-process the printed products, transfer, package, and ship them out accordingly. Factories will operate on a net-zero emission approach, where production processes and material utilization are efficient and sustainable enough to produce near-zero waste to the environment.

1.5.12 Sustainability in manufacturing

Demands for manufactured consumer goods compounded as the world population expanded because of improved economy and living standards. The manufacturing industry became one of the key sectors within the modern socioeconomic systems as it contributed substantially to wealth generation and job creation, especially in developing economies [44]. Regrettably, such heavy industrial activities took their toll on the well-being of our planet.

An industrial system is assumed unsustainable when the rate of resource consumption and waste production by the society surpasses the ability of nature to transform the wastes back into nutrients and resources for the environment [44]. In response to widespread environmental degradation, limited natural resources and rapid climate change, the UN World Commission on Environment and Development held an urgent meeting in 1987 and outlined a series of guidelines in an effort to minimize the repercussions that manufacturing had on the environment. Titled *Our Common Future*, the report detailed recommendations to encourage industrial operations to lessen nonrenewable natural resources consumption and reduce wastes and pollutions production where possible. In addition, manufacturing activities should strive to minimize the irreversible negative impacts on human health and the environment [45]. Consequently, manufacturing companies started to retrofit their factories and industrial machines with incrementally cleaner and more efficient "engines" to make them cleaner, faster, and quieter to appease political pressure from their government while keeping pace with ongoing market demands. Despite such efforts to push for "greener" manufacturing, the perception of sustainability did not transpire well enough in practice.

Going forward, in 1992 the Business Council for Sustainable Development, comprising a group of 48 industrial companies, decided to come together to readdress pressing issues that had resurfaced in manufacturing and the environment. They ended up adopting the buzzword "eco-efficient" and later advocated the concept of eco-efficiency among representatives of 167 countries at the Earth Summit in Rio de Janeiro [45].

1.5.12.1 Eco-efficiency

In conventional industrial systems the flow of materials is typically linear and single directional. According to the cradle-to-grave economic model, raw materials are extracted from the environment, converted into products, and disposed of eventually, hence turning nature's resources into wastes and the environment into a graveyard. The eco-efficiency strategy is to recover as much value from the used materials and repurpose them into other products before they are discarded to landfills or incinerators [46], hence getting more from less through recycling; more products or services with less use of resources, wastes produced and toxicity to the environment. As a result, companies that embraced eco-efficiency benefitted substantially in cost savings and reduced environmental impact.

Although the eco-efficiency approach may potentially lead us to a healthier and sustainable planet, doing less harm to the environment does not rectify a deep-rooted problem [45]. We cannot negate the fundamentally destructive nature of the cradle-to-grave model. Reusable materials are

44 Digital Manufacturing

laboriously recovered and recycled as an end-to-pipe solution. However, these materials are rarely designed to be 100% recyclable from the start. By contrast, the entire process is, in fact, downcycling. Downcycled materials are typically lower in quality and have limited repurpose use [46]. As a result, products created from these "recycled" materials are often inferior and have lower economic value. Moreover, overly processed materials are harmful and do not regenerate back to support the ecological systems of the environment when dumped in landfills.

Granted that the cradle-to-grave model may seem limited and flawed, it still holds certain merits in the contemporary manufacturing industry. Rather than operating on a "take, make, and dispose" linear approach, the properties of materials can be redefined and their flow pathways remodelled to create a cyclical, cradle-to-cradle "metabolism" system [46]. Materials flowing through the system can be designed to maintain their status as resources, feeding through different processes as nutrients while gathering intelligence after some time. Hence, the materials will be upcycled rather than downcycled to become waste. Developing healthy and environmentally benign products and industrial systems is central to the concept of eco-effectiveness and, within the cradle-to-cradle system, lies the eco-effective strategies [47]. It focuses on developing and redesigning products and their associated material flows to create a synergistic relationship between the ecological and economic systems [46].

1.5.12.2 Eco-effectiveness

The eco-effective approach to sustainable manufacturing calls for a fundamental conceptual shift away from the conventional cradle-to-grave linear material flow system to a cradle-to-cradle cyclical system as materials flow in regenerative, closed-loop cycles driven by renewable energy.

The Ellen MacArthur foundation proposed a circular economic model in the late 1900s to counter and alleviate the negative impacts brought on by the linear model [48]. The goal is to guide the manufacturing industry in building and rebuilding the overall system health of its economic activities cyclically. The model is founded on three fundamental principles: to design out waste and pollution, keep products and materials at their highest value, and regenerate the natural system. Therefore it represents a systemic shift that will eventually build on long-term resilience, generate business and economic opportunities, and provide environmental and societal benefits [48]. In addition, the concept of the model is supported by several other schools of thought, such as industrial symbiosis, performance economy, sharing economy, biomimicry, and cradle-to-cradle design.

Cradle-to-cradle design is a biomimetic approach to designing products and systems that are inspired by nature's processes. German chemist Michael Braungart and American architect William McDonough, together with

A historical perspective on industrial production and outlook Chapter | 1   45

Environment Protection Encouragement Agency (EPEA) Hamburg, developed the cradle-to-cradle design framework in the late 1900s. The goal was to provide a conceptual reference for design and manufacturing engineers to "redesign" products with a minimal negative impact and leave a positive ecological footprint on the environment [47].

The framework proposes a set of design principles for redesigning products and industrial processes. Materials are transformed into "nutrients" at various stages of a product life cycle while flowing perpetually within two discrete metabolisms: biological metabolism and technical metabolism [46]. Fig. 1.10 shows a schematic diagram of a cradle-to-cradle design framework where materials flow as biological and technical nutrients between the biosphere and technosphere, respectively.

Biological nutrients are biodegradable materials or by-products of biodegradation processes that flow through the biological metabolic cycle. They are nonhazardous to living systems and can be used on humans safely or returned to the environment to feed other biological processes. Products designed as biological nutrients are called products of consumption [46]. They can be "consumed" through decay or attrition from daily uses and returned to the natural environment to feed other living systems. Examples of products as biological nutrients can be rubber tires, organic cotton textiles, concrete fillers made from pulverized ash, etc.

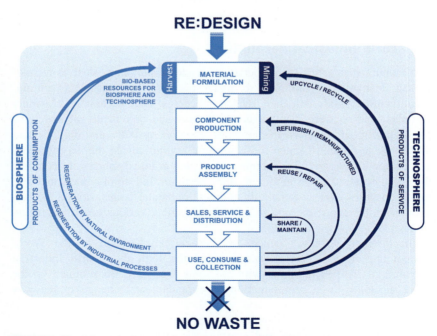

FIGURE 1.10 Schematic diagram of a cradle-to-cradle design framework [47].

46 Digital Manufacturing

Technical nutrients can be mineral or synthetic materials flowing through the technical metabolic cycle. They can remain useful within a closed-loop system of manufacture, recovery, and reuse and are kept in good condition at their highest value through many product lifecycles. Products conceived as technical nutrients can be durable goods rendered to consumers as a service. Hence, they are also referred to as products of service [46]. The manufacturer or commercial vendor owns the material assets and maintains them for continual reuse. They will lease them out to customers as a service for a defined duration. The customers will pay for the service and return the goods to the owner when the loan period is up. Thus the same product can be shared and used repeatedly by different customers whenever they need it. Examples of technical nutrients can be the hired vehicles, leased accommodations, rented tools and machines, etc.

The shift in approaches from eco-efficiency to eco-effectiveness can draw companies closer to fostering sustainability in manufacturing. The cradle-to-cradle design presents a broad framework for manufacturers to develop eco-effective industrial systems. It provides concepts and guidelines for redesigning products and systems where industrial materials flow and circulate endlessly without generating any waste. Moving forward, to put the cradle-to-cradle design framework into practice, business proprietors will need to acquire the right technologies, human resources, and business model and strategies.

1.6 Forth Industrial Revolution

The manufacturing industry experienced a global paradigm shift as technological innovations continued to make progress into the 21st century. Manufactured goods and their related processes soon became increasingly complex, and so were manufacturing systems and organizational structures [39]. The rise of advanced technologies, complex systems, and digital infrastructures vastly transformed the world and revolutionized the way we conceived new values in the economic, social, and political spheres. In contrast, they also fostered uncertainties in our consumer and employment markets, raised concerns over sustainability and environmental issues, shifted the balance of global economic power, and brought about various unsettling megatrends [49].

In the early 2000s, founder and executive chairman of the World Economic Forum, Klaus Schwab, coined the term the Fourth Industrial Revolution (IR4) in his book with the same title [50]. He summed up the label succinctly as a representation of a fundamental change in the way humans interact. As our physical, digital, and biological worlds converge with the aid of advanced technologies, it is pertinent that individuals around the world learn to harness and leverage the benefits brought on by these opportunities to build an inclusive and human-centric future for themselves.

A historical perspective on industrial production and outlook Chapter | 1   47

Like all industrial revolutions that took place previously, IR4 has the potential to lift the overall global income levels and improve everyone's quality of life. However, this time Schwab warned that the ones gaining the most from the rise will very likely be those who can afford and have access to the digital world [50].

Economists Erik Brynjolfsson and Andrew McAfee added that intellectual and physical capital providers, such as the innovators, shareholders, and investors, will most likely make up the largest group of beneficiaries against those who supplied labor instead [51]. Emerging technologies such as artificial intelligence, machine learning, automation, and autonomous robots will inevitably replace human labor in the highly connected technology-driven economy. They will also bring in better returns and create a safer and more efficient work environment.

Schwab also pointed that demands for highly skilled workers will go up while those with less education and lower skills will drop [50]. Hence, income inequality between the rich and poor will likely persist and worsen. Besides exerting influence on society, IR4 will also cause significant changes to the economy, existing business models, national and global security, and the workforce. Therefore it is critical and pertinent that the government bodies, big corporations, and individuals worldwide start developing a comprehensive and globally shared understanding of what IR4 is. They need to identify the different drivers and potential impacts to seize the opportunities it creates in the future.

Schwab concluded his book by stating that change is unavoidable as technology advances and shapes our future generations [50]. Although IR4 may potentially "robotize" humanity and turn us into unemotional beings, it can empower humankind as a new collective and heighten our moral consciousness built on a shared sense of destiny. Therefore we must enable ourselves, including our children, with the necessary tools for everyone to navigate and discover new meanings from this change.

1.6.1   Industrie 4.0

The concept and objectives of Industrie 4.0 (I4.0) can be easily confused with IR4 as both terms were coined slightly around the same time. Many in the manufacturing industry often use them interchangeably to describe the adoption of advanced digitalization and the integration of industrial manufacturing and logistics processes [52]. In theory, I4.0 only makes up one of the components framed under the greater umbrella of IR4. It focuses extensively on the engagement of digitalization, organizational transformation and productivity enhancement in manufacturing processes and production systems of organizations.

I4.0 is a strategic program established by the German federal government, universities, and private enterprises in the early 2000s. It aims to help

48 Digital Manufacturing

organizations realign their current and future capabilities so that they can engage more intelligently in the competitive markets [53]. The initiative advocates the digitalization and integration of manufacturing systems and processes with emerging technologies. By fostering human−machine interactions and converging them with the Internet of Things (IoT) technology, new ICT infrastructures, and a worldwide network connection, a plethora of user data can be harnessed, shared, and analyzed along the entire global chain. Hence, the information can be inferred and leveraged to improve the overall consumer experience and life cycle of manufactured products.

In addition, I4.0 will greatly influence and disrupt the way businesses are conducted and reshape existing business models. The effects of this change will eventually cascade down to impact society, our natural resources, and the environment. Notably, other production strategies and initiatives similar to I4.0 were also championed by other industrial countries such as the United States, United Kingdom, France, Japan, Korea, and China around the same time to help them stay uniquely competitive in the global manufacturing market [54].

I4.0 aims to achieve several objectives in response to the developing megatrends [52]. It promises to help factories quicken their "time-to-market" process of manufactured goods by offering them higher and faster innovation capabilities through emerging technologies. With the help of faster and more versatile production processes such as AM, these factories will possess the flexibility and capabilities to manufacture quality customized products cost-effectively in smaller quantities to cater to the individual consumer market. In addition, by adopting sustainable practices and cradle-to-cradle design, factories are encouraged to take on a more environment friendly and eco-effective approach to resource management and material utilization in their production line.

With I4.0, the hierarchical structure in traditional organizations will be flattened, condensed, and better interconnected. It can decentralize and democratize the task of decision-making to share responsibilities and empower individuals. Factory operators will be equipped with the necessary skills and knowledge to work collaboratively with cobots. With the rise of social networks, digital markets, and e-business practices, the new disruptive business models in I4.0 will focus not only on manufacturing products but also on creating new provisions for services to deliver a complete consumer experience [55].

Despite the many perceived benefits and supportive policies granted by I4.0, businesses and organizations can still encounter tough challenges when trying to adopt the program into their factories [52]. Implementing I4.0 and a successful digital transformation strategy will require a strong leadership, willingness to acquire collaborative operation models, confidence and trust in the digital workflows, availability of relevant digital skills and knowledge,

A historical perspective on industrial production and outlook Chapter | 1   49

strong support from the labor force, and progressive government policies, rules and regulations.

1.6.2 Cyber-physical system

When Germany was pushing its I4.0 program to business proprietors and factories worldwide, the US government proposed an alternative approach to champion a similar cause. Defined as cyber-physical system (CPS), it is a complex engineering system that converges and integrates the physical, computation and networking, and communication processes together. It aims to improve the overall performance, quality, and controllability of manufacturing processes for smart digital factories [56].

In a CPS, functioning physical devices, components, and machines are translated into cyberspace as virtual models composed of data generated from sensor networks and manual input. In response, these virtual replicas or "digital-twins" will act as feedback agents to control and monitor the physical counterparts via the system's vast computer network [56].

During its normal operations, built-in sensors on physical machines and devices will collect and send feedback data to their "digital-twins" for ongoing updates and monitoring. These abstract data are processed in cyberspace using smart algorithms so that factual information about the machines' health status, performance, and risk can be calculated and synchronized in real time. With real-time information feeding back concurrently to corresponding virtual machines in the system, they can gain self-awareness by assessing their performance history and diagnosing potential component failures, predicting future breakdowns, and eventually prompting maintenance of their physical counterpart [56].

By comparing and analyzing information in a connected network system, discrete virtual machines can self-compare and adjust their settings and performance through referencing a knowledge data library gained from their working history. The system can autoconfigure and autoprogram customized manufacturing criteria of individual physical machines based on their current operating condition and output indicators. As a result, the manufacturing process of products can be tailored according to every physical machine integrated into the production line, ultimately leading to a more flexible and higher quality output with optimum operation cost.

Intelligent functions and autonomous capabilities are indeed unique attributes of a smart digital factory. Intelligent machines and systems will operate in tailored programs based on any production rate and customer demands [56]. The CPS can present the readiness and flexibility required to cater to on-demand services in a manufacturing system that is nowhere near achievable in traditional factories. For this reason, factories of the future will take on a system of systems approach to manufacturing, where resources and capabilities of various OSs used for monitoring and feedback of events,

50 Digital Manufacturing

processes, and devices are integrated with information systems used for data-centric computing [57].

1.6.3 Factory of the future

Information and operational technologies will converge seamlessly in the factory of the future. Meaningful data together with emerging technologies such as AM, autonomous robots, closed-loop control systems, artificial intelligence and machine learning, virtual prototyping, and simulations will present extensive opportunities for established factories to reengineer their existing mechanized and automated processes toward smart manufacturing [58].

Production operators and manufacturing engineers can use advanced ICT to maneuver seamlessly between the physical and virtual worlds through CPS. By using simulation technology and virtual prototyping of physical machines, their "digital-twins" can aid operators to predict potential failures, customize processes, and review designs without having to stop the ongoing production line. Manufacturing engineers can make use of virtual and augmented reality technologies to train their operators in complex processes, maintenance, and troubleshooting tasks. Autonomous machines and cobots are equipped with communication devices, computer vision, machine learning, and predictive analytics capabilities so that they can collaborate intelligently with one another and react swiftly to any production crisis without human supervision.

Multidisciplinary experts such as product designers, materials scientists, and industrial engineers can take advantage of biomimicry and biomimetic abstractions to resolve challenges presented in highly complex technical systems. By studying the fundamental principles in nature and leveraging the power of big data analytics, algorithm-driven generative design, virtual prototyping, simulation, and optimization software, they can construct an array of design iterations digitally, analyze them, and select one that is esthetically pleasing, structurally balanced, and mechanically sound [39]. User-centric and consumer-oriented designs will take center stage in the new product design and development process. Ethnographic observations and user experience feedback from various consumers groups are collected using a contextual and experiential approach in the design process. Collaborative work among multidisciplinary teams in the product development process can contribute enormous value to the overall product life cycle of future consumer products [59].

IoT, cloud computing, big data analytics, and new ICT infrastructures such as 5G connectivity, data centers, and cloud servers will connect and control various operations and processes, hence improving communications in digital factories and between external parties within the global network [60]. Intelligent sensors are applied to raw materials and finished products in warehouses to provide real-time updates that can help optimize inventory control and facilitate internal traceability. Horizontal and vertical systems are effectively interconnected and integrated to enable fast data exchange, communication, and distribution of valuable information among the management teams, factory operators, and material suppliers. Information is digitally stored, processed, and distributed whenever needed within a comprehensive computer network system consisting of computers, servers, enterprise resource systems, inventory management programs, business-related applications, etc. Advanced ICT, 5G technology, edge computing, and systems can boost communication and provide interconnectivity without latency. Cyber-physical security will play a critical role in protecting, detecting, and responding to cyberattacks when important data and information flowing within the network are compromised.

Digital factories will be eco-effectively designed to embrace cradle-to-cradle design and net-zero manufacturing strategies. In this context, as shown in Fig. 1.11, a holistic perspective is to use available energy (exergy from the second law of thermodynamics) in Joules as the key measure on the supply side [4].

Given the pool of available energy, the objective will be to provide services such as transportation and manufacturing based on their discrete needs on the demand side. As the operational demands of digital factories are flexible, they can be "shaped" based on the available energy on the supply side throughout the day. A job or work request can be executed when the sun rises, where solar power is available, or when night falls with other energy sources. Hence, we can schedule these jobs according to the service-level agreement (SLA) with the customers. An SLA that needs to be executed right away may choose to use available energy drawn from natural gas, while jobs under a more flexible SLA can kick in only when power from the sun comes online. In addition, an integrated supply–demand management system, described in simple terms previously, can ensure the most effective utilization of energy available from a range of sources, including waste streams [4].

Although humans will no longer be required to perform manual labor as autonomous machines and intelligent programs can take over the monotonous and repetitive tasks, their association with the digital factory of the future remains crucial. Humans possess unique experiences, insights, and imaginations that enable them the ability to provide valuable creative inputs at various stages and nodes in the production process [61]. In addition, collaboraton is critical to positive innovation and is reinforced through interacting, enhancing, and augmenting the power of technology. Personnel ranging from top management to business executives to shop floor operators will adopt different collaborative problem-solving approaches and methods such as creative design thinking to discover, execute, and validate design strategies and innovative business solutions.

52 Digital Manufacturing

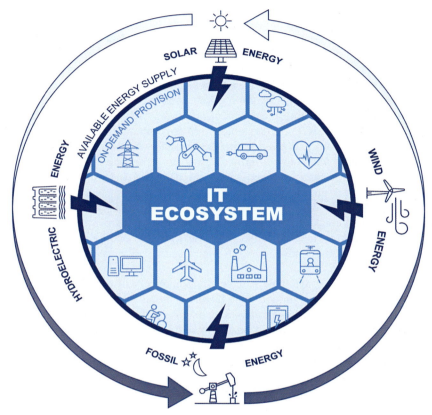

FIGURE 1.11 A holistic perspective on the constant production and on-demand provision of available energy within an information technology or IT ecosystem [4].

Indeed, state-of-art technologies tomorrow can present boundless possibilities to motivate companies and empower individuals. Nevertheless, we must be willing to embrace them with an open heart and a curious mind to sow the seeds of knowledge and opportunities today to reap the fruits of innovation and success tomorrow.

1.7 Summary

Advances in technology have dramatically changed the way humans create objects and interact with one another over the last 300 years. From the agrarian society that used primitive agricultural techniques and manual handicraft production in cottage industries, society transformed drastically after steam engines were invented. Heavy machinery flooded the factory shop floor as the manufacturing industries boomed. Coal became the new currency but

A historical perspective on industrial production and outlook **Chapter | 1** **53**

sickened the environment and the people. As the world progressed and shifted its attention toward science and innovation, society experienced sudden enlightenment when electricity, automation, robots, and computers became readily accessible. They brought new economic opportunities to the manufacturing industry and changed the way people approached work and play. The pace of life quickened as more people adopted a sophisticated and individualistic lifestyle with the freedom to pick and choose things they fancy. The advent of modern computers and the World Wide Web transformed how knowledge and information were created, stored, and shared.

Factories evolved over the centuries from a simple operation running on manual labor and steam engines to state-of-the-art interconnected manufacturing systems with minimal human operators today. In the future, artificial intelligence, 3D printers, and autonomous machines will take over the shop floor from handling operations to managing supply chains and monitoring cyber-physical security. The factory of the future is a self-organizing entity of networked manufacturing equipment with the flexibility and capability to respond to on-demand production in the unpredictable and volatile market economy.

Despite such great innovations and promises of technology, we cannot take them for granted now and expect them to provide future needs for our society. Moreover, we cannot simply extend our current infrastructures and conduct business as usual with a reactive stance toward resource and environmental concerns. By learning valuable lessons from the past and leveraging the benefits of technological advances, we can create sustainable cyber-physical solutions for the future. The role of technology will be to reinvent the consumption—production model and manufacturing processes to mitigate the impact on the environment and slow down the consumption of the finite pool of global natural resources.

Is technology innovation a boon or a bane to humanity? It will undoubtedly be a boon when we engage it positively with a curious mind. We need to be conscious, recognize the inevitable changes around us and learn to embrace state-of-the-art technologies. They will not scourge humanity but rather enable us the means to collaborate, adapt, and weather competitively through the unpredictable and fast-evolving economy of tomorrow. Indeed, only the active participation of creative individuals eager to plan for tomorrow, advocate for sustainability, and make creative decisions in precarious situations is imperative to the future success of a sustainable, digital manufacturing economy.

References

[1] A.H. Maslow, A theory of human motivation, Psychological Review 50 (4) (1943) 370–396.

[2] United Nations Food and Agriculture Organization, Global agriculture towards 2050. <http://www.fao.org/fileadmin/templates/wsfs/docs/Issues_papers/HLEF2050_Global_Agriculture.pdf>, 2009 (cited December 2019).

54 Digital Manufacturing

[3] P.K. Wright, Twenty-First Century Manufacturing, Prentice-Hall, Upper Saddle River, NJ, 2001.

[4] C.D. Patel, Sustainable ecosystems: enabled by supply and demand management, Distributed Computing and Networking, Springer Berlin Heidelberg, Berlin, Heidelberg, 2011.

[5] L. Patriquin, The agrarian origins of the industrial revolution in England, Review of Radical Political Economics 36 (2) (2004) 196−216.

[6] F.A. Sharman, An introduction to the enclosure acts, Journal of Legal History 10 (1) (1989) 45−70.

[7] R.C. Allen, The Industrial Revolution: A Very Short Introduction, Oxford University Press, Oxford, 2017.

[8] A. Smith, An Inquiry into the Nature and Causes of the Wealth of Nations, W. Strahan and T. Cadell, London, 1776.

[9] K. Marx, Das kapital, in: N. Capaldi, G. Lloyd (Eds.), The Two Narratives of Political Economy, John Wiley & Sons, Inc, Hoboken, NJ, USA, 2011, pp. 409−446.

[10] M.P. Groover, Automation, Production Systems, and Computer-Integrated Manufacturing, 3 (ed.), Prentice-Hall, Upper Saddle River, NJ, 2008.

[11] F.W. Taylor, The Principles of Scientific Management. Norton Library, W.W. Norton, New York, 1967.

[12] C. Clarke, The history of production systems in the automotive industry, Automotive Production Systems and Standardisation: From Ford to the Case of Mercedes-Benz, Physica-Verlag HD, Heidelberg, 2005, pp. 71−125.

[13] D.A. Hounshell, From the American System to Mass Production, 1800−1932: The Development of Manufacturing Technology in the United States. Studies in Industry and Aociety, Johns Hopkins University Press, Baltimore, 1985.

[14] D. Mourtzis, M. Doukas, The evolution of manufacturing systems: from craftmanship to the era of customisation, in: V. Modrák, P. Semančo (Eds.), Handbook of Research on Design and Management of Lean Production Systems, IGI Global, Hershey, PA, 2014, pp. 01−29.

[15] G. Halev, Expectations and Disappointments of Industrial Innovations, Springer, Cham, Switzerland, 2017.

[16] S.A. Marglin, Lessons of the golden age: an overview, in: S.A. Marglin, J. Schor (Eds.), The Golden Age of Capitalism: Reinterpreting the Postwar Experience, Clarendon Press, Oxford, 1990.

[17] Y.P. Tang, Advanced Manufacturing Technology, third ed., Science Press; Alpha Science International Ltd., Beijing, 2015.

[18] J.H. Gilmore, B.J. Pine-II, The four faces of mass customization, Harvard Business Review 75 (1997) 91.

[19] M. Hamalainen, J. Karjalainen, Social manufacturing: when the maker movement meets interfirm production networks, Business Horizons 60 (6) (2017) 795−805.

[20] A.P.M. Velenturf, et al., *A New Perspective on a Global Circular Economy*, In *Resource Recovery From Wastes: Towards A Circular Economy*, The Royal Society of Chemistry, 2020, pp. 1−22.

[21] T.C. Chang, R.A. Wysk, H.P. Wang, Computer-Aided Manufacturing, 3 (ed.), Pearson Prentice Hall, Upper Saddle River, NJ, 2006.

[22] M.P. Groover, Groover's Principles of Modern Manufacturing: Materials, Processes, and Systems, John Wiley & Sons, Inc, Hoboken, New Jersey, 2017.

[23] S.K. Vajpayee, Principles of Computer-Integrated Manufacturing, Prentice-Hall, Inc, Englewood Cliffs, NJ, 1995.

[24] F.M.L. Amirouche, Historical perspective on digital computers, Principles of Computer-Aided Design and Manufacturing, Pearson Prentice-Hall, Upper Saddle River, NJ, 2004.

[25] S.P. Mitrofanov, Scientific Principles of Group Technology, National Lending Library for Science and Technology, Boston, 1966.

[26] C.D. Patel, The case of art to part: the rise of systemic intent based design with 3D digital manufacturing, 2019. Available from: <https://www.engineering.com/story/the-case-of-art-to-part-the-rise-of-systemic-intent-based-design-with-3d-digital-manufacturing> (cited November 2020).

[27] S. Krish, A practical generative design method, Computer-Aided Design 43 (1) (2011) 88−100.

[28] D.D. Bedworth, M.R. Henderson, P. Wolfe, Computer-Integrated Design and Manufacturing, McGraw-Hill, New York, 1991.

[29] J. Harrington, Computer Integrated Manufacturing, R. E. Krieger Pub. Co., Huntington, N.Y, 1979.

[30] R. Palaniswamy, T. Frank, Enhancing manufacturing performance with ERP systems, Information Systems Management 17 (3) (2000) 43−55.

[31] K.T. Ulrich, S.D. Eppinger, Product Design and Development, 5 (ed.), McGraw-Hill, New York, 2012.

[32] T.A. Salomone, Concurrent engineering background, What Every Engineer Should Know About Concurrent Engineering, Routledge, New York, 1995, pp. 1−31.

[33] A.K. Kamrani, Collaborative design approach in product design and development, in: A. K. Kamrani, E.S.A. Nasr (Eds.), Collaborative Engineering: Theory and Practice, Springer US, Boston, MA, 2008, pp. 1−17.

[34] S.S.A. Willaert, Rd Graaf, S. Minderhoud, Collaborative engineering: a case study of concurrent engineering in a wider context, Journal of Engineering and Technology Management (JET-M) 15 (1) (1998) 87.

[35] T. Sandler, Collective Action: Theory and Applications, University of Michigan Press, Ann Arbor, 1992.

[36] S.C.Y. Lu, et al., A scientific foundation of collaborative engineering, CIRP Annals— Manufacturing Technology 56 (2) (2007) 605−634.

[37] M. Borsato, M. Peruzzini, Collaborative engineering, in: J. Stjepandic, N. Wognum, W. Verhagen (Eds.), Concurrent Engineering in the 21st Century Foundations, Developments and Challenges, 2015, pp. 165−196.

[38] A. du Plessis, et al., Beautiful and functional: a review of biomimetic design in additive manufacturing, Additive Manufacturing 27 (2019) 408−427.

[39] G. Byrne, et al., Biologicalisation: biological transformation in manufacturing, CIRP Journal of Manufacturing Science and Technology 21 (2018) 1−32.

[40] J. Chang, et al., Advanced material strategies for next-generation additive manufacturing, Materials 11 (1) (2018).

[41] S. Ford, M. Despeisse, Additive manufacturing and sustainability: an exploratory study of the advantages and challenges, Journal of Cleaner Production 137 (2016) 1573−1587.

[42] I.J. Petrick, T.W. Simpson, 3D printing disrupts manufacturing: how economies of one create new rules of competition, Research-Technology Management 56 (6) (2013) 12−16.

[43] M. Kumke, H. Watschke, T. Vietor, A new methodological framework for design for additive manufacturing, Virtual and Physical Prototyping 11 (1) (2016) 3−19.

56 Digital Manufacturing

[44] K.R. Haapala, et al., A review of engineering research in sustainable manufacturing, Journal of Manufacturing Science and Engineering 135 (4) (2013).

[45] G.P. Nassos, N. Avlonas, Practical Sustainability Strategies: How to gain a Competitive Advantage, John Wiley & Sons, Inc, Hoboken, New Jersey, 2020.

[46] M. Braungart, W. McDonough, A. Bollinger, Cradle-to-cradle design: creating healthy emissions − a strategy for eco-effective product and system design, Journal of Cleaner Production 15 (13) (2007) 1337−1348.

[47] EPEA GmbH, Cradle to cradle. <https://epea.com/en/about-us/cradle-to-cradle>, 2020, (cited June 2021).

[48] E. MacArthur, Towards the circular economy, Journal of Industrial Ecology 2 (2013) 23−44.

[49] P.T. Kidd, A 21st century paradigm, Agile Manufacturing: Forging New Frontiers, Addison-Wesley, Wokingham, England, 1994.

[50] K. Schwab, The Fourth Industrial Revolution, World Economic Forum, Geneva, Switzerland, 2016.

[51] E. Brynjolfsson, A. McAfee, in: A. McAfee (Ed.), The Second Machine Age: Work, Progress, and Prosperity in a Time of Brilliant Technologies, W. W. Norton & Company, New York, 2014.

[52] L.M. Fonseca, Industry 4.0 and the digital society: concepts, dimensions and envisioned benefits, Proceedings of the International Conference on Business Excellence 12 (1) (2018) 386−397.

[53] Digital Transformation Monitor, Germany: industrie 4.0. 2017.

[54] J. Zhou, et al., Toward new-generation intelligent manufacturing, Engineering 4 (1) (2018) 11−20.

[55] D. Ibarra, J. Ganzarain, J.I. Igartua, Business model innovation through industry 4.0: a review, Procedia Manufacturing 22 (2018) 4−10.

[56] J. Lee, Smart factory systems, Informatik-Spektrum 38 (3) (2015) 230−235.

[57] International Electrotechnical Commission, Factory of the future. 2015.

[58] S. Mittal, et al., Smart manufacturing: characteristics, technologies and enabling factors, Proceedings of the Institution of Mechanical Engineers, Part B: Journal of Engineering Manufacture 233 (5) (2017) 1342−1361.

[59] C.H. Chen, L.P. Khoo, N.F. Chen, Consumer goods, in: J. Stjepandić, N. Wognum, W.J. C. Verhagen (Eds.), Concurrent engineering in the 21st century: foundations, developments and challenges, Springer International Publishing, Cham, 2015, pp. 701−733.

[60] V. Alcácer, V. Cruz-Machado, Scanning the industry 4.0: a literature review on technologies for manufacturing systems, Engineering Science and Technology, an International Journal 22 (3) (2019) 899−919.

[61] A. Gilchrist, Smart Factories, in Industry 4.0: The Industrial Internet of Things, Apress, Berkeley, CA, 2016.

Chapter 2

Digital product design and engineering analysis techniques

Tianyu Zhou[1], Weidan Xiong[2], Yuki Obata[3], Carlos Lange[1] and Yongsheng Ma[4]

[1]Department of Mechanical Engineering, University of Alberta, Edmonton, AB, Canada, [2]Guangdong Laboratory of Artificial Intelligence and Digital Economy (SZ), Shenzhen University, Shenzhen, China, [3]Division of Mechanical and Aerospace Engineering, Faculty of Engineering, Hokkaido University, Sapporo, Hokkaido, Japan, [4]Department of Mechanical and Energy Engineering, Southern University of Science and Technology, Shenzhen, China

Abbreviations

0D	0-dimensional
2D	2-dimension
3D	3-dimensional
AD	axiomatic design
AI	artificial intelligence
AM	additive manufacturing
API	application programming interface
B-Reps	boundary representations
CAD	computer-aided design
CAE	computer-aided engineering
CAX	computer-aided technologies
CFD	computational fluid dynamics
DSM	design structure matrix
FEA	finite element analysis
FEM	finite element method
FVM	finite volume method
GIS	geographic information system
IGES	initial graphics exchange specification
LiDAR	light detection and ranging
MBD	multibody dynamics
NURBS	nonuniform rational B-spline
PLM	product lifecycle management
QFD	quality function deployment
STEP	standard for the exchange of product data
UI	user interface

Digital Manufacturing. DOI: https://doi.org/10.1016/B978-0-323-95062-6.00003-6
© 2022 Elsevier Inc. All rights reserved.

2.1 Introduction

With the application of modern computer-aided design (CAD) and computer-aided engineering (CAE) tools and advanced design methodologies, digital product design, simulation, and optimization largely reduce product development time, making it possible for new products to come into the market within just a few months or days. Rapid digital form creation, parametric modeling, and digital engineering simulation make it possible for the product digital model to be rapidly built, modified, verified, and optimized.

This chapter aims at introducing the digital product design process and modern CAE simulation techniques aided by artificial intelligence (AI). It is divided into seven sections. Section 2.2 gives a general introduction of the product design process, where different design stages and their focus are introduced. Section 2.3 presents the creation of 3D digital forms in a CAD environment, which is fundamental for rapidly build product digital models and realizing generative product design. Section 2.4 provides readers with intent-based systemic design approaches, which is a combination of the functional feature design approach, feature-based CAD modeling techniques, and decision-making systems for designers. In Section 2.5, computational fluid dynamics (CFD) simulation will be presented as a typical CAE tool for product performance validation. A detailed case study of a car body design, aerodynamic simulation, and redesign optimization will be shown as a detailed demonstration. Section 2.6 will introduce current challenges and future work for modern CAD and CAE technology. The last Section 2.7, provides a summary of this chapter.

2.2 Product design process

Design, modeling, performance validation, and redesign are product development processes that form an iterative design cycle (see Fig. 2.1). In the face of a brutal product competitive situation nowadays, reducing development time for a new product (such as a car and an airplane) is of great importance for its marketing success. AI has been gradually applied to product design areas, such as intelligent decision-making systems for design solution generation in recent years. With feature-based modern product design and

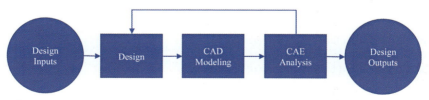

FIGURE 2.1 Iterative design cycle.

modeling methods in CAD, modernized digital product design is of great importance for realizing fast and reusable parametric product design solutions. Advanced engineering analysis approaches, such as finite element analysis (FEA) and CFD in CAE are powerful tools for product performance validation, design verification, and optimization.

There are several stages in the product design process, and each of them has its focus and level of design information abstraction and representation. The product design process can be described in the following steps:

Initial design intents or customer requirements are analyzed and decomposed into design parameters, geometric constraints, assembly relationships, and other design rules. In this stage, the design is mainly conceptual, and it involves an understanding of people's needs and how to meet them with products.

Based on the functional requirements, AI-based design algorithms and preferred design strategies, an initial design scheme for every design module will be automatically generated with the help of modern decision-making design systems. In this stage, design solutions for functional modules and their relationships have been generated.

With the interaction between the intelligent design systems and the product designers, the digital product design CAD model will be created to realize rapid initial design scheme generation. In this stage, detailed design models for parts and assemblies are created.

After getting the digital CAD model of the initial design scheme, product function and performance analysis and validation will be implemented through engineering analysis. Advanced CAE systems play an important role in simulation verification. Commonly used CAE approaches include FEA, CFD, and multibody dynamics (MBD), etc. In this chapter, CFD will be introduced in detail in Section 2.5 as a powerful tool for aerodynamics simulation. Design optimization or design pattern selection will then be performed based on the engineering analysis results until all requirements are satisfied, which will gradually generate the final digital product design outputs.

The generative design (GD) and engineering analysis form the entire digital product design process (see Fig. 2.2), where the design phase and engineering phase go on concurrently and collaboratively.

2.3   3D digital form creation

CAD is a term to describe the use of computers to aid the creation, modification, analysis, or optimization in product design cycles. Various CAD software are developed to increase the productivity of the designers or engineers, to improve communications through documentation, to improve the quality of design and create a database for manufacturing. Designers are able to use CAD systems to view digital objects in various representations,

60 Digital Manufacturing

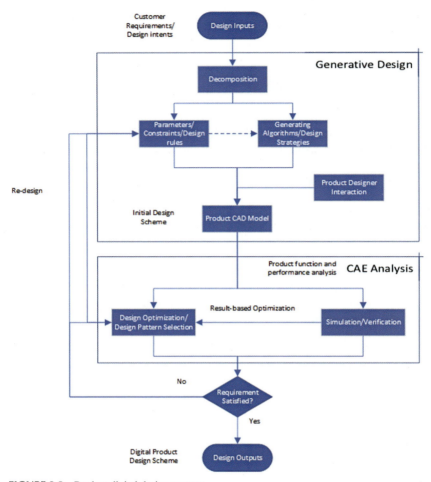

FIGURE 2.2 Product digital design process.

create complex objects with modeling approaches, modify objects with model processing techniques, test these objects by simulating real-world conditions, etc.

In the very early-stage of CAD technologies, the designers would define an 3D object by drawings of its front, top, and side views. In this stage, CAD is referred as computer-aided drawing or drafting, assisted designers in creating digital 2D drawings of an 3D object from the three orthogonal views.

In the 1960s, wireframe modeling was developed to represent an object by all its defining edges, curves and vertices. Fig 2.16 (E) shows a typical wireframe model. However, this approach could still bring ambiguity to complex objects. The first breakthrough in CAD was surface modeling and solid modeling, which transformed the designer's understanding from 2D drawing into 3D representation [1].

This section will introduce fundamental mainstream techniques in 3D geometric modeling. It will begin with an introduction to common 3D object representations in Section 2.3.1, while approaches that are widely used in 3D digital form construction and manipulation are touched on next. In Section 2.3.3, a case study in creating the symbolic architectural design will be explored and presented.

2.3.1 3D digital forms

Modeling is the process of describing a shape to the computer. A digital object in the virtual world is represented by geometric elements, such as points, edges, surfaces, and solids. The geometry of a digital object could be seen as the spatial or topological relations of these geometric elements [2].

The types of primitives and feasible operations vary from one representation to another. Before commencing the modeling activity, the first task to perform is to choose a proper representation to describe the target object.

2.3.1.1 Form representations

Figs. 2.3–2.7 show five types of common data representations in the computer science community.

A point cloud approximates the external surface or envelop of a 3D shape by a set of unstructured points. The essential information of each point is spatial coordination. Point clouds are commonly obtained through photogrammetry, depth sensing, LiDAR (mobile mapping, terrestrial laser scanning, and aerial LiDAR), etc. This representation is widely used in scenarios involving scanning, including 3D-perception in autodriving, heritage digitalization, and object digitization. However, due to the lack of topological relationships among data points, as shown in Fig. 2.3, it is unhandy to

FIGURE 2.3 An angel sculpture in the form of point cloud. *From angel [3D model] GNU LGPL V3. https://github.com/bldeng/GuidedDenoising/tree/master/models/RealData/angel.*

62 Digital Manufacturing

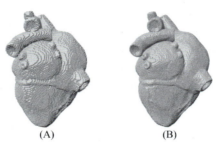

FIGURE 2.4 A human heart in the form of voxels with (A) medium and (B) high resolutions. *From Human heart [3D model] by thunk3d.scanner, 2021, Sketchfab, CC by 4.0. https://sketchfab.com/3d-models/human-heart-94505ec969cf468684d4393a3de99478.*

FIGURE 2.5 A fan disk in the form of triangular mesh. *From fan disk [3D model] GNU LGPL V3. https://github.com/bldeng/GuidedDenoising/tree/master/models/Fandisk0.3.*

FIGURE 2.6 An easy stool design in the form of surface. *Courtesy of C.H. Frankie Tee (2008), HP-NTU Digital Manufacturing Corporate Laboratory.*

FIGURE 2.7 Section view of a rotational joint in the form of solid. *Courtesy Y.B. Tian (2021), Beijing University of Aeronautics and Astronautics.*

manipulate a point cloud. Engineers often convert point clouds into other representations, which provide topological information for further editing and manipulation.

A voxel-based model is a volume model, which approximates a 3D solid by a set of voxels. A voxel represents a single sample, or data point, which includes spatial coordination, color information, etc., on a regular volumetric grid in three-dimensional space. Voxel model is frequently used in the visualization and analysis of medical data (e.g., volumetric imaging) [3], scientific data [e.g., geographic information system (GIS)], and physical simulation. Voxel model is often used in modeling tasks whose input is volumetric data. However, when describing a model in high resolution, such as the right model (B) in Fig. 2.4, they are typically large data sets and require a large amount of memory.

To model a shape, instead of describing the volume, simply describing the shape boundary can reduce the need for storage space while preserving the resolution and improving the overall smoothness. For example, a voxel-based model can be converted to a smoother mesh surface via marching cubes [4].

A mesh is a discretization of a geometric domain with smaller discrete elements. A mesh surface is a discretization of a smooth surface with a collection of connected faces. These faces are often convex polygons, such as triangles and quadrilaterals. Each face relates to its neighboring faces by common edges and vertices, as shown in Fig. 2.5. The mesh representation simplifies the rendering process of visualizing the 3D model onto the computer screen. Thus, it is widely used in the entertainment industry, such as in game engines. A volumetric 3D mesh grid is a discretization of a shape volume with a collection of connected cells. These basic 3D cells are tetrahedron, quadrilateral pyramid, triangular prism, and hexahedron. The mesh grid is widely used in applications involving physical simulations, such as the finite element method (FEM).

A surface model uses curvy boundary edges and surfaces to approximate the exterior shape of an object, as shown in Fig. 2.6. A curvy edge can be represented in two forms: (1) parametric forms, that the coordinates of points locating on the curve are related by a parametric variable; (2) or nonparametric

64 Digital Manufacturing

forms, that the coordinates of points locating on the curve are related by a function explicitly or implicitly. Classic types of curves representations include Bezier Curve, B-Spline Curve, Nonuniform Rational B-Spline (NURBS) Curve, T-Spline Curve, etc. In computer-aided geometric modeling, curves are commonly used to define 3D surfaces. Depending on surface features, 3D surfaces can be classified into swept, ruled-generated, or free-formed surfaces [5]. The surface representation is widely used in creating industrial products because of its mathematical precision.

A solid model is a representation of the geometric shape of a physical object. A solid can be created with two methods: (1) boundary representations (B-Reps), where designers may specify geometric elements to define the mathematical representation of boundary surfaces and knit these surfaces together as a finite volume; (2) designers may select a set of primitive shapes (e.g., blocks and cylinders) and combine them using logic operations (e.g., union, intersection, and difference). This process is called constructive solid geometry. The resultant solid model is an unambiguous, complete, smooth, and highly detailed digital approximation of an object or an assembly of objects (such as an engine or an airplane) [6]. Fig. 2.7 shows a section view of a rotational joint that consists of detailed parts.

2.3.1.2 Properties

Different representations have their strength and weakness. Thus they are suitable for different applications. Table 2.1 summarizes several properties of these representations: the availability of spatial data, the availability of topological structure, the definition of volume, the level of smoothness, the level of memory consumption for a model with a satisfying resolution, and several common scenarios. Designers and engineers can choose the target representation based on the requirement of the target representation.

2.3.2 Form modeling

Besides considering the primitives of representation, selecting suitable modeling operations is crucial as well. Software with direct modeling methods allow designers to manually create highly detailed models from scratch and modify existing models through interactive digital sculpting methods, such as SketchUp and ZBrush. Creating a model purely by manual sculpting is labor-intensive work. This section will introduce four useful approaches in chronological order. These approaches are widely used in 3D digital form construction and manipulation, which largely increase the level of automation.

2.3.2.1 Modeling approaches

- Parametric modeling provides an alternative to generate an accurate model via parameter editing. For example, designers can create a cylinder

TABLE 2.1 Properties of model representations.

Representation	Spatial data	Topological structure	Volume	Smoothness	Memory requirement	Common scenarios
Point cloud	Y	N	N	Low	High	3D scanning
Voxel	Y	Y	Y	Low	High	Medical volumetric imagingPhysical simulation
Surface	Y	Y	N	High	Low	Manufacturing industryIndustrial design
Mesh surface	Y	Y	N	Medium	Medium	Entertainment industryAnimation industry
Mesh volume	Y	Y	Y	Medium	High	Manufacturing industryPhysical simulation
Solid	Y	Y	Y	Medium/high	Low/medium	Manufacturing industryIndustrial design

66 Digital Manufacturing

by editing several parameters (top diameter, bottom diameter, and height) instead of manually adding primitives from scratch or drag and pull from a base shape.

- Procedural modeling creates model/scene algorithmically. A set of rules or grammars are defined to allow designers to create complex model/scene with a few parameters [7]. For example, the rule can use feature attribute information stored in GIS data, such as the number of floors, roof type, and wall material type, to generate a series of alternate 3D models that accurately represent the properties of each feature. Procedural modeling techniques are also widely used in describing the natural 3D phenomenon, such as plants, water, and terrains [8,9].
- Parts digitalization is a reverse engineering process that starts from a concrete object in the physical world and obtains a digital 3D form that describes its shape in a virtual world. Engineers first collect the shape information with 3D scanning tools in contact, transmissive, or reflective ways. The collected raw data, often a point cloud, can then be used to construct a digital 3D model in more viable representations through postprocessing.
- Generative Design (GD) is a process that uses learning algorithms to explore the variants of a design beyond what is currently possible using the traditional design process. Designers or engineers begin by inputting design goals, along with parameters or constraints, to the algorithm. The algorithm then explores a wide range of possible permutations of an initial solution, quickly generating design alternatives to find the possibly best solution. The GD mimics the natural evolutionary approach and adds randomness to the design process. The created designs could exist neither in the database nor in the designers' minds before. Current techniques often employ unsupervised evolutional-algorithms or generative adversarial networks.

Companies investing in the technology of GD aim to reduce component mass, improve the performance of their designs, minimize manufacturing process time, and create new products suited to the next generation of customers. They are concerned more than ever with customization and uniqueness (Table 2.2).

2.3.3 Case study

This section introduces a modeling process of an architectural concept design, which is an Egypt theme museum. The design is created via an interactive design system with AI algorithms. This system assists designers in generating concept designs of buildings, whose silhouettes mimic user inputs profiles from certain angles [10]. Fig. 2.8 shows the overview of this system.

Digital product design and engineering analysis techniques Chapter | 2   67

TABLE 2.2 Properties of modeling approaches and common software.

Approaches	Input	Goal	Notable software
Parametric Modeling	• Parameters	Create model with several defining parameters	• SOLIDWORKS • CATIA • AutoCAD • Blender • Rhino
Procedural Modeling	• Parameters	Create model/scene algorithmically with a set of rules or grammar	• 3ds Max • Houdini • OpenSCAD • Grasshopper 3D • Blender
Part digitalization	• Physical object	Replicate the geometric shape of physical object to virtual space	• 3DF Zephyr • Canoma • SketchUp
Generative design	• Goals • Parameters • Constrains	Create model (models) automatically with AI algorithms	• FUSION 360

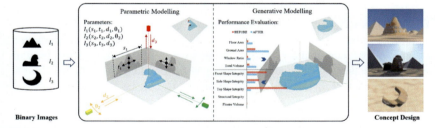

FIGURE 2.8   Overview of the shape-inspired architectural concept design system.

An architectural design, especially a symbolic one, should satisfy several fundamental requirements: firmness, commodity, and delight [11]. Most important of all, a feasible architectural design should be structurally sound under the force of gravity. The structural performance of a building model can be measured with the FEM, which subdivides a large system into small and simple elements then numerically construct and solve differential equations [12]. Fig. 2.9 shows two designs with feasible stability (A) and problematic over-hanging volume (B).

Several essential requirements were considered when choosing a proper representation for this design system:

68 Digital Manufacturing

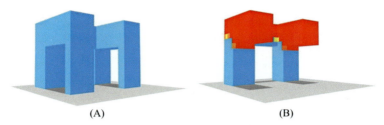

FIGURE 2.9 Two architecural design with (A) feasible stability, and (B) overhanging volume. The strain of each voxel is denoted from *yellow* (small) to *red* (large).

The model should contain volumetric information.

The model should have a suitable structure for physical simulation.

The elements of the model should be structured such that the algorithm can evaluate the quality of the design conveniently.

The resolution of the design is acceptable such that the design could be converted into a smoother version later.

The process of creating a 3D model from 2D silhouette images should be fast such that the designer can use the system as an interactive design tool.

After considering all the primary factors, the system employs the voxel-based representation with acceptable resolutions over the design processes.

2.3.3.1 Initialization and parametric modeling

The basic input to the system is three binary images, I_1, I_2, and I_3. Each silhouette template is associated with a viewpoint or camera. Two cameras, associating with template I_1 and I_2 respectively, are located on the ground level. While the top camera, associating with template I_3, is located overhead. The camera i, associate with template I_i, can be defined by several parameters:

- Image scale, s_i.
- Image translation, t_i.
- Camera rotation, θ_i.
- Camera distance, d_i.

Given the three silhouette templates, the system first creates an initial design with default parameter values, as shown in Fig. 2.8. Then, the designer can interactively manipulate any of the mentioned parameters to modify the voxel-based model.

2.3.3.2 Design evaluation

The system employs several structural and aesthetical criterions to measure the quality of the created designs. These include floor area, ground area, window ratio, total volume, shape integrities, structural integrity, and floater volume. These criterions have varying importance for different types of building. A set of numbers, called weights, are given by the designers to represent the relative importance of all the criterions. An extra function combining the weights and criterions is employed to evaluate the overall performance of the created design.

2.3.3.3 Generative modeling and result

Creating a design by manual modeling, such that the design has satisfying performance overall the mentioned requirements, is challenging and labor-intensive. With the evolutional algorithms, this system generates and evaluates thousands of designs and returns a best-found design without the interference of the designer. He or she can then polish the design according to the requirements. After finalizing the design, the designer can convert the voxel-based design into other representations for further usages.

Fig. 2.10 shows the resultant design in the form of voxels (A) and smoother mesh surface (B). The mesh surface is used to generate the rendering views in Fig. 2.8.

2.4 Intent-based systemic design

The intent-based systemic design aims at modeling large and complex systems in a collaborative environment with relatively high efficiency, where robust, consistent, and efficient information flow throughout the entire systemic design is important. How to decompose and map design inputs, such as customer requirements and/or design intents, accurately with product functional modules? How to build a reusable and editable CAD model which can clearly represent the geometric and nongeometric relationship between parts/faces/interfaces? How to make the final design solution for each functional unit and mechanism? The answers to these questions are in the following sections.

(A)   (B)

FIGURE 2.10 Two representations of the building design, Egypt theme museum: (A) voxel model and (B) mesh model.

2.4.1 Functional feature approach

As mentioned in Section 2.1, design inputs should be analyzed and decomposed into detailed design elements, such as design parameters, geometric constraints, assembly associations, and other design rules, such as physical laws, logical relationships, and parametric dimensional restrictions. Traditionally, the design intents cannot be automatically fused into the CAD modeling process because different levels of semantic representation, information abstraction, and feature granularity cannot be associated in a unified modeling method, so it is important to decompose and map design inputs with product functional modules. Functional feature is a feature type that generically integrates the functional design intent, engineering physics, and product geometric model in a consistent functional-physical modeling approach. Cheng et al. proposed a templated framework of the functional feature design process, namely functional feature modeling cube [13], to connect functional design, behavior modeling, and structure design with functional features (see Fig. 2.11).

This modeling procedure [14] can be described in the following stages.

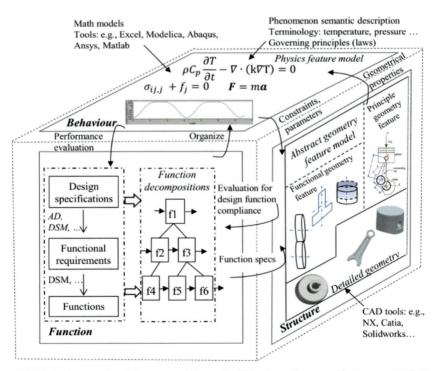

FIGURE 2.11 Functional feature modeling cube [13]. (https://www.tandfonline.com/doi/full/10.1080/09544828.2017.1291920, Taylor & Francis Ltd).

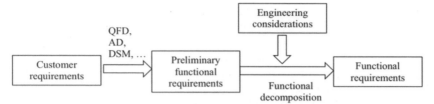

FIGURE 2.12 Information mapping from customer requirements to functional requirements [14].

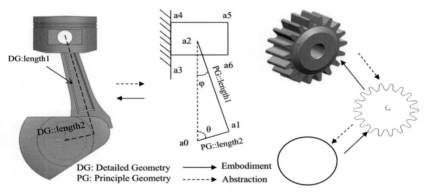

FIGURE 2.13 Design abstraction and embodiment of geometries of different levels [14].

2.4.1.1 Functional analysis of the design

A few approaches, especially in the system design level, such as axiomatic design (AD) [15], design structure matrix (DSM) [16], and quality function deployment (QFD) [17], are available and applicable to correlate and/or integrate functions with other design aspects such as customer requirements, design parameters, product structures, and design tasks. These approaches have been applied by building up the mapping relations among any two of those aspects in the form of an associated matrix. The conceptual, functional design approach helps the mental flows of the organize engineers and the existing correlation approaches (AD, DSM, QFD, etc.) are helpful to model relations between functions and other design aspects (see Fig. 2.12).

2.4.1.2 Abstract geometry features modeling

Design intents could be supported by abstract geometry features, which could be extracted or abstracted from parts. From 0D (a mass point) to 3D (detailed spatial model), abstract geometry features can represent different levels of design details (see Fig. 2.13). Design parameters within the abstract geometry feature should be well organized accordingly, such as references, constraints, and parametric relations.

Next, the relations among those abstract geometries should be identified spatially. For example, a single solid part may contain a few abstract geometry features. In this case, the spatial relations among different abstract geometry features need to be considered carefully, which could be done by adjusting their references to make them well constrained and parameterized. Parameters within abstract geometry features should be named meaningfully and nonrepetitively, and the constructed abstract geometry features could be placed into a feature library and could be reused for future design activities.

2.4.1.3 Detailed computer-aided design part modeling

In this stage, modeling activities are performed to construct the CAD model for detailed design. After identifying the abstract geometry features, it is often not easy to automatically generate a detailed CAD model with current CAD systems. Ideally, the synthesis process could be realized by combining the abstract geometry features with proper positioning and additional feature operations. Instead, designers manually make use of available modeling operations to construct the geometry to embed the abstract geometry features and their spatial relations.

In summary, the functional feature modeling procedure includes three simple steps (see Fig. 2.14).

1. Functional analysis, which correlates the functions with design inputs such as design intents and customer requirements.
2. Abstract geometry features modeling, which enables the design functionalities by organizing references, constraints, and parameterizations with the abstract geometry features.

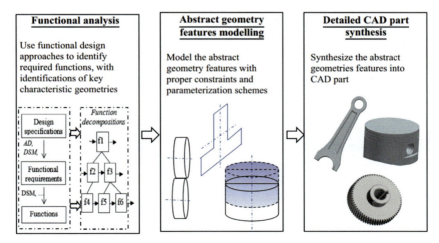

FIGURE 2.14 Schematic overview of the functional feature modeling procedure [14].

3. Detailed CAD part modeling, which is implanted by embedding the abstract geometry features and their spatial relations in a designer constructed geometry.

2.4.2 Feature-based computer-aided design modeling

There are many modeling methods in CAD and other engineering analysis systems, such as wireframe modeling, surface modeling, and solid modeling. GD is an iterative design process, which means the digital product model itself should be easy to build, control, edit, and reuse. Also, the gap among models concerning different stages of the product lifecycle management (PLM) needs sufficient information flow to ensure information consistency. In the design stage of a car, for example, a CAD model was used for detailed design and assembly. However, in the performance validation stage, applications such as CFD simulation for its aerodynamic behavior, detailed design and assembly will amass enormous extra computational cost. Small gaps between design parts may also cause unstable or nonconvergence aerodynamic simulation results. In this case, we need to defeature design details to ensure convergence and computational economic. Therefore, a CAE model needs to be built based on the original CAD model. When it comes to design optimization, a proper way of managing change propagation between the CAD model and the generated CAE model is vital for keeping the high efficiency of the iterative design-analysis-redesign cycle.

From the above introduction, we can conclude that for fulfilling the efficient iterative product design process, the model should be parametric, which is easy to be built, controlled, edited, and reused. Also, there should be some useful link between different stages of the PLM for sufficient information transfer availability, information consistency guarantee, and change propagation control. In this case, the feature-based modeling method would be a good candidate for meeting the above requirements.

Ontology speaking, feature is defined as an object class with rich and associated properties of product and process engineering from both geometric and nongeometric aspects. Such features could act as information transformation links among different stages of a product lifecycle engineering. In the authors' and other research groups' research outcomes, many derived concepts, such as associative assembly feature, complex machining feature, CAD/CAE integration feature, have been proposed and adopted in applications of CAD and other computer-aided technologies (CAX) under this approach. Feature-based modeling, also referred to as "parametric modeling" in the context of geometric modeling, is a practical and versatile method that regards constraints, dimensions, and many other quantified properties to realize rapid model generation and modification.

The feature-based modeling method has the following, but not limited to, advantages for GD.

2.4.2.1 Object-oriented

Compared with modeling with steps (process-oriented modeling), feature-based modeling is object-oriented, which has better encapsulation, reusability, and specialized operation methods design for each part or specific design unit. For example, a "hole" is a specific feature. Editing the parameters of the hole would not have much influence on other parts or the whole assembly. A "hole" can be duplicated, mirrored, arrayed, and parametrically modified for reuse. Lastly, a "hole" has its subtypes and operation methods for editing (see Figs. 2.15 and 2.16).

2.4.2.2 Easy to represent relationships among parts/faces/interfaces

There are many kinds of relationships among different object classes. Inheritance/abstraction and association are the two most important relationships among different features. Inheritance means that the child class keeps most of the attributes from the parent class while deriving its unique attributes and operation methods. As for CAD modeling, the feature-based method allows subtype parts inherited from general parts while developing its own parameterization, controlling, and editing methods. As such, it will be a good advantage for building up a specialized design library and for customization design, parameterization, and optimization. Fig. 2.15 shows an example of subtypes of precision stepped center pins and their customized parameter editing user interfaces (UIs). It demonstrates how feature-based modeling methods can be used for parametric modeling and design library generation for standard components. Apart from standard components, most

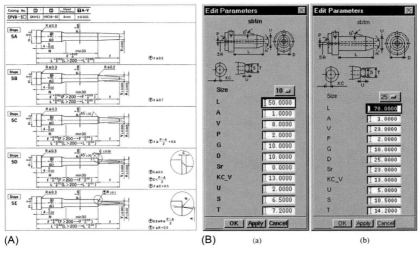

FIGURE 2.15 (A) Standard part subtypes and (B) customized parameter editing UIs [18].

Digital product design and engineering analysis techniques Chapter | 2 75

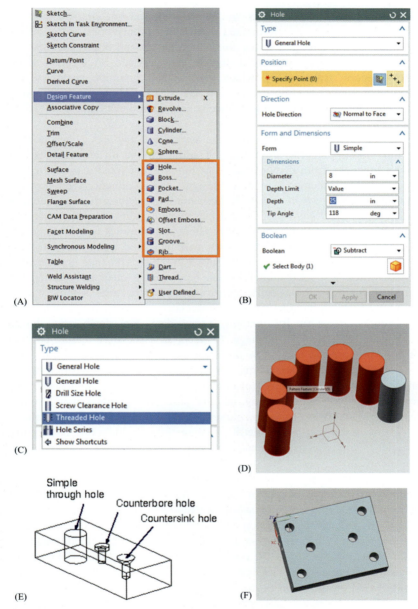

FIGURE 2.16 Some hole feature operations in Siemens NX. (A) Feature type selection, (B) hole parameters operation interface, (C) hole type selection interface, (D) hole array operation, (E) different types of hole creation, and (F) hole mirror operation [19].

76 Digital Manufacturing

customized parts can also use a similar feature-based modeling method for parametric modeling and customized part design library construction. They are important for mass product customization design and digital manufacturing. In reverse, many paralleled classes can generate an abstract class, namely generic feature [20], which extract their common attributes and operation methods. The generic feature acts as a modeling template and is capable of defining many kinds of mechanical components and application-oriented models.

Another important relationship is the association that acts as an information link among parts/faces/interfaces. Traditionally, at different stages of a product lifecycle, from documentation of requirement specifications to conceptual design, detailed structure design, and production, engineering knowledge is stripped off except the bare minimum geometrical and control data, such as CAD solid models and cutting tool paths [21]. For example, when we export a model from a CAD system and import it to a CAE system for analysis, the model in most cases will be saved as standard for the exchange of product data (STEP) or initial graphics exchange specification (IGES), which is a solid model format. However, a lot of information has been lost, such as the specific constructive feature tree, assembly constraints, and design parameterization. Associative relations among engineering features are normally ignored; hence data consistency and design changes are challenging to manage. As such, an associative feature concept has been proposed to tackle this problem. An associative feature is defined as a set of semantic relationships among product geometric entities, which can be defined as a single object entity in an engineering application. The product geometric entities may include assemblies, components, solids, faces, edges, vertices, surfaces, curves, points, vectors, datum references, etc. The relations can be geometric or nongeometric, including constraints, dependencies, equations, memberships, part-whole relations, coupling, patterns, etc. [22].

Fig. 2.17 shows an example of a cooling circuit design for an injection molding part. After injection, the injected high-temperature plastic or other infills will go through a fluxion-solidification-shrinkage process. The final part may be smaller in dimension compared with the mold cavity, and stress concentration may happen for an unbalanced cooling effect. Traditionally, it is hard to estimate precisely the cooling effect on the final part. Hence, mold design was a tedious job. The mold cavity and core need to be modified frequently based on the result of the cooling simulation and final part behavior for satisfying dimension accuracy and residual stress requirements. After modifying the mold cavity and core, the cooling channels need to be changed manually in order to cater to the modified design pattern. With the help of the associative feature, this modification could be realized automatically by user-defined parametric constraints and change propagation rules to guarantee information consistency. In summary, associative features act as an

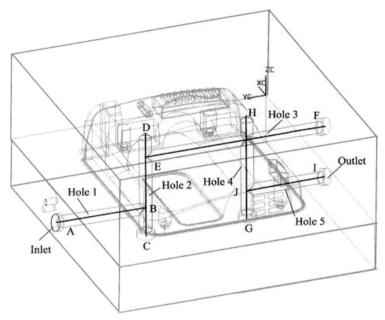

FIGURE 2.17 An example of cooling circuit [21].

important role in information flow among different parts/faces/interfaces and different stages of PLM.

2.4.2.3 A good way to manage information flow in concurrent engineering

As discussed previously, a feature is an object class with its unique data structure, attributes, and operation methods. Most features are application-driven and act as different roles in different product design and analysis stages. Information or knowledge might be stripped off during the model translation between CAD and CAX, which is a handicap for product model interoperability. Some research efforts, such as multifaceted feature [23], common data model [24], and CAD/CFD integration feature [25], have been made to manage design change propagation, enhance the information consistency robustness, and realize CAD/CAX integration. For example, a multifaceted feature design method considers parametric design from many different views and aspects, such as design, analysis, and manufacturing.

A database management system is developed, typically stored and operated by a powerful workstation as the parent model for managing model information related to different aspects. When we need the model for a specific application, an application-driven child model will be automatically created by information extraction from the parent model. As the designer,

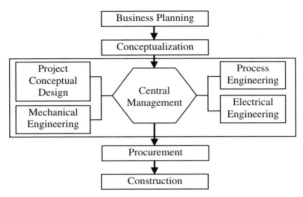

FIGURE 2.18 Collaborative engineering processes framework [20].

engineer or manufacturer modifies the child model, timely communication between the child model and the parent model will be recorded, and the changes will be propagated to the parent model. In this way, concurrent and collaborative engineering will be possible because of the efficient data management method. Fig. 2.18 shows a templated framework for the collaborative engineering processes, where the multifaceted feature behaves as a useful information management tool in the central management workstation.

2.4.2.4 Friendly to manufacturing analysis and additive manufacturing optimization

In traditional product design methods, design, analysis, and prototyping were done separately, which is time-consuming and the nonconcurrence made it difficult for consistent product optimization. Manufacturing analysis, such as toolpath planning and interference inspection, has made the design and optimization to be more considerate and critical for manufacturability friendly. Moreover, the growth of 3D digital manufacturing further changes the paradigm since the convergence of rapid prototyping and manufacturing expanded the original design space to a larger field of exploration [26].

The feature-based method for intent-based systematic modeling provides sufficient information for manufacturing analysis, such as toolpath analysis. It is because manufacturing information, such as tool selection and volume generation methods, will be defined by designers when creating the manufacturing model. The simulation outcome will be useful guidance for selecting the specific manufacturing strategies. Fig. 2.19 demonstrates the outcome of the simulation result, which can be used as toolpath optimization.

Given the abilities in additive manufacturing (AM), the designer can optimize shapes to drive intent-based systemic design even further since the design pattern may greatly influence the simulation and optimization

Toolpath Strategies

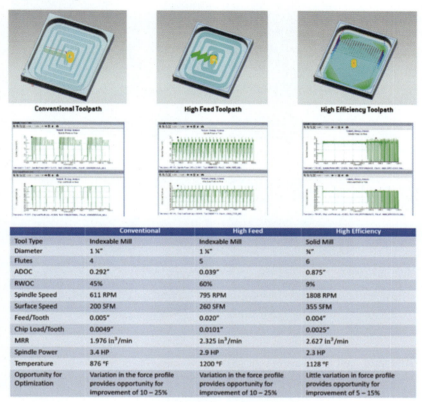

FIGURE 2.19 Comparison of three different toolpath strategies [27].

processes in AM. Fig. 2.20 shows the simulation results of a single design structure from the design stage to a prototype fully cooled down to room temperature after AM. Designers may further optimize the design pattern after getting the simulation result. In some hybrid manufacturing systems, such as conformal cooling system design and AM for injection molds [28], where AM and subtractive manufacturing may be applied at the same time, the association among added volume entities and subtracted volume entities should be treated carefully. In a multipurpose product design process, where different product performance factors, such as product weight, cost, and manufacturing cycle time, are considered simultaneously, the intent-based systematic design may be a good choice for systematically fulfilling the integration of design intent, CAD modeling, and AM simulation [26]. Chapter 5 will provide a more detailed introduction about simulation and optimization for AM.

80 Digital Manufacturing

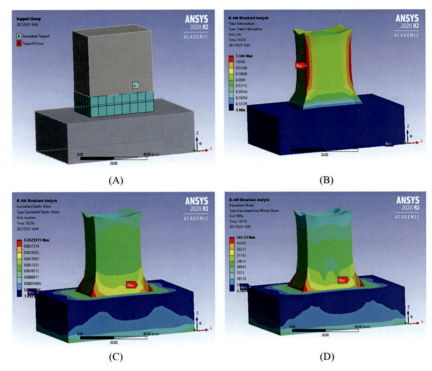

FIGURE 2.20 AM simulation of a single design structure. The original design and support structure (A), deformation after AM (B), equivalent elastic strain after AM (C), and equivalent stress after AM (D).

2.4.3 Two typical decision-making types: retrieval and inspirational

GD is an iterative design process, and it is important to make a proper decision for every iteration. For example, topology optimization by solid isotropic material with penalization method approach for structural design needs an optimization algorithm for deciding whether to keep or abandon mesh-based design variables for each design optimization iteration [29]. As for digital product design, many advanced decision-making designer expert systems have been developed based on AI algorithms. These advanced expert systems are growing fast. In recent years, many new systems based on artificial neural network came into being with great advantages, such as reinforcement learning. It will make the designer expert system itself behave more and more superior. However, there is still some uncertainty about the upcoming development trend for these modern decision-making systems. Some traditional decision-making systems have already been widely used nowadays

Digital product design and engineering analysis techniques **Chapter | 2** **81**

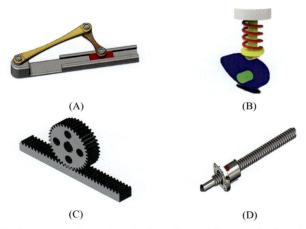

FIGURE 2.21 Four commonly used mechanisms that transform rotational motion into linear reciprocating motion. Crank-slider mechanism (A), Cam-follower mechanism (B), Pinion-rack mechanism (C), and Screw-nut mechanism (D) [30].

for designers. Two of the most common types among them are retrieval and inspirational systems.

The retrieval decision-making system behaves in a search-decide-verify mode. The rule-based retrieval system will search from the existing design libraries for potential design candidates based on the design requirements. After that, the designer will choose an option and will verify its feasibility. For example, when designing a device or mechanism that can transform rotational motion into linear reciprocating motion, four common candidate mechanisms, namely crank-slider, cam-follower, pinion-rack, and screw-nut, are automatically sorted out from the existing design library database (see Fig. 2.21). The designer will choose the best option according to his knowledge and verify its feasibility.

The inspirational decision-making system contains an AI solution module, where design knowledge is fused, and recommendation algorithms are implemented. The system will list available solutions and automatically show the user its initial recommendations. Fig. 2.22 shows a hole-type recommendation solution for a customized threaded adapter, where threaded holes are sorted from available hole types as the suggested design solution for the joint.

With the help of the decision-making systems, the designer will select a design pattern accordingly. After that, the designer will choose a model from the parametric design library. Customized dimensions, constraints, and parametric relationships will be defined and edited by the designer to realize a rapid design. The designer can decide whether or not to make the library adapt to the update if new design patterns are created, which can also be

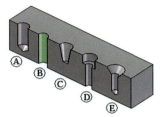

FIGURE 2.22 Recommended solution for hole type in threaded adapters [31].

reused for future similar design requirements. In this way, the system itself will have more and more available design candidates, and software developers can update relevant selection algorithms to make the system gain design "experience."

2.5 Engineering analysis

After getting the design model, functional verification or performance validation will be implemented by engineering analysis. Generally speaking, there are three methods for solving an engineering problem, analytical, experimental, and numerical methods (see Fig. 2.23).

The analytical approach is to solve the engineering problem by mathematical modeling and get the analytic solution of the unknown relationships. However, it might be limited if the problem is complex and the equations are hard to find analytical solutions. The experimental method is to solve the problem by creating a real model or a scale model and do the experiment, such as conducting a wind tunnel testing for the car body-air resistance coefficient. But it can be expensive and time-consuming. The numerical method is to solve the problem by numerical analysis. The results might be interpolated by certain points or elements, such as CAE simulation, where domain discretization and numerical solution iteration methods are commonly used. The numerical method is expanding fast in recent years due to the drastic improvement of hard devices' computing power, the advancement of discretization algorithms, and the application of new iteration and optimization methods.

Engineering simulation is a valuable tool for checking whether the design model satisfies functional requirements and generating some supportive results for design optimization in digital product design. Commonly used CAE approaches include FEA or FEM, CFD, MBD, etc.

The FEM is a widely used method for numerically solving differential equations arising in engineering and mathematical modeling [32]. Such differential equations exist as governing equations in many engineering fields, such as structural analysis and mass transport. The FEM subdivides a large system into smaller, simpler mesh-based parts named finite elements by

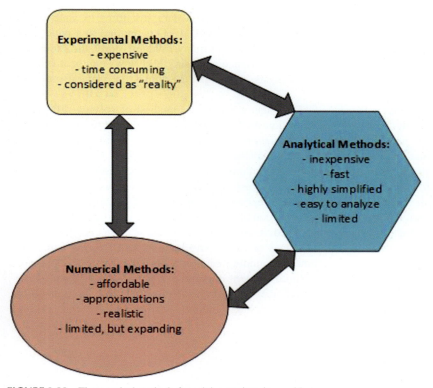

FIGURE 2.23 Three typical methods for solving engineering problems.

space discretization. The unknown function over the discretized domain is approximated by interpolating the numerical solutions of a finite number of points, which will transform the original differential equations into a system of algebraic equations. The simple equations that model these finite elements are then assembled into a more extensive system of equations that models the entire problem called the global equations. These equations will then be solved iteratively after applying initial and boundary conditions.

MBD is the study of the dynamic behavior of interconnected rigid or flexible bodies, each of which may undergo large translational and rotational displacements [33]. For example, the Robotic mechanism design and trajectory interference analysis will depend on the result of MBD simulation to a great extent.

2.5.1 Computational fluid dynamics simulation

This section will present a detailed introduction to CFD. The demonstration case in Section 2.5 is based on CFD simulation. CFD is a branch of fluid mechanics that uses numerical analysis and data structures to analyze and solve problems that involve fluid flows [34]. CFD is applied to a wide range of research and engineering problems in many fields of study and industries, including aerodynamics and aerospace analysis, fluid flows and heat transfer, engine and combustion analysis, etc. Fig. 2.24 shows the general processes of CFD analysis.

Similar to FEA, CFD is also a numerical method of solving differential problems, where discretization methods and solution algorithms are two critical parts of the whole simulation. In CFD, the most popular discretization method is the finite volume method (FVM). Like FEA, FVM also needs space discretization (meshing), numerical result interpolation, and a method for representing and evaluating partial differential equations in the form of algebraic equations. The difference is that in the FVM, volume integrals in a partial differential equation containing a divergence term are converted to surface integrals using the divergence theorem. These terms are then evaluated as fluxes at the surfaces of each finite volume. In CFD, the finite volume is the small volume surrounding each node point on a mesh. Compared with FEA, where each finite element may have several nodes for approximation, FVM is a good balance between result accuracy and computational cost. It is also conservative, making it more suitable for large and complex

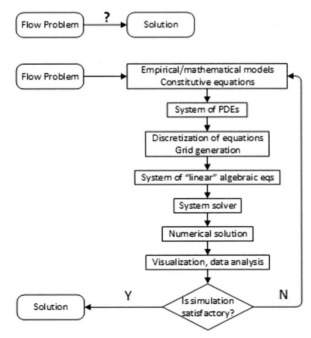

FIGURE 2.24 General processes of CFD analysis.

problems. Furthermore, advanced CFD software such as ANSYS CFX provides the users with specialized models and solvers for different types of flow. For example, there are customized advection scheme models such as the UDS model and turbulence models such as the k-epsilon model. The users can choose models suitable for their specific type of flow simulation.

As a powerful engineering analysis tool, CFD plays a vital role in evaluating certain products' aerodynamic behavior and outer profile design, such as the profile body of a car, warship, and missile. CAD/CFD integration has become a popular research topic for realizing rapid product body optimization to improve its aerodynamic behavior, and such design optimization might be multiobjective. For example, car body designers wish to decrease the car body's drag coefficient C_d without increasing its lifting coefficient C_l. For such application, rapid modification of the product model and consistent information flow between CAD and CFD environment is vital to realize the iterative design efficiently; simulation-based analysis; optimization-based redesign processes. Lei et al. proposed an AI-aided CFD analysis in the feature-based cyclic CAD/CFD interaction process, where an association between product geometric feature and CAE boundary features was built up, and a rule-based automatic CFD model selection was proposed for an expert flow simulation system [35]. Fig. 2.25 shows how this system works.

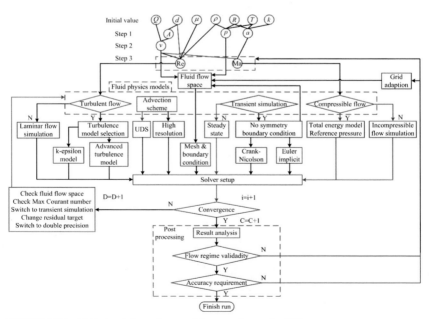

FIGURE 2.25 CFD simulation robust model generation cycles [25].

2.5.2 Case study

A case study of a car body design, analysis, and optimization processes is demonstrated in this section. It goes through a step-by-step product digital design process, as mentioned previously. First, the parametric car body model of Range Rover Velar (see Fig. 2.26 for its dimension) is built, and the CFD simulation is performed for calculating its aerodynamic drag coefficient (C_d) and lifting coefficient (C_l). Next, shape optimization is implemented by changing its design parameters, and CFD verification is done for getting a convincible optimization result. Siemens NX is used for parametric CAD modeling, while CFD simulation is carried out using ANSYS CFX.

2.5.2.1 Parametric computer-aided design model build-up

The parametric enclosed outer profile of the car is required for aerodynamic performance simulation of the car body. The model is simplified by defeaturing some design details. The process is also applied for actual aerodynamic tests for clay models (see Figs. 2.27–2.29).

2.5.2.2 Computational fluid dynamics simulation

Velocity is assumed to be 100 km/h, and the flow is turbulence with the calculated Reynolds number (Re) as 4.4×10^6. The aerodynamic drag coefficient (C_d) and lifting coefficient (C_l) are calculated numerically, which are

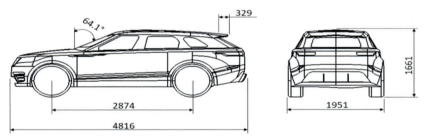

FIGURE 2.26 Dimensions of simplified Range Rover Velar model (unit: mm, degree).

FIGURE 2.27 Comparing models-perspective view. Left: original car model (front) [36] Right: simplified model (front).

Digital product design and engineering analysis techniques Chapter | 2 **87**

FIGURE 2.28 Comparing models-back view. Left: original car model (rear) [37] Right:simplified model (rear).

FIGURE 2.29 Bottom view of the simplified model.

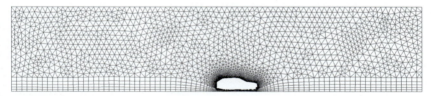

FIGURE 2.30 Initial mesh.

0.31 and −0.013, respectively. Compared with the dragging coefficient (C_d) provided officially of the actual car as 0.32, we can conclude that our model is accurate enough and the result of CFD simulation convincible. The detailed CFD simulation includes the following steps:

1. Initial mesh, where the computational domain is meshed by properly sized elements (see Fig. 2.30).
2. Simulation setup, where the computational domain, boundary conditions, and domain properties and simulation settings are setup (see Figs. 2.30 and 2.31, Tables 2.3 and 2.4).
3. Initial simulation and mesh refinement, where CFD simulation is implemented, and local mesh of some critical areas are refined (see Figs. 2.32 and 2.33).
4. Decision of final mesh and mesh independence test, where different mesh refinement locations are attempted for many cases, and the best case is

88 Digital Manufacturing

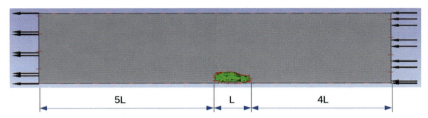

FIGURE 2.31 Computation domain.

TABLE 2.3 Boundary conditions.

Boundary name	Condition
Inlet	Uniform velocity 27.778 (m/s)
Outlet	Relative pressure 0 (Pa)
Symmetry	Symmetry
Ground	Moving wall
Wall	Free-slip
Ceiling	
Car body	No-slip

TABLE 2.4 Domain properties and simulation settings.

Fluid	Air @ 25°C
Density ρ (kg/m^3)	1.185
Dynamic viscosity ν [kg/(m s)]	$1.831 \cdot 10^{-5}$
Advection scheme	UDS (upwind differencing scheme)
Turbulence numeric	First order
Convergence criteria	RMS $1 \cdot 10^{-4}$
Advection scheme for refinement runs	CDS (central differencing scheme)
Turbulence numeric for refinement runs	High resolution
Convergence criteria for refinement runs	RMS $1 \cdot 10^{-4}$
Flow regime	Turbulence
Turbulence model	k-Omega with shear stress transport
Blockage ratio (%)	0.86

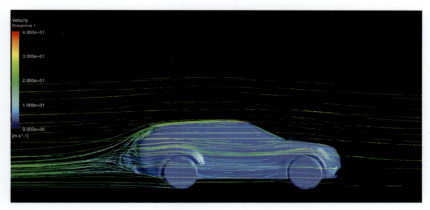

FIGURE 2.32 Streamline of initial simulation.

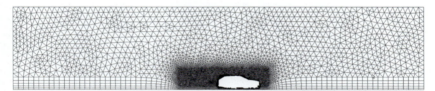

FIGURE 2.33 Local mesh refinement.

selected. After that, further changes to the mesh element size are made to assess whether the mesh size is good enough for the simulation result.

2.5.2.3 Shape optimization

Several design parameters are considered editable for the model. In this case, two design parameters are modified, windshield angle θ, and tail plate length l. For nondimensionalizing, l is divided with the wheelbase length W to generate a nondimensionalized number l/W. Simulation is repeated to obtain the results shown in Table 2.5. Result comparison when changing design parameters.

By slightly increasing the windshield angle (from 64.1 to 65.6 degrees) and tail plate length (l/W from 0.114 to 0.137), the car body will have better aerodynamic performance, where drag coefficient (C_d) and lifting coefficient (C_l) are both slightly decreased (see Figs. 2.34–2.36).

2.6 Current challenges and future work

Although CAD and CAE technologies have made significant progress in the past few decades, there are still some challenges today that hinder the realization of a fast digital product design process.

TABLE 2.5 Result comparison when changing design parameters.

	C_d	C_l	Change of C_d (%)	Change of C_l (%)
Basic (64.1;0.114)	0.3063	−0.01301	–	–
$\theta = 65.6$	0.3091	−0.01286	0.927	−1.49
$\theta = 67.1$	0.3088	−0.01282	0.817	−1.82
$l/W = 0.137$	0.3038	−0.01266	−0.792	−3.00
$l/W = 0.164$	0.3071	−0.01627	0.279	24.6

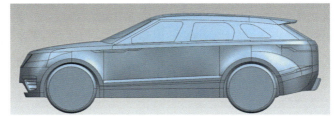

FIGURE 2.34   Basic model.

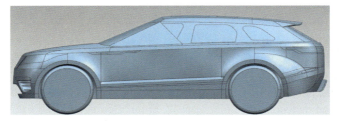

FIGURE 2.35   The model with windshield angle of 65.6 degrees.

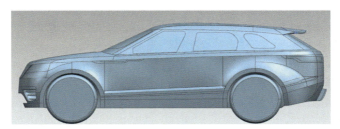

FIGURE 2.36   The model with $l/W = 0.137$.

2.6.1 Current challenges

The authors would like to point out some current challenges according to their own knowledge.

2.6.1.1 Interoperability among different design and engineering analysis stages

The development history of CAD and CAE systems is relatively separate at the beginning, and features are mostly application-driven. Thus, a robust and unified data structure and universal operation methods have not been realized yet. Andreas proposed different levels of interoperability model between webs for data exchange [38], which is also suitable for describing interoperability levels among CAD and CAE systems (see Fig. 2.37). Nowadays, most CAD and CAE systems stay at the level of syntactic interoperability. The reason is that each CAD and CAE software develops its library functions and operation algorithms separately, which can be checked on their API. Some researchers/technical engineers develop plug-ins as converters among different file formats [39], which is the current level of integration among most of the different CAD and CAE software. This method is relatively straightforward. However, it is not robust enough and is hard to maintain with the version update of the CAD/CAE software because it's not solving the interoperability problem from the root (data structure, develop language,

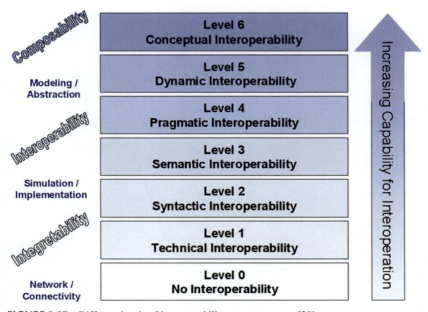

FIGURE 2.37 Different levels of interoperability among systems [38].

and operation methods). Besides, missing information such as geometry and topology may cause translation problems [40].

2.6.1.2 Change propagation and information consistency management

The collaborative engineering environment involves many participants in today's systemic PLM companies (see Fig. 2.38). Other involvers may influence the design solutions. Thus, information consistency among different sectors is vital for making the "correct design decision." In many companies, the database is managed by a central workstation, which may cause information delay as information changes cannot be updated simultaneously. Furthermore, the maintenance of such a powerful workstation is long and tedious, which may cause a "chain reaction" in the product planning. It has become a handicap for realizing the application of the multifaceted feature mentioned before because the centralized information management structure makes the reliability of the central workstation so important for efficiently supporting the whole PLM process.

2.6.2 Expected future work

The authors would like to point out some future research work expectations according to their own knowledge.

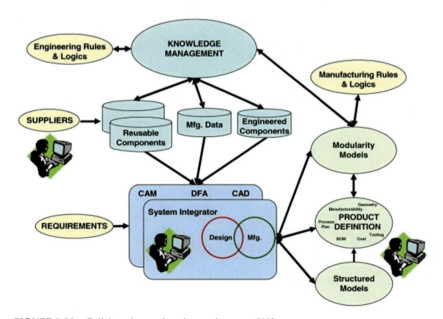

FIGURE 2.38 Collaborative engineering environment [41].

2.6.2.1 Standardization of model formats in a unified approach

STEP and IGES files were outstanding achievements in the past for data exchange among different CAD and CAE systems. However, these kinds of data exchange are far away from adequate. Information may be stripped off from the original file, which is hard to recover. Manually importing or exporting these files and adding necessary information for a specific application may also cause a semantic conflict. The interoperability on different engineering software platforms requires information extraction from a master model. The information would be stripped off until only the common data remains. From a top-down point of view, the proposed standard model should be an object class that contains all the attributes and methods throughout the entire product design and even manufacturing stages. The master model should be read and edited by all CAD and CAE software. Information is exchanged rather than stripped off and then added separately for a particular application.

2.6.2.2 Decentralized computer-aided design/computer-aided engineering network and cloud processing algorithms

As mentioned before, the PLM platforms in many companies are centralized frameworks. Such distribution depends largely on the central database management system. However, the design process itself and the decision-making rules are not managed by or based on the minority of senior supervisors but are engaged by most of the participants (see Fig. 2.39). Thus, a decentralized, distributed design and manufacturing system would be more suitable for the future product design process. Models are stored, edited, and optimized in a cloud system. As a result, cloud processing algorithm design

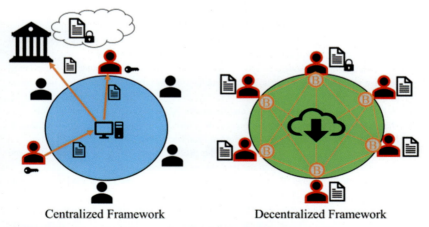

FIGURE 2.39 Centralized framework versus decentralized framework [42].

94 Digital Manufacturing

for CAD and CAE might eventually become a popular trend for the future development of CAD and CAE systems.

2.7 Summary

This chapter introduces advanced digital product design and engineering analysis techniques. CAD for 3D form creation, intent-based systemic design, and modern engineering analysis have significantly reduced the product development time. 3D form creation is fundamental for constructing and manipulating digital products' geometric models. The intent-based systemic design aims to model large and complex systems in a collaborative environment with sufficient information (geometric and nongeometric). The information flow throughout the entire systemic design cycle is managed in a consistent and efficient way. Engineering analysis is important for product performance validation after the product design stage through digital simulation. CFD simulation and shape optimization are demonstration in detail in a case study of a car body design. The current challenges of CAD and CAE include low level of interoperability and poor information consistency management remain. Finally, proposed future development trends of the CAD/CAE systems are pointed out.

References

[1] Y. Bar-Cohen (Ed.), Advances in Manufacturing and Processing of Materials and Structures, CRC Press, 2018, pp. 15−50.

[2] Z. Bi, X. Wang, Computer Aided Design and Manufacturing, John Wiley & Sons., 2020, pp. 38−53.

[3] C. Lee, A. Badal, Y.S. Yeom, K.T. Griffin, D. McMillan, Dosimetric impact of voxel resolutions of computational human phantoms for external photon exposure, Biomedical Physics & Engineering Express 5 (6) (2019) 065002.

[4] W.E. Lorensen, H.E. Cline, Marching cubes: a high resolution 3D surface construction algorithm, ACM Siggraph Computer Graphics 21 (4) (1987) 163−169.

[5] B. Khoda, Computer-aided design of additive manufacturing components, Laser-based additive manufacturing of metal parts: modeling, optimization, and control of mechanical properties; 2017.

[6] M.K. Agoston, Computer Graphics and Geometric Modeling, Springer, London, 2005.

[7] D.S. Ebert, F.K. Musgrave, D. Peachey, K. Perlin, J.C. Hart, S. Worley, Texturing & Modeling: A Procedural Approach, Morgan Kaufmann, 2003.

[8] M. Saldana, An integrated approach to the procedural modeling of ancient cities and buildings, Digital Scholarship in the Humanities 30 (Suppl. 1) (2015) 48−63.

[9] R.M. Smelik, T. Tutenel, R. Bidarra, B. Benes, A survey on procedural modelling for virtual worlds, Computer Graphics Forum 33 (6) (2014) 31−50.

[10] W. Xiong, P. Zhang, P.V. Sander, A Joneja, ShapeArchit: shape-inspired architecture design with space planning, Computer-Aided Design 142 (2022). Available from: https://doi.org/10.1016/j.cad.2021.103120.

Digital product design and engineering analysis techniques Chapter | 2   95

[11]   R. Venturi, D.S. Brown, S. Izenour, Learning from Las Vegas, MIT Press, Cambridge, MA, 1967.

[12]   J.N. Reddy, An Introduction to the Finite Element Method, McGraw-Hill, New York, 2010.

[13]   Z. Cheng, Y. Ma, Explicit function-based design modelling methodology with features, Journal of Engineering Design 28 (3) (2017) 205−231. Available from: https://doi.org/10.1080/09544828.2017.1291920. Available from: www.tandfonline.com.

[14]   Z. Cheng, Y. Ma, A functional feature modeling method, Advanced Engineering Informatics 33 (December) (2017) 1−15. Available from: https://doi.org/10.1016/j.aei.2017.04.003.

[15]   N.P.S. Amro, M. Farid, Axiomatic Design in Large Systems, Springer, 2016.

[16]   S.D. Eppinger, T.R. Browning, Design structure matrix methods and applications, Design Structure Matrix Methods and Applications (2018). Available from: https://doi.org/10.7551/mitpress/8896.001.0001.

[17]   V.H. Torres, J. Ríos, A. Vizán, J.M. Pérez, Integration of design tools and knowledge capture into a CAD system: a case study, Concurrent Engineering Research and Applications 18 (4) (2010) 311−324. Available from: https://doi.org/10.1177/1063293X10389788.

[18]   Y.S. Ma, S.B. Tor, G.A. Britton, The development of a standard component library for plastic injection mould design using an object-oriented approach, The International Journal of Advanced Manufacturing Technology 22 (2003) 611−618. Available from: https://doi.org/10.1007/s00170-003-1555-8.

[19]   M.C. Leu, W. Tao, A. Armani, K. Kolan, NX 12 for engineering design; December 2017. pp. 54−74.

[20]   Y.S. Ma, Semantic Modeling and Interoperability in Product and Process Engineering, Springer Series in Advanced Manufacturing; 2013, pp. 89−142.

[21]   Y. Ma, T. Tong, Associative feature modeling for concurrent engineering integration, Computers in Industry 51 (2003) 51−71. Available from: https://doi.org/10.1016/S0166-3615(03)00025-3.

[22]   Y. Ma, G.A. Britton, S.B. Tor, L.Y. Jin, Associative assembly design features: concept, implementation and application, The International Journal of Advanced Manufacturing Technology (2007) 434−444. Available from: https://doi.org/10.1007/s00170-005-0371-8.

[23]   C.M. Vahid Salehi, Methodological integration of parametric associative CAD systems in product lifecycle management environment, in: ASME 2009 International Design Engineering Technical Conferences and Computers and Information in Engineering Conference, 2009, pp. 505−514.

[24]   G.P. Gujarathi, Y.S. Ma, Parametric CAD/CAE integration using a common data model, Journal of Manufacturing Systems 30 (3) (2011) 118−132. Available from: https://doi.org/10.1016/j.jmsy.2011.01.002.

[25]   L. Li, Y. Ma, C.F. Lange, Association of design and simulation intent in CAD/CFD integration, Procedia CIRP 56 (2016) 1−6. Available from: https://doi.org/10.1016/j.procir.2016.10.006.

[26]   C. Patel, The case of art to part: the rise of systemic intent based design with 3D digital manufacturing; 2019 [Online]. <https://www.engineering.com/story/the-case-of-art-to-part-the-rise-of-systemic-intent-based-design-with-3d-digital-manufacturing>.

[27]   N. Saini, A. Lam, Simplify machining complexity using integrated toolpath analysis. Third Wave Systems. NX MANUFACTURING. [Online]. <www.thirdwavesys.com> <https://blogs.sw.siemens.com/nx-manufacturing/simplify-machining-complexity-using-integrated-toolpath-analysis/ >.

96 Digital Manufacturing

[28] Z. Wei, J. Wu, N. Shi, L. Li, Review of conformal cooling system design and additive manufacturing for injection molds, 17 (2020) 5414−5431. doi:10.3934/mbe.2020292.

[29] O. Sigmund, A 99 line topology optimization code written in matlab, Structural and Multidisciplinary Optimization 21 (2) (2001) 120−127. Available from: https://doi.org/10.1007/s001580050176.

[30] CAD Models. GrabCAD Community Library [Online]. <https://grabcad.com/library>.

[31] Hole Types. LingHunt Engineering Inc. [Online]. <https://linghunt.com/GUIDES/CustomThread/CustomGuide.html>.

[32] Finite Element Method − Wikipedia [Online]. <https://en.wikipedia.org/wiki/Finite_element_method>.

[33] Multibody System − Wikipedia [Online]. <https://en.wikipedia.org/wiki/Multibody_system>.

[34] Computational Fluid Dynamics, Wikipedia [Online]. <https://en.wikipedia.org/wiki/Computational_fluid_dynamics>.

[35] L. Li, C.F. Lange, Y. Ma, Artificial intelligence aided CFD analysis regime validation and selection in feature-based cyclic CAD/CFD interaction process, Computer-Aided Design and Applications 15 (5) (2018) 643−652. Available from: https://doi.org/10.1080/16864360.2018.1441230.

[36] Vauxford, "File:2017 Land Rover Range Rover Velar First Edition D3 3.0 Front." [Online]. < https://commons.wikimedia.org/wiki/File:2017_Land_Rover_Range_Rover_Velar_First_Edition_D3_3.0_Front.jpg

[37] Vauxford, "File:2017 Land Rover Range Rover Velar First Edition D3 3.0 Rear." [Online]. < https://en.wikipedia.org/wiki/File:2017_Land_Rover_Range_Rover_Velar_First_Edition_D3_3.0_Rear.jpg

[38] A. Tolk, What comes after the semantic web - PADS implications for the dynamic web, 20th Workshop on Principles of Advanced and Distributed Simulation (PADS'06) (2006). Available from: https://doi.org/10.1109/PADS.2006.39.

[39] K. Shea, R. Aish, M. Gourtovaia, Towards integrated performance-driven generative design tools, 14 (2005) 253−264. Available from : https://doi.org/10.1016/j.autcon.2004.07.002.

[40] M. El-mekawy, A. Östman, K. Shahzad, Geospatial interoperability for IFC and CityGML: challenges of existing building information databases geospatial interoperability for IFC and CityGML: challenges of existing building information databases, IEEE-Innovations in Information Technology (2008).

[41] Ali K. Kamrani, Collaborative Design in Product Design and Development, Collaborative Engineering, Springer, Boston, MA, 2008. Available from: https://doi.org/10.1007/978-0-387-47321-5_1.

[42] D. Puthal, N. Malik, S.P. Mohanty, E. Kougianos, C. Yang, The blockchain as a decentralized security framework [Future Directions], IEEE Consumer Electronics Magazine 7 (2) (2018) 18−21. Available from: https://doi.org/10.1109/MCE.2017.2776459.

Chapter 3

Design methodologies for conventional and additive manufacturing

Xue Ting Song[1], Jo-Yu Kuo[2] and Chun-Hsien Chen[1]
[1]*HP-NTU Digital Manufacturing Corporate Lab, School of Mechanical and Aerospace Engineering, Nanyang Technological University, Singapore,* [2]*Department of Industrial Design, National Taipei University of Technology, Taipei, Taiwan*

List of abbreviations

AM	Additive Manufacturing
AEM	Assemblability Evaluation Method
ASF	Assembly Sequence Flowchart
BATC	Beijing Automotive Technology Center
CAD	Computer-Aided Design
CM	Conventional Manufacturing
DFA	Design for Assembly
DFAM	Design for Additive Manufacturing
DFM	Design for Manufacturing
DFMA	Design for Manufacturing and Assembly
DFX	Design for X/Excellence
HDB	Housing and Development Board
HP	Hewlett-Packard
ITM	Industry Transformation Map
MAT	Minimum Assembly Time
NM	Theoretical minimum number of parts
TM	Assembly Time
TO	Topology Optimization
VR	Virtual Reality

3.1 Introduction

Design methodologies within the context of manufacturing refer to the study of principles, practices, and procedures related to the creation of a mechanical design. These methodologies provide guidance and inform the product

---

Digital Manufacturing. DOI: https://doi.org/10.1016/B978-0-323-95062-6.00007-3
© 2022 Elsevier Inc. All rights reserved.

98 Digital Manufacturing

design process. The primary focus is to develop a deep and practical understanding of the design process and aid design engineering teams in achieving design outcomes effectively. The first reported design research dates back to 1861, when Reuleaux, a German mechanical engineer, published a handbook on machine design, *The Constructor*, and a second book, *The Kinematics of Machinery*, which details classifications of mechanisms and machine motions for the machine design process [1,2]. The earliest design theories and methods centralize on modeling the design process and the theoretical abstraction of design knowledge.

Design methodologies that evolved from the introduction of new systematic design methods were first introduced in the 1960s. These methods were developed based on a structure of analysis and testing to help designers cope with complex problems and to encourage innovation via productive brainstorming and collaborative thinking amongst multidisciplinary teams. Each proposed idea can be evaluated based on user-defined needs-based goals. They are frequently employed in fields of design practices, such as engineering, industrial, architectural, and urban design. As manufacturing technologies continue to advance, design methodologies have been intensively studied to update and better inform designers and engineers about how to design mechanical products.

3.1.1 Design for Assembly

The Design for Assembly (DFA) methodology was developed in efforts to reduce assembly time (TM) and product life cycle cost by reducing the number of parts and assembly operations. It was not until 1983 when Boothroyd and Dewhurst developed a software tool for DFA, which was then widely used in industries to design manual, automated, and printed circuit board assemblies [3].

DFA focuses on the minimization and standardization of parts, subassemblies, and assemblies. The goal is to reduce the cost and time of assembly by consolidating single-function parts into multifunctional designs and combining noncritical parts into critical parts, resulting in improved assemblies. DFA also helps teams streamline assemblies and complex models by simplifying attachment directions, hence reducing attachment time [4]. While part manufacturing costs are not brought into analysis yet, the efficiency rating and estimated TMs from a DFA analysis serve to provide benchmarks against which future design iterations, previous designs, or competitors' products can be compared [3].

3.1.2 Design for Manufacturing

The widespread adoption of DFA led to the development of the Design for Manufacturing (DFM) methodology to help more businesses optimize the design and manufacture of an assembly. Insights from a DFM analysis can be gathered to complete a product's analysis postassembly and premanufacture. DFM aims to simplify product design by reducing part count via basic rules, analytic formulas, or finite element process simulations [4]. It focuses on optimizing individual parts to reduce or eliminate the need for expensive, complex, and unnecessary features that would be difficult to manufacture. It can also help teams select cost-effective materials and processes during the manufacturing stage, thereby reducing overall product cost and achieving better time-to-market efficiency.

Fig. 3.1 shows the assimilation of DFA and DFM in the design process proposed by Pahl and Beitz [5]. DFA serves its purpose effectively in the conceptual design stage where concept generation occurs, while DFM is most useful in the embodiment and detail design stages. The combined

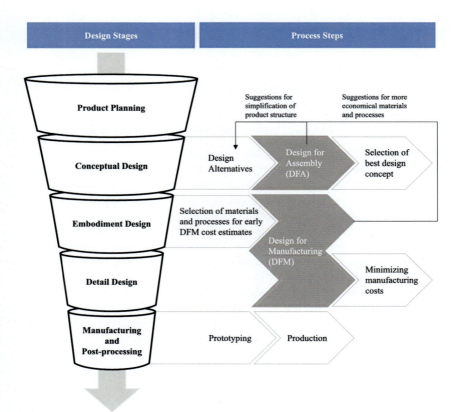

FIGURE 3.1 DFA and DFM in a modified product design process. *DFA*, Design for Assembly; *DFM*, Design for Manufacturing.

100 Digital Manufacturing

application of DFA and DFM in the product design process is better known as the Design for Manufacturing and Assembly (DFMA) methodology.

3.2 Design methodologies for conventional manufacturing

Conventional manufacturing (CM) refers to traditional manufacturing processes such as machining, joining, sheet metalworking, thermoforming, blow molding, die casting, injection molding, and more. To design and manufacture successful products with CM, designers, engineers, and manufacturers have applied different design methodologies in the product design process. There are three main classification of approaches:

1. Qualitative approaches include general guidelines to manage the decision-making process for a diversity of products, manufacturing processes, and materials.
2. Quantitative approaches include widely accepted and tested methods by Boothroyd-Dewhurst, Hitachi, Lucas-Hull, and other computer-aided DFA and DFM methodologies.
3. Mixed approaches include software tools for qualitative and quantitative analysis. One such example is CONDENSE—a concurrent design evaluation system to determine design specifications and facilitate design selection [6].

3.2.1 Design for Manufacturing and Assembly guidelines

DFMA is useful for improving piece-functional designs by inspiring system-functional designs. As a qualitative approach, it can be viewed as a knowledge-based technique that invokes a series of guidelines, principles, recommendations, or rules of thumb for designing a product to ease manufacturing and assembly tasks while lowering the manufacturing costs. Empirical data based on experimentation and successful case studies can provide guidelines and metrics to improve the assemblability of a design. These guidelines inform design directions and set the fundamental constraints in conceptualizing product design. Table 3.1 presents a set of key DFMA guidelines.

In general, minimization, standardization, and modularization of components in an assembly are the key characteristics associated with the DFMA guidelines [10]. These guidelines generally lead to the benefits of an improved design with simplified assemblies (in terms of unambiguous part orientation, less handling, fewer adjustments, etc.), fewer manufacturing operations, lower costs from eliminating complex tooling and dedicated fixtures, greater modularity, and easier repair and maintenance. The main drawback of DFMA guidelines is the nonspecificity of the recommendations as they are meant to inform design decisions on a general level. Hence, the expertise of a human designer or engineer is required to interpret and evaluate appropriate decisions before applying them to specific products and applications.

Design methodologies for conventional and additive manufacturing Chapter | 3 **101**

TABLE 3.1 A set of key Design for Manufacturing and Assembly (DFMA) guidelines [7–10].

No.	DFMA guidelines	Remarks
With respect to individual parts		
1	Reduce part count and types of parts	By redesigning parts to be multifunctional or multi-use in a product or product family and minimizing the number of different types of parts, e.g., the number of flexible and inflexible parts in an assembly
2	Use standardized off-the-shelf parts	By using industry-standard parts to avoid additional costs
3	Leverage on part symmetry	By deliberately designing the parts to be symmetrical to lower production costs, or asymmetrical to improve part insertion
4	Ensure ease of fabrication	By using common manufacturing methods and as few dissimilar materials as possible
With respect to subassemblies and assemblies		
5	Design for modularity	By minimizing assembly directions for handling, insertions, or reorientations and developing a layered top-down assembly or robotic assembly to reduce costs over manual methods
6	Design for error-proof assembly	By eliminating the possibilities for errors due to incorrect assemblies, and ensuring that each assembly step is clearly visible and accessible
7	Minimize or eliminate the use of separate fasteners	By designing for efficient joining and fastening where parts are self-aligning and easy to insert, e.g., incorporate snap-fits on the cover of an enclosure instead of using screws for easy fastening and removal for maintenance
8	Ensure that the product fits in standard packaging	By ensuring easy disassembly and assembly of the product

Source: Adapted from Refs [7–10].

3.2.2 Design for Manufacturing and Assembly procedures

Over the years, several quantitative evaluation frameworks were proposed to help businesses adopt DFMA in their practices. The three most prominent DFMA methods used are the Boothroyd-Dewhurst DFMA, the Hitachi Assemblability Evaluation Method (AEM), and the Lucas-Hull DFA. These methods share the same philosophy but use different approaches to

102 Digital Manufacturing

quantitatively score and assess the assemblability of a product design in the early stages of the design process.

3.2.2.1 Boothroyd-Dewhurst Design for Manufacturing and Assembly

Boothroyd and Dewhurst first coined the term DFMA in 1983, and it subsequently became the most widely recognized design methodology adopted in the manufacturing industry. As the foremost developers of the DFMA strategies and software, Dewhurst described that "DFMA makes one look critically at the structure of products, the relationship between parts and the number of parts, the securing methods, and try to design the assembly content out of it while keeping the design flexible" [7]. The goal of DFMA is to provide manufacturing analysis during the conceptualization of product designs in a sequential and organized fashion. By analyzing every part and operation in a basic assembly, engineers can rigorously confront the complexity of the design and work toward simplifying it by implementing DFMA in their practices.

There are two key measures to rate a design's assemblability: (1) by directly comparing the part count in an assembly and how each part is ideally assembled, and (2) by using a rating system to compare the part count and assemblability for alternative designs. After scoring each part based on its assemblability, the values can be summed up for the entire design and used as an indicator for the overall design quality. By providing measurements of assemblability, objective comparisons can be made directly across design alternatives via scorekeeping [7,10].

When designing for assemblies, it is essential to begin by choosing an appropriate assembly system. The design engineering team must be aware of the product's assembly procedure and whether manual, automated, or special purpose assembly machines will be involved [8]. In any case, a manual assembly analysis is usually performed first, followed by an analysis for high-speed automatic or robotic assembly [8]. The ease of assembly reflects the efficiency of the design, which further informs other design and business-related decisions later [3]. The cost of assembly is related to the manufacture of the product, its method of assembly, and the associated tooling and labor costs. A simple analysis of the expected production volume, payback period required, part count in the overall assembly, and variations of product styles is also conducted to select the assembly system.

1. Determining the necessity of each component

   The ease and cost of handling and insertion are first evaluated to determine the necessity of each component. The findings are subsequently compared to a database, which estimates the time and cost of assembly for that component. Three DFA criteria must be examined for each component that is added to the assembly [3]. Fig. 3.2 shows the decision flowchart of the examination to determine if these components are critical to the product for fundamental reasons.

Under these three criteria, areas where the product assembly can be simplified are identified as each specified part and its associated operations are assessed. If the answers to all the questions in Fig. 3.2 are "yes," the component is scored "1" and deemed an essential part of the product that is necessary to the product's structure for functionality. If any of the questions were answered "no," the component would be scored "0" and deemed unnecessary. Hence, it can be eliminated or combined with other critical parts of the product. The summation of all critical parts is known as the theoretical minimum number of parts (NM).

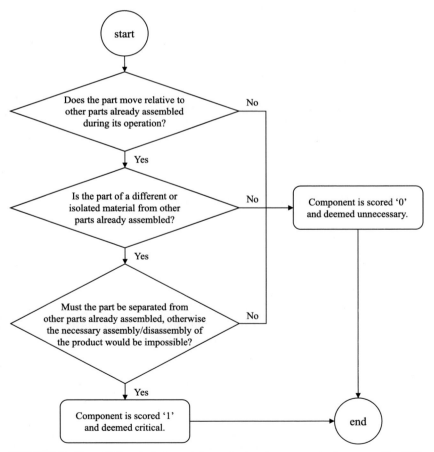

FIGURE 3.2 Three DFA criteria to examine against each component of an assembly. *DFA*, Design for Assembly.

104 Digital Manufacturing

Less significant parts can either be merged into these critical parts or eliminated totally, hence improving the part's manufacturability.

For instance, closed-end springs have a shorter handling time than open-end springs as the increased symmetry introduces redundancy and reduces the handling time. Incorporating the washer seat into the screwhead and increasing the length of the shaft eliminates the need for holding a separate washer, making alignment easier and reducing insertion time. Proper trade-off decisions concerning the product's assembly and manufacturability can be carried out by quantifying the effects of design changes in terms of the TM, assembly costs, and individual part costs.

2. Estimating the handling and insertion times and costs of each component

The product assembly is then further evaluated in terms of its difficulty and duration of operations [3]. The difficulty of an assembly operation is examined based on two considerations: (1) how the part is handled (grasped, oriented, manipulated, etc.) and prepared for insertion, and (2) how the part is inserted or fastened into the assembly. The duration of an assembly operation will be evaluated by summing up the time taken for handling and inserting the part into the assembly.

In a typical manual assembly process, a two-digit handling code and a two-digit insertion code will be identified from industry-tested data tables. Data features from these tables include metrics such as the size and weight of the part, orientation for handling, part alignment, part securing method, and more. These codes are cross-referenced to estimate the time taken to assemble a product. These information are then used to determine the following metrics [3,8]:

a. TM: estimated by the summation of handling and insertion times with its actual part count from industry-tested data tables,

b. minimum assembly time (MAT): estimated by multiplying 3 seconds by NM to derive the ideal TM (assuming all critical parts require an average of 3 seconds for assembly—1.5 seconds of handling and 1.5 seconds of insertion),

c. assembly efficiency percentage: calculated by dividing MAT by TM, then multiplying by 100,

d. assembly cost: calculated by multiplying the total TM with a factor accounting for standard labor rates and overheads,

e. design efficiency: calculated by dividing the ideal TM by the actual TM.

It is also necessary to estimate assembly costs so that any possible savings can be considered in the design iteration. Utilizing knowledge bases of DFMA time standards and cost models, it is possible to make these estimates, including the individual part cost and the associated tooling, without having the detailed design of the parts yet. As actual TM decreases through part reduction efforts, the DFA index increases to indicate product improvement. These metrics encourage the iterations of

designs to reduce handling and insertion difficulties and durations by pointing to specific changes and expressing the penalty associated with poor design.
3. Redesigning the product to improve its assembly and manufacturing operations.

After the standard durations and difficulties of all assembly operations have been estimated, the identification of noncritical parts for elimination is prioritized before further iterations or design alternatives are considered. Due to business, economic, and technical considerations of manufacturing individual components, there may be other constraints preventing such eliminations. Subsequently, the focus on design iterations can be shifted to the improvement of any assembly and manufacturing operation that results in excessive TM.

Upgraded from paper evaluation forms, the Boothroyd-Dewhurst Inc. has developed its own set of computer-aided evaluation tools, called the DFA Product Simplification program and DFM Concurrent Costing module, to help companies employ DFMA with better efficiency. The DFA Product Simplification program helps design engineering teams identify areas for substantial cost reduction by consolidating or eliminating parts while maintaining the full functionality of the product. The DFM Concurrent Costing module offers an early concurrent cost analysis of the primary cost drivers for manufacturing a product by benchmarking the design with industry-tested data tables. Integration with computer-aided design (CAD) tools such as Pro/Engineer is also offered. The automated evaluation process provides a quantitative analysis based on an expert system that analyzes the design, repeats the design optimization by applying appropriate design rules, and evaluates its quality after each iteration [10]. An analysis report can then be generated for client quotations, providing a basis for planning and decision-making. With a reliable cost analysis backed by industry estimates, teams can also challenge suppliers' estimates and make more informed decisions in the assembly and manufacture of the product. Certainly, the use of software enforces structure and adds rigor to the overall project analysis.

DFMA provides manufacturing engineers a reliable quantitative analysis for discussions, holding similarity to the capability of CAD for design engineers. Engineers can quickly compare design proposals, choose alternative materials or manufacturing methods, modify part features, and consider necessary trade-off decisions between parts consolidation and production process costs to optimize the product design. It serves as a communication tool to push the concurrent engineering movement (refer to Section 3.3) and presents perspectives from team members of different departments, avoiding the interorganizational and interdisciplinary problems that commonly plague new product teams. DFMA also encourages early collaborations among teams as it allows for a nonthreatening way to discuss design issues without the animosity of criticism [7]. The main benefit of DFMA lies in its ability to lead teams to successful part consolidation, fater developments, and reduced costs. With fewer parts to wear out, the product's function and lifetime quality improve as well. Hence, this is a significant reason why companies adopt DFMA to slash product development time and lower assembly and manufacturing costs [8].

3.2.2.2 *Hitachi Assemblability Evaluation Method*

First developed by Hitachi Corp. in 1976, the AEM aims to identify design "weaknesses" and facilitate design improvements at the earliest evaluation, by simplifying the automatic insertion of parts. The AEM assesses the assemblability of a product design by evaluating the assemblability evaluation score E and the assembly cost ratio K. E is the sum of all assemblability scores for the individual tasks divided by the total number of tasks, and K is the total assembly cost of a new design divided by that of the previous design [9]. In short, the E-score assesses the design quality by evaluating the difficulty of assembly operations while the K-score estimates the assembly cost improvements. Fig. 3.3 outlines the procedure of an AEM analysis.

First, the procedure defines the operations necessary for each insertion of the product, such as fixturing, forming, rotating, and joining. For every motion different from a simple downward motion (the most elementary assembly operation performed by a human or machine), penalty points will be assigned [9]. The E-score is measured as the design efficiency with a perfect design representing a score of 100. Hitachi accepts an overall E-score of 80 and above for automatic assembly. An overall K-score of 0.7 or less typically achieves 30% cost savings or more. By achieving the target values of E and K, the AEM helps teams focus on problem areas to be improved, where part count can be reduced and where assembly operations can be made simpler.

Interestingly, the AEM does not include direct analysis for handling as these are sensitive to part configurations and are difficult to determine at an early design stage [9]. However, the estimated assembly operation time includes both the part handling and insertion times. Hitachi continued to improve the functionality and accuracy of the method by developing an

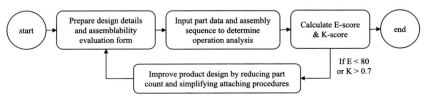

FIGURE 3.3 Procedure of AEM analysis [7]. *AEM,* Assemblability Evaluation Method. *Adapted from S. Ashley, Cutting costs and time with DFMA, Mechanical Engineering, 117 (3) (1995) 74−77.*

improved version of the AEM in 2002, known as the Extended Assemblability Evaluation Method [11].

3.2.2.3 Lucas-Hull Design for Assembly

The Lucas-Hull DFA method originated from the United Kingdom as a result of a collaboration between Lucas Corp. and the University of Hull [9]. Unlike the Boothroyd-Dewhurst DFMA and the Hitachi AEM methods, cost analysis is not included in the Lucas-Hull DFA. Similar to Hitachi's AEM, penalty points are assigned and summed for each design issue identified. This method evaluates a product based on its handling/feeding and insertion/fitting processes using an Assembly Sequence Flowchart (ASF). Fig. 3.4 illustrates an example of an ASF generated using the Lucas-Hull DFA method.

The Lucas-Hull DFA generates three steps of analysis:

1. The design efficiency (the ratio of essential parts to the total number of parts) is determined using a functional analysis to separate parts into essential and non-essential categories.
2. Each part is examined by cross-referencing a knowledge base in a handling analysis, where the handling ratio (the index total divided by the number of essential parts) is determined.
3. An insertion index and an insertion ratio will be determined based on a knowledge base to produce the insertion analysis.

Each symbol in the ASF contains a penalty factor, and the respective sums of these factors make up the handling index in the handling ratio and the insertion index in the insertion ratio. These scores are then compared to predetermined thresholds or referenced to values established in previous designs, followed by design iterations to improve the product's assemblability and manufacturability.

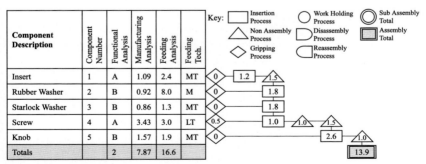

FIGURE 3.4 An ASF for a headlamp trim screw before redesign [12]. *ASF*, Assembly Sequence Flowchart. *Adapted from C. Barnes, et al., Assembly sequence structures in design for assembly, in: Proceedings of the 1997 IEEE International Symposium on Assembly and Task Planning (ISATP'97)-Towards Flexible and Agile Assembly and Manufacturing, 1997, IEEE.*

108 Digital Manufacturing

3.2.3 Applications of Design for Manufacturing and Assembly

The success of DFMA can be attributed to its systematic procedure in analyzing the assemblability and manufacturability of a product. The reduction in part count produced a snowball effect in time and cost reduction on assembly, manufacturing, labor, inventory, documentation, and administration [8]. This results in simpler and more reliable products produced at lower costs. These factors also heavily reduce the overheads of many businesses, which usually form the largest fraction of the total product cost.

The fundamental concept of DFMA has enabled the creation of new design tools, improving the productivity of global teams and businesses. These digitalized DFMA tools also enable closer collaborations between designers, engineers, and manufacturers during the early stages of the product design process. Throughout its development, various companies have reaped the benefits of concurrent engineering by utilizing DFMA to cut costs, improve assembly and manufacturing efficiencies, create new business opportunities, produce better product quality, and conduct less repair due to fewer manufacturing failures [7,8]. The following subsections highlight case studies of DFMA applications in different industries, proving its foundational logic and effectiveness as a robust design methodology.

3.2.3.1 Consumer products

This section presents some case studies of the application of DFMA for consumer products.

3.2.3.1.1 Hasbro (1993)

In preparation for the 1993 Christmas season, Hasbro implemented DFMA to analyze the design and identify cost improvements for its Talk n' Play Fire Truck. Toy retailers were constantly looking into ways to minimize inventories and maximize productivity as timing is critical in their businesses [7]. By employing DFMA, Hasbro was able to compare trade-off information and develop a high-quality yet profitable product within their hectic schedule constraints. The metal toy truck was eventually redesigned in plastic, while the total part count was reduced from 33 to 5 within a TM of 22 seconds instead of 198 seconds [3].

3.2.3.1.2 Hewlett-Packard (1994)

In 1994, Hewlett-Packard (HP) adopted DFMA in the development of its digital multimeter, 34401A. As a result, the input connection scheme and a front panel were redesigned to be assembled without screws. After the analysis, the redesigned multimeter had 18 parts instead of 45 parts and can be manually assembled by one person in under 6 minutes, compared to 20 minutes before [3].

Design methodologies for conventional and additive manufacturing Chapter | 3 **109**

3.2.3.1.3 Motorola Solutions (2012)

With 25 years of successful implementation, DFMA has proven to be effective in helping Motorola Solutions classify and improve product designs. In 2012, a global DFA team was formed to realign the company's separate engineering teams with a unified DFA procedure. Engineers divided the company's extensive portfolio into a series of product families and calculated the range and average DFA index for each product grouping, which serve as benchmarks for the company's future designs [13]. This facilitated the improvement of products as teams gained a quantified target to pursue. Global teams can also better identify how close new designs are to the company's predetermined goals.

3.2.3.1.4 Endress + Hauser (2016)

Endress + Hauser, a global provider of laboratory measurement instrumentation, automation solutions, and services, first adopted DFMA in its practices in 2016. Within the first two years, they reported a cost savings potential of more than US $1 million generated per annum [14]. In-house DFMA project leaders have also consistently requested DFMA evaluations for each new product development, with employees actively volunteering to participate in their DFMA workshops.

3.2.3.2 Aerospace and automotive industries

This section presents some case studies of DFMA implementation within the aerospace and automotive industries.

3.2.3.2.1 McDonnell Douglas Corporation (now Boeing) (1994)

The primary application of DFMA in the aerospace industry is to improve the efficiency of the plane's systems by ensuring lightweight structures. The then-company applied DFMA in their process and saw an average of 37% reduction in part count and 46% decrease in fastener count [3].

3.2.3.2.2 Beijing Automotive Technology Center (2017)

In 2017, an engineer of Beijing Automotive Technology Center's (BATC) design cost engineering department realized major design defects in the front bumper system of their new model vehicle [15]. The mounting structure was not well designed and could trigger potential breakage. Poor positioning of the mounting points also resulted in weak stability of the assembly. The engineer realized that the problem arose from the adoption of an old body structure and the poor positioning of assembly points. To reduce design risks and manufacturing costs, BATC redesigned the middle pole structure of the vehicle's front end and applied the design changes to all existing and new models under the same platform. Consequently, it resulted in a 66%

110 Digital Manufacturing

reduction in part count, 76% reduction in manufacturing time, and 30% reduction in manufacturing costs [15]. The weight of the part was also reduced by 24% as plastic was used to replace steel.

3.2.3.3 Construction in Singapore

This section presents a case study of Singapore's construction industry. The Housing and Development Board (HDB) of Singapore has long adopted precast technology since the 1980s. To achieve the productivity goals set by the government, the adoption of DFMA was mandated in 2014 for developments on Government Land Sales sites, sparking a widespread adoption of prefab through DFMA [10]. As of 2017, approximately 70% of a typical HDB block's concrete structure was built using prefab methods with mass timber and structured steel, which were manufactured offsite and assembled onsite [10]. Singapore continues to focus on its buildability framework and encourage buildable features such as prefabricated bathrooms and standardized precast staircases. To further boost the economy and open opportunities for innovationto improve productivity, the Building and Construction Authority of Singapore charted specific Industry Transformation Maps (ITM) for key sectors of growth in 2017 [16]. As part of the Construction ITM, DFMA remains a widely supported strategy for developers and contractors when building in Singapore. While applying DFMA technologies contributes to lean buildings and a cleaner built environment, a reduction in construction time is one of the most important factors driving the construction industry to adopt DFMA [17]. However, the lack of adoption in the private sector continues to limit the industry's ability to achieve higher productivity performance. Hence, more government initiatives are in progress to further encourage an industry-wide adoption of DFMA strategies [10].

3.2.4 Limitations of Design for Manufacturing and Assembly

Since the proliferation of mass consumption in the mid-1900s, companies have been constantly pressured to rush and produce an unprecedented amount of goods and services, gradually flooding global supply chains to answer high levels of demand. Despite the benefits of cost- and time-savings in assembly design and manufacturing process planning, there are several challenges in employing DFMA that limit its impact in professional practices.

3.2.4.1 Getting the management on board

To reap the potential benefits of DFMA, companies need to initiate a top-down approach to train employees and actively employ DFMA in their practices. This means setting aside manpower, time, and capital resources for the process changes it demands to establish a company-wide adoption. Some companies are reluctant to change due to the lack of resources and the fact

Design methodologies for conventional and additive manufacturing Chapter | 3 **111**

that significant market share has been gained in the past without employing DFMA.

The macro goal of improving the product design process within the company is often secondary to the immediate outcomes on specific products that can be produced by engaging external DFMA experts. Yet, DFMA consultants are usually outsourced as a last resort to upgrade the company's product design capabilities to fight its competition, or when the company has specific cost objectives for a product [7].

3.2.4.2 Difficulties in the integration of design engineering teams

To successfully implement DFMA within teams' practices, significant culture changes and mindset shifts are necessary. The success of employing DFMA lies in the strength of the collaboration between designers, engineers, and manufacturers. + Designers and engineers ought to submit and discuss their designs for a preliminary critical analysis within the team, while manufacturers should involve themselves early from the conceptual design stage as well.

Assembling and managing an interdisciplinary team is also tricky as the management often does not give such teams enough authority in decision-making [7]. It is critical that all team members admit the limits of their capabilities and work together effectively to implement DFMA in their practices successfully. When mutual understanding of one another's requirements is established, the subjectivity and rivalry from the decision-making process can be minimized [7]. Successful integration and collaboration within the team will ensure a harmonious team dynamic critical to reaping the benefits of the DFMA methodology.

3.2.4.3 Lack of resilience in the Design for Manufacturing and Assembly software

As manufacturing technologies continue to advance, the DFMA software needs to be updated with the latest manufacturing methods such as additive manufacturing (AM) technologies for engineers to make more informed decisions when choosing manufacturing processes. The resilience of the DFMA software also depends on the constant update of existing case studies and accurate representations of the specific manufacturing methods, to provide as reference values or benchmarks. This remains a tall order as a large number of examples are required to build a reliable database within the DFMA software.

3.2.4.4 Different objectives and project-specific constraints for each product

The issue of choosing metrics for different objectives is also important as various kinds of scorekeeping will lead to different results. Without other references or benchmarks, teams rely on the cost and time estimates for each

112   Digital Manufacturing

design choice made by the DFMA software. Hence, these estimates must have sufficient reliability and creditability for engineers to buy into the program. However, since not all teams focus on the product's cycle times and costs, DFMA is inherently limited in its ability to calibrate its analysis to the specific purposes of each unique product.

The effective utilization of DFMA is also dependent on project-specific requirements. The benefits of DFMA increase exponentially with mass production. When mass production is not required, the adoption of DFMA may not be as useful for the project teams. For construction-related projects, teams often struggle to balance the constraints of restricted site layouts and mass production. Hence, just-in-time delivery is often adopted, especially in small-scale projects, due to the lack of on-site storage space [10]. Employing DFMA in these situations may not be suitable and it may cause companies to incur higher costs instead.

3.3   A paradigm shift

Due to the limitations of CM technologies, current products designed with conventional design methodologies may be compromised in their functionalities and performance. As manufacturing technologies continue to advance, more comprehensive design methodologies are emerging to synthesize state-of-the-art design and manufacturing knowledge in an increasingly accessible manner. A growing number of objectives also need to be met simultaneously within a single part or product beyond the optimization of a single objective. As a result, Design for X (DFX) was initiated to help teams meet specific design goals in a more structured and guided approach. With the rise of AM, new design methodologies considering the opportunities offered and restrictions imposed by different AM processes need to be developed and standardized. Pioneering efforts from academia coupled with the industry-led demands of practical applications subsequently gave rise to the concept of Design for Additive Manufacturing (DFAM).

3.3.1   Design for X

The concept of concurrent engineering was proposed in the 1990s to substitute the sequential engineering and design approach as more design objectives were made necessary with the development of more complex products. The parallel processing of activities occurred to facilitate the concurrent considerations within the design and manufacturing process [18]. Simultaneously, the concept of collaborative engineering appeared, and cross-disciplinary teams started to adopt a collaborative approach to facilitate concurrent engineering. This resulted in a new design support system, which aims to cover the whole product life cycle, known as collaborative product development.

Design methodologies for conventional and additive manufacturing Chapter | 3 113

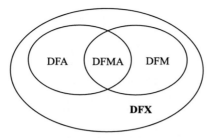

FIGURE 3.5 Classification of DFX, DFA, DFM, and DFMA. *DFA*, Design for Assembly; *DFM*, Design for Manufacturing; *DFX*, Design for X/Excellence; *DFMA*, Design for Manufacturing and Assembly.

Following the success of DFA and DFM, new design methodologies oriented toward specific objectives emerged to achieve greater customer satisfaction by maximizing all desirable characteristics identified in the product and minimizing lifetime costs. These methods are collectively known under the generic term DFX. DFX refers to a generic name for members of a family of design methodologies adopted to improve product design by addressing the product's full life cycle needs [19,20]. The "X" in DFX represents all desirable characteristics (cost, safety, user-friendliness, sustainability, performance, maintainability, etc.) or excellence in general [20]. With this development, DFA, DFM, and DFMA became design methodologies considered under the umbrella term DFX. Fig. 3.5 shows a Venn diagram classifying the aforementioned terms.

The industrial adoption of DFX in the product design process grew in popularity due to its easy deployment and ability to improve competitiveness (product quality, time-to-market, etc.), rationalize decision-making, and increase the productivity of design engineering teams [21,22]. Today, DFX is recognized as the earliest consideration of design objectives and constraints, enabling the capitalization and dissemination of design knowledge. The more complex the product, the greater the value of DFX as designers need to consider all worthwhile objectives to produce successful products.

3.3.2 Design for Additive Manufacturing

With recent advancements in manufacturing technologies, AM has become a viable option for manufacturing end-use products beyond rapid prototyping and low-volume production samples. CM constraints are largely alleviated with greater design freedoms enhanced by AM. For instance, conventional design limitations such as uniform wall thickness, the avoidance of sharp corners, weld lines in injection molding, and more can be overcome using AM technologies. Chapter 4 elaborates more on the different types of AM

114 Digital Manufacturing

technologies, their process considerations, and the industry drivers for AM adoption.

To facilitate the utilization and exploitation of AM potentials, Bourell et al. recommended developing new design methodologies dedicated to the AM paradigm, known as DFAM [23]. Compared to CM technologies, AM distinguishes itself with the different levels of interdependence among design considerations and manufacturing process variables, allowing designers to leverage the enhanced degrees of design freedoms enabled by AM. There is a need to rethink the conventional design process given the specific capabilities of AM. These new ways of making also necessitate the definition of new design parameters. New and existing designs can be optimized to better fit these new parameters as 3D geometries become more complex. Hence, efforts to define DFAM can help overcome previous cognitive barriers when designing with conventional process restrictions while exploiting the design potentials of AM to optimize multiple design objectives.

3.3.2.1 Definition of Design for Additive Manufacturing

DFAM is most commonly defined as a methodology that aims to "maximize product performance through the synthesis of shapes, sizes, hierarchical structures, and material compositions, subject to the capabilities of AM technologies" [21,24,25]. It aids product conceptualization and helps designers develop new design possibilities by exploring the unique capabilities of AM. DFAM can also be redefined as a set of design methodologies and tools, which supports DFAM approaches, that helps designers consider the specificities of AM throughout the full product design process [21].

3.3.2.2 Design freedoms afforded by additive manufacturing

Designers can freely explore designs by leveraging the material, geometrical, hierarchical, and functional complexities enabled by AM, allowing teams to better exploit the unique capabilities of AM to reduce weight, part count, materials, labor time, and costs. Considering the manufacturing process limitations of AM (available processes, variable settings of various AM machines, usable materials, etc.), designers will work within feasible means to achieve the desired life cycle objectives. Despite the complexities of process planning with AM, there are greater benefits to be reaped due to the ability to customize geometries and the economical fabrication that comes with early investments in hard tooling (molds, dies, fixtures, etc.) [26]. Considering both design freedoms afforded and restrictions imposed by AM provides a realistic representation of the product, resulting in product innovation. Table 3.2 lists the design freedoms afforded by AM, which can become design opportunities or restrictions depending on applications. Section 3.4.2 will elaborate on these significant factors as a set of design considerations in the various product design stages.

TABLE 3.2 Design freedoms afforded by additive manufacturing (AM).

Material complexity	Functionally graded material parts and custom metallurgy (alloys, laminates, other composites, etc.) can be fabricated by varying material compositions distributed throughout the part to achieve desired mechanical and chemical properties. A single material can also be blended with void space to create variable porosity within a single part. Different process parameters can also modify material percentage throughout the part via point-by-point material variation control or discrete control within and between layer transitions to achieve desired qualities with property gradients for a given set of load, constraints, and objectives (ultralight and stiff structures, minimal material usage, increased strength, uniform stress distribution, etc.). This will affect the mechanical strength and physical properties of the part. Hence, process variability may expand or constrain the freedom of design depending on applications
	The wide choices of materials expand beyond polymers and metals (Chapters 6 and 7) to include variable mixes such as edible ingredients, multimaterial complex parts, and embedded electronics (Chapter 8). New fusing agents are also constantly under development to improve material properties, such as conductivity, translucency, and flexibility. The flexibility of material composition and property control capability within AM will thus continue to increase
Geometrical complexity	It is possible to directly fabricate organic and freeform geometries with AM by varying micro- and meso-structures to optimize structural properties. Unique and esthetic form factors can now be emphasized for end-use products, not just prototypes. Products can also be fabricated with varying degrees of customization, personalizing consumer products to specific preferences and enabling customer-driven design and mass customization with infinite variations
Hierarchical complexity	Within the part, complex multiscale internal structures (honeycombs, lattices, trusses, foam internals, cellular and sandwich structures, etc.), both homogeneous and uniform or heterogeneous and periodic with spatial variations, can also be created via topology and shape optimization to increase part functionality and improve product performance. The size, type, orientation, and boundary conditions of the internal structures can also affect the overall material and structural properties (porosity, deformation, failure, etc.) of resulting geometries. Other complex surfaces, textures, and porosities can also be created by AM processes with micro- or nanoscale resolution for improved functionality. Metamaterials can be produced on demand by varying internal lattice structures to produce useful application-specific properties
Functional complexity	AM enables parts to be integrated with multiple features to realize multiple functionalities. Operational mechanisms and embedded properties such as compliant mechanisms and living hinges can be fabricated directly to achieve multifunctional parts

Source: Adapted from Refs [24,26,27].

3.3.3 Trend of hybrid manufacturing production

It is crucial to note that the advent of DFAM does not imply the irrelevance of DFMA. More so, the advances of AM technologies and DFAM bring about opportunities for AM to complement current CM technologies and DFMA methodologies, while helping fulfill specific objectives listed under DFX simultaneously. While AM represents the latest evolution of manufacturing technologies, the conventional DFA, DFM, and DFMA rules are also applicable in AM. AM can overcome certain process limitations specific to CM technologies by removing material, geometrical, hierarchical, and functional limitations to produce complex components without a drastic increase in cost. This is in line with the trend of hybrid AM−CM production, which is the integration of AM technologies and supplementary CM processes to extend the technological capability of manufacturing by exploiting the advantages of each discrete process synergistically [28−30].

The most common form of hybrid AM−CM production involves building a near-net shape via AM technologies, then postprocessing it to achieve the final part specifications via CM processes [28,30−32]. In today's world of digital manufacturing, where AM is well-positioned to drastically impact conventional operations and time-sensitive production scenarios, hybrid manufacturing demonstrates immense potential for growth in the production of more complex parts that requires greater flexibility while maintaining high accuracy in a relatively short production time [30,32]. The fusion of multiaxis subtractive technologies in machining quality-formed surfaces and the superiority of additive technologies in building intricate parts opens up new ways to explore their consolidated advantages [28]. From a product planning perspective, this benefits product development, technology management, and operative production planning [29]. From an economic perspective, it may be more cost-efficient for companies to adopt hybrid manufacturing due to shorter production time, higher production rate, and more accessible processes capabilities, enhancing the competitiveness and the speed of innovation in industrial companies [29−31].

AM-fabricated parts often encounter poor repeatability, inconsistent dimensional accuracy, rough surface finish, slow process speed, and more as "an inevitable outcome of their layered nature" [28,31]. Hence, hybrid manufacturing is employed to achieve improved geometric dimensional tolerances, better surface integrity, increased material removal rate, and desired mechanical and material properties with extended application areas [30,32]. Several efforts have been initiated to incorporate design tools and systems that facilitate hybrid manufacturing. To encourage the adoption of direct digital manufacturing beyond product research and development, Strong et al. developed a system of strategically located AM hubs to links existing traditional manufacturers with evolving AM technologies through classical facility location models [32]. These AM hubs would act as suppliers to traditional manufacturers

Design methodologies for conventional and additive manufacturing Chapter | 3   **117**

who possess the demand for hybrid AM production, while leveraging the current capabilities and excess capacity of multiple traditional manufacturing facilities. Dilberoglu et al. also built a virtual manufacturing simulator for a hybrid workstation employing concurrent AM and CM technologies, enabling the integration of hybrid AM−CM production and expanding the existing direct digital manufacturing supply chain [28].

With these developments, it has become increasingly challenging to determine a clear distinction between the terms "designers," "design engineers," and "engineers," as their skillsets overlap in the product design process. From here on, the authors use these terms interchangeably to refer to engineers who actively participate in the product design process by designing parts and controlling design parameters in the design and manufacturing processes.

3.4 Design methodologies for additive manufacturing

Extensive studies have been conducted in attempts to propose a framework or methodology for DFAM [21,22,24,25,27,33−39]. These works include a range of specific design guidelines, rules, and frameworks to inform different actors of the product design process.

3.4.1 Notable Design for Additive Manufacturing research works

Over the last decade, researchers have proposed various design methodologies to generalize the adoption of DFAM. Boyard et al. proposed a design methodology to optimize parts by arranging a modular graph of functions and linking segments representing their corresponding functional and spatial organization [35]. The initial working structure and product layout are created by reusing or adapting similar sets of functions based on available geometries in a 3D-model database. Adam and Zimmer introduced a method to reconstruct AM parts' designs by its standard geometrical elements, namely, basic elements, element transitions, and aggregated structures [33]. Using function- and process-dependent design rules, suitable ranges for attribute values are recommended to ensure the manufacturability of standard elements in high qualities. Salonitis proposed an axiomatic design approach to scientifically assess good design ideas from the conceptual design stage, in an extended zigzag decomposition of the design space into domains via the independence axiom and the information axiom [38]. The decomposition is terminated when all functional requirements based on customer needs are fully satisfied by the selected set of design parameters. Bikas et al. proposed a first-time right design framework to reduce design iterations by meeting part's specifications and achieving optimum manufacturability [34]. It aims to extract manufacturability knowledge required to morph an optimal part design by bridging the gap between generic DFAM knowledge and the computational approach with AM factors'

118 Digital Manufacturing

parametrization, using multiple data inputs from real-time manufacturing and improved shape algorithms [34].

In addition, there were multiple attempts focused on reviewing and classifying existing DFAM methodologies to inform and guide designers on various levels of decision-making. Laverne et al. evaluated DFAM for concept assessment and decision-making, classified DFAM into component-based and assembly-based DFAM, and developed an assembly-based DFAM method focused on early design stages [21,22]. Similarly, Yang and Zhao reviewed and classified representative design methods related to AM and proposed a new design method that integrates function integration and structure optimization to realize less part count and better performance, while complying with the process constraints of manufacturing, assembly, and standardization [39]. To better synthesize functional requirements and process knowledge simultaneously, Yang and Zhao also suggested developing an analytic model for design rationalization and multifunctional optimization to mathematically represent design decisions [39]. Kumke et al. classified DFAM approaches and proposed a modular DFAM framework by integrating existing DFAM methods and tools into various design stages, providing specific and continuous support throughout the product design process [25]. The framework facilitates a goal-oriented utilization of AM potentials and the development of AM-conformal designs by considering the users' level of AM knowledge. Gibson et al. categorized DFAM into two broad considerations: (1) opportunistic design, where designers exploit the design opportunities afforded by AM, and (2) restrictive design, where designers focus on the restrictions imposed by AM [24]. The former perspective helps to identify AM characteristics to leverage and optimize the design space, while the latter perspective focuses on understanding the process constraints and determining how to satisfy the product requirements within these constraints.

Some researchers also worked on developing more specific DFAM methods catered to specific key actors of the product design process. Maidin et al. proposed a design feature database, which serves as a knowledge repository for inspiration and provides design recommendations to aid industrial designers in the creative process of AM parts redesign [36]. To demonstrate the major opportunities and constraints when designing for AM, systematic reviews of the current DFAM knowledge have been conducted by Thompson et al. and Pradel et al., highlighting issues related to design and redesign for direct and indirect AM production [27,37]. Pradel et al. also mapped the current DFAM knowledge into a coherent framework based on the generic design process, which assumes AM as a manufacturing process for the sequential production of end-use components [37].

In 2018, the ISO/ASTM 52910 standard was published to support general guidance and identify AM issues by providing guidelines and recommendations when using AM in product design [26]. The standard includes a general process for designing mechanical parts for structural applications based on different design objectives and a process for identifying general potentials for AM utilization. With increasing demand from industries for a standardized set of design guidelines and process adherence, particularly to facilitate future certifications of AM-fabricated parts, this standard serves as a critical starting point to optimize operations and processes for the widespread adoption of AM technologies in digital manufacturing. The standard also discusses design freedoms that AM offers and various important considerations related to geometries, materials, postprocessing, and more when designing parts for AM.

Many other proposed frameworks and methodologies based on past literature are often specific to certain design stages or AM processes, machines, and materials. There is a need to consolidate DFAM knowledge supporting different key actors at each critical design stage for consistency in language and presentation. Furthermore, the effectiveness and applicability of existing DFAM methods and tools need to be further examined, validated, and updated [25,33,36]. To fully exploit AM-specific potentials for new product generation and to facilitate designs optimized for AM, a combination of general DFAM approaches independent of applications, processes, machines, and materials is required [25]. Developing and defining DFAM knowledge remains a major challenge to systematically exploit AM's potentials and opportunities in product innovation and manufacturing. All AM practitioners are encouraged to do supplementary readings on relevant resources to properly assess the gaps within the latest literature before embarking on DFAM research or a new project that heavily involves DFAM.

3.4.2 Design stages of a general Design for Additive Manufacturing framework

Design considerations from product planning to manufacturing and postprocessing can differ significantly for different AM processes, machines, and materials. A general framework that informs designers of critical design considerations when employing AM technologies can provide early guidance in the product design process. Specifically, to bridge this paradigm shift from CM to AM and encourage the adoption of AM in facilitating hybrid manufacturing, a general DFAM framework adapted from the conventional design process is necessitated. Incorporating design recommendations independent of AM processes, machines, materials, industries, and applications, the framework synthesizes the opportunities and restrictions of utilizing AM as design considerations, built upon knowledge from the relevant established DFAM approaches in Section 3.4.1.

Fig. 3.6 outlines the proposed DFAM framework according to the main design stages listed on the left and key design considerations of the respective design stages listed on the right. It is adapted to the distinctive characteristics of AM by providing continuous coverage of the product design

120 Digital Manufacturing

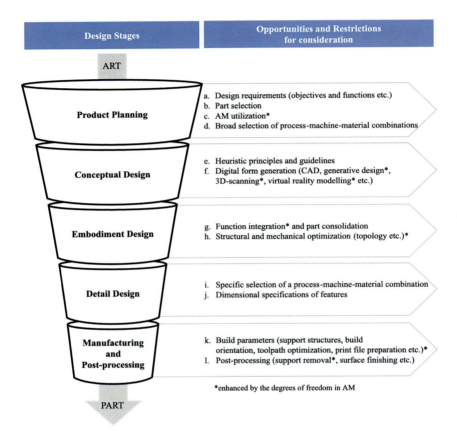

FIGURE 3.6 Proposed DFAM framework with key considerations at various design stages. *DFAM*, Design for Additive Manufacturing.

process, with detailed AM-specific considerations at each design stage to utilize AM design potentials systematically. The modular framework also allows for easy integration with various existing DFAM approaches and adaptations to promote advancements in AM technologies with redefined demands. It is crucial to build upon past research efforts and extract the different abstraction levels of DFAM into a holistic framework. The framework is proposed as a design-oriented tool, providing general key DFAM considerations rather than specific technical information in the following five design stages adapted from Pahl and Beitz's design process [5,25,37]:

1. Product planning: Design requirements (objectives and functions, etc.), part selection, AM utilization, broad selection of process−machine−material combinations.

2. Conceptual design: Heuristic principles and guidelines, digital form generation [CAD, generative design, 3D-scanning, virtual reality (VR) modeling etc.]
3. Embodiment design: Functional integration and part consolidation, structural and mechanical optimization (topology, etc.)
4. Detail design: Specific selection of a process—machine—material combination, dimensional specifications of features.
5. Manufacturing and postprocessing: Build parameters (support structures, build orientation, toolpath optimization, print file preparation, etc.), postprocessing (support removal, surface finishing, etc.)

The following section aims to inform design engineers of the key design considerations when designing for AM to optimize the conventional design process from the product planning stage to the manufacturing stage and beyond. As AM technologies stabilize, more designers and engineers are using AM processes and AM-compatible materials in the design and manufacture of products. There are also more methods to explore and create digital forms, integrate functions, and optimize structural and mechanical performance. Different build parameters and postprocessing considerations specific to AM can also expand design possibilities in terms of the part's material, geometrical, hierarchical, and functional complexities, enhanced by the increased degrees of design freedom with AM. Note that this framework does not specify the technical, economic, and manufacturing feasibility of the product design in detail. Other business, cost, and environmental considerations that should be considered predesign or those that will surface postdesign will not be discussed here, but should be considered in formal product design analysis.

3.4.2.1 Product planning

In the product planning stage, the design requirements of the overall product and its individual parts are first determined to start the product design process. Next, evaluations on the selected parts to be fabricated via AM, the availability of AM technologies, and the feasibility of AM production will be conducted. A broad combination of suitable processes, machines, and materials will then be preselected, allowing designers to focus on designs that narrow down process limitations in the later manufacturing stage.

3.4.2.1.1 Design requirements

Design requirements, such as the objectives (most lightweight design, least material costs, fastest time-to-market, etc.) and functions of the part, should be defined in the product planning stage. Common systematic engineering design approaches, such as quality function deployment, function analysis, morphological chart, and more, can also be utilized to translate these design requirements into engineering design characteristics. Design engineers ought to consider essential design features that need to be included in the part. Consulting an interactive design feature database or online repository with access to a comprehensive collection of physical AM artifacts and applications can provide insights to preliminary design solutions based on shared experiences [36]. The effectiveness of such databases can serve as a starting point for a new design or redesign.

3.4.2.1.2 Part selection

Next, selecting suitable parts for AM fabrication will define subsequent process planning parameters. Examples of good part candidates for AM fabrication include highly complex geometries that require dimensional precision, specific customization, unique combinations of material properties, integrated internal structures that require supports, and more. In addition, parts with low anticipated production volumes are also suitable for AM fabrication when there are significant improvements in manufacturing time and process costs. Fig. 3.7 shows a step stool with a simple loft profile, which can be manufactured faster and cheaper via CM (A), compared to one with complex supporting geometries, which will be a good part candidate for AM fabrication (B).

3.4.2.1.3 Additive manufacturing utilization

After selecting the parts, it is critical to identify whether the part can be fabricated via AM or not. This decision depends on factors such as the availability of AM materials, access to AM processes and machines, stipulated project duration, and cost. Reasons to employ AM fabrication should also be justified at various management levels based on the technical and economic feasibility of the production.

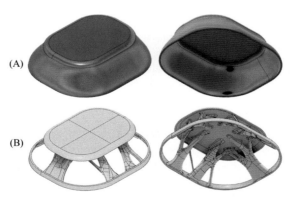

FIGURE 3.7 A poor part candidate (A) and a good part candidate (B) for AM fabrication. *AM*, Additive manufacturing.

Design methodologies for conventional and additive manufacturing Chapter | 3 **123**

If a part can be fabricated economically using CM technologies, or if it involves frequent production of simple part shapes in large production volumes, AM is not necessary. As AM provides a higher degree of design freedom and different levels of interdependence among the manufacturing variables, there may be limitations imposed due to the complexities of process planning [26]. AM should be considered, especially when flexibility in fabrication is required, including the production of complex or customized geometries, heterogeneous material properties or distributions, and especially obsolete parts. Nonetheless, the capabilities of available manufacturing technologies should be compared with the product requirements to determine technologically feasible production alternatives [29]. One must think additively and balance the costs, value delivered, and associated risks when deciding to pursue AM as a manufacturing technology in the design process.

3.4.2.1.4 Broad selection of process–machine–material combinations

Integrating the design and manufacturing requirements of the part early in its product planning stage is critical as more than 80% of the life cycle costs will be determined at this stage, even though only 10% of the expenses are incurred [40]. Using available methods and design support tools, designers can narrow down the list of appropriate process–machine–material combinations by comparing between AM and CM processes or amongst AM technologies. The suitability of a specific process for a part should be evaluated before being employed. The technical elements of the part, such as its size, surface, geometrical, physical properties, or other highly prioritized design features, can form a basis to identify and inform suitable process–machine–material combinations. Different criteria used to select amongst the AM technologies include process capability, machine accuracy, material properties, material cost, build time, production volume, and more.

According to the classifications under ISO/ASTM 52900:2021, there are currently seven classifications of AM technologies [41]. Each of these seven technologies includes a list of processes with different considerations explicitly associated with the type of process–machine–material variable settings. Table 3.3 lists the key advantages and process parameters of the seven AM technologies for reference (refer to Chapter 4 for more definitions of the seven AM technologies).

To facilitate the concurrent considerations of design and manufacturing constraints, a core AM technology and material class or type (polymer, metals, etc.) can be preselected before designing the part as this will affect the subsequent part quality and properties of the part to be manufactured. Potentials for AM fabrication can be identified as teams can evaluate the technical viability of a part via a particular AM technology and its associated

124 Digital Manufacturing

TABLE 3.3 Key advantages and process parameters of the seven additive manufacturing (AM) technologies.

Process	Key advantages	Key process parameters
Vat polymerization	Very good surface finish but support structures required	Energy beam power, scan pattern, etc.
Material extrusion	Anisotropic properties weaker in z-direction due to limitations of interlayer bonding. Support structures required for overhanging features	Material composition, nozzle diameter, material feed rate, scan speed, extrusion temperature, build chamber atmosphere, temperature, etc.
Material jetting	Able to locally control material composition by depositing different materials in different areas, resulting in fully dense material with good surface finish due to small size of droplets. Support structures required	Ink formulation, deposition pattern, stand-off distance, etc.
Sheet lamination	To be melted and consolidated at elevated temperature and pressures. Postprocessing is required to remove excess material	Sheet material composition, sheet thickness, bonding mechanism, cutting process selection, etc.
Powder bed fusion	Require support to maintain part accuracy and prevent warpage. Results in fully dense parts due to full melting. May not need support if the process produces some porosity to maintain dimensional accuracy	Laser/e-beam power, scan speed, powder composition, powder bed temperature, etc.
Binder jetting	Able to control pore characteristics with repeatability and reproducibility. Self-supporting	Powder selection, binder selection and formulation, powder-binder selection, etc.
Direct energy deposition	Able to fabricate and repair metal parts. Does not utilize support structure	Laser power, material feed rate, scan speed, atmosphere, etc.

machine and material availability. Depending on the part's application and operational environment, design engineers can specify the desired material properties early in the product design process to narrow down the selection of AM technologies. From these preliminary selections, they can then decide whether to take advantage of the layered built part's anisotropies, design to avoid the weakness of potential anisotropies, or choose another AM process that ensures isotropic material properties during print.

The emphasis on enhanced design freedom and achievable complexity of AM-fabricated parts may be diminished by other manufacturing parameters that limit the quality of such parts. While AM offers new possibilities in product design and manufacture, the advantages and drawbacks of employing AM are partially dependent on the upstream and downstream process steps [31]. Hence, considering a broad selection of process−machine−material combinations in the early phase of product design can facilitate the implementation of AM in the process chain and fulfill product requirements.

3.4.2.2 Conceptual design

Designers can leverage the design freedoms that AM offers in the conceptual design stage to exploit AM potentials fully. Taking advantage of AM process capabilities and the enhanced degrees of freedom, complex designs that satisfy multiple design objectives can be constructed within the available design space. At this stage, a plethora of design alternatives are created and evaluated upon the predefined set of design requirements. Innovation can coexist with other design objectives, such as increasing the product complexity while cutting costs. As such, heuristic principles, guidelines, and multiple approaches for digital form generation, which aid in concept generation, exploration, and selection, are discussed in the following subsections.

3.4.2.2.1 Heuristic principles and guidelines

There has been extensive development of DFAM guidelines, principles, and rules that are independent of AM processes, machines, and materials (refer to Section 3.4.1). Despite being generic and abstract in nature, these DFAM heuristic principles can provide a starting point for conceptual designs and foster creativity throughout the product design process [37]. However, it may be challenging to apply them pragmatically due to the nonspecificity of these heuristic principles. Several research efforts based on industrial case studies aim to develop DFAM rules that are process-, machine-, and material-specific, which may inspire future conceptual designs for similar applications. New design opportunities can also be systematically exploited by considering the material, geometrical, hierarchical, and functional complexities of the part, as listed in Table 3.2.

It is also beneficial for designers to consider the general design rules for CM technologies when designing for AM as it will be easier to transit from AM to CM when required. Applying conventional design rules in DFAM also offers advantages in improving overall part quality while reducing efforts in redesigning parts for AM fabrication. More importantly, this will prevent designers from creating novel forms enabled only by AM, but could not be otherwise produced by CM [37].

3.4.2.2.2 Digital form generation

To create complete digital models of a component, designers need to consider the 3D modeling capabilities of CAD tools used. There are various ways to generate and explore design alternatives, using solid 3D modeling software, generative design methods, 3D-scanning, VR modeling, and more.

Beyond pen and paper, designers heavily rely on modern CAD tools to conceptualize designs by creating solid or mesh models with boundary representations and nonuniform rational basis splines (NURBS). Fig. 3.8 presents a chair modeled in Rhinoceros 3D, a popular NURBS-based CAD software program. Some software programs such as OpenSCAD also allow designers to generate 3D models using code. With increasing part complexity, the conventional parametric system may become inadequate for modeling organic shapes and complex multiscale geometries. Moreover, general solid 3D modeling systems cannot easily generate multiscale cellular structures or indicate material variations. Data structures to store such geometric, structural, and material properties will be vital for future applications [27].

As a result, new methods utilizing data and artificial intelligence to generate algorithm-driven designs are on the rise. One such development is Grasshopper 3D, which is a visual programming system that runs within Rhinoceros 3D. Besides enabling easy parameterization of intricate elements, such as the size of spacings, the thickness of connectors, and the number of faces in XYZ directions, it allows for the integration of Python scripts to build algorithms for generating designs iteratively. Other generative design algorithms such as shape grammars, biomimicry, cellular automata, space syntax, and artificial neural networks also support form exploration. In biomimicry, the understanding of nature inspires genetic algorithms to generate complex and organic shapes based on the natural evolution of biological

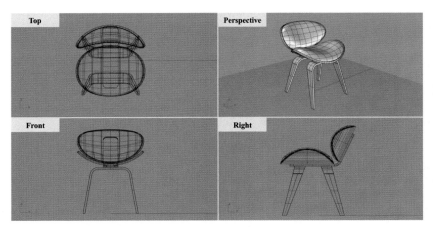

FIGURE 3.8 A chair modeled in Rhinoceros 3D.

systems. With deep learning, artificial neural networks can iteratively test and generate optimal designs that fulfill the product requirements. Fig. 3.9 shows multiple design outcomes of a bracket generated by Autodesk Fusion 360's generative design algorithm.

Designers can also use 3D-scanning to reverse engineer the design of parts by capturing and converting physical objects into digital models. A surface mesh will be first generated by scanning the surface topology of an object, which can later be modified into a CAD model to create a design. Specifically, parts can be highly customized by referencing specific 3D shapes, thereby allowing customization parameters to be tailor-made to parts with specific purposes. Fig. 3.10 gives an example of how customized sports equipment, such as a bike saddle, can be reverse engineered and modified to suit riders' comfort. Personalized prosthetics and medical aids, such as the finger/hand brace as seen in Fig. 3.11, can also be produced via 3D-scanning of the patient's hand, which can then be fabricated via AM and used for functional purposes.

VR modeling via haptic devices is another method to generate and explore form factors digitally. Using VR software and tools, such as HTC Vive and Oculus Rift, designers can sculpt freeform shapes in space, import existing models for reference, and interact with a 3D CAD model in real time. This enables design teams to brainstorm novel 3D designs while

FIGURE 3.9 Generative design outcomes of a bracket.

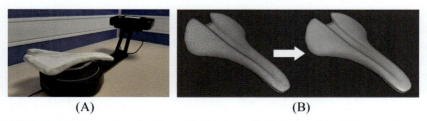

FIGURE 3.10 3D-scanning of a bike saddle (A); its meshed and smoothened 3D scans (B).

128 Digital Manufacturing

FIGURE 3.11 Finger/hand brace printed via HP Multi Jet Fusion. *HP*, Hewlett-Packard. *Courtesy HP Smart Manufacturing Application and Research Center (SMARC) Singapore, HP Inc., 2021.*

collaborating and iterating faster, encouraging cocreation in virtual environments. The gesture-based forms can then be refined and constructed in a CAD software program with more geometric details. Newer developments also allow users to generate complex 3D models based on 2D images or sketches via profile recognition and interactive interfaces powered by deep neural networks [42,43].

After all design alternatives have been produced and compared with one another, design concepts that satisfy the design requirements listed during the product planning stage will undergo further evaluation and design development.

3.4.2.3 Embodiment design

The selected design concepts are further reviewed and optimized at the embodiment design stage in terms of its product architecture and design configuration. The focus of this stage is to realize the design objectives set out in the product planning stage. The product architecture can be improved via function integration and part consolidation, given the advantages of AM. Furthermore, the design configuration can be optimized via different structural and mechanical optimization techniques focusing on the shape, size, topology, and mechanical properties of the part to be produced.

3.4.2.3.1 Function integration and part consolidation

Due to the fundamental limitations of CM technologies, designers may not be able to achieve a globally optimal consolidated structure for manufacturing. AM technologies, on the other hand, can enable combinations of forms and integrate complementary functions into a smaller number of critical parts. Integrating functions optimizes the product architecture to achieve its desired properties and performance, improving the design effectiveness and efficiency. Designers can incorporate assembly features into the parts to enable easy insertion and fixation during assembly operations for necessary assemblies. Parts of an existing assembly can also be consolidated into a single printable object, reducing the overall part count. As a result, TM, tooling, and inventory costs can be reduced or eliminated while improving the product's functionalities and performance [27]. Designers have the opportunity to rethink the design of assembled products as assemblies that can be directly fabricated, such as articulated joints and discontinuous interlinked structures with movable parts. Interlocking features to create a range of intricate fabric-like structures, such as chainmails as seen in Fig. 3.12, are also possible when designed with sufficient gaps between the interconnected chain links to prevent accidental fusing during the manufacture.

Compliant mechanisms with kinematic joints that enable relative motion between parts can also be incorporated into the part design. It is achieved by designing bending patterns into working mechanisms or structural elements that bend with the desired input—output behavior. Thin and folded sections can perform like a living hinge, allowing parts to collapse and expand to a certain degree and achieve different equilibrium states. With these hinges, products can be printed as a single assembly with interlocking parts. Fig. 3.13 demonstrates examples of a diamond flexible hinged structure and a collapsible array of living hinges.

The structural geometry of a part can also be redesigned to control its mechanical behavior under given sets of loading conditions. Different folding tessellations can create specific effects, such as a springy tension. These configurations can be printed in 2D and then folded into 3D to form the final shape. The shape and tightness of the folding pattern will directly impact the

FIGURE 3.12 Chainmails made from interlocked chain links [44].

FIGURE 3.13 A diamond flexible hinged structure (A) and an array of living hinges forming a compliant mechanism collapsible by hand (B) [44].

130 Digital Manufacturing

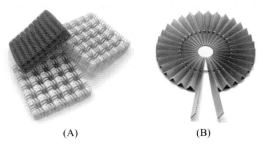

FIGURE 3.14 A spring lattice structure (A) and a 3D folding fan (B) [44].

part's movement. Fig. 3.14 presents a spring lattice structure and a 2D foldable sheet that can transform into a 3D folding fan.

The use of AM presents an opportunity to consolidate parts and print assemblies as a whole, as long as it does not result in a loss of functionality. The three critical questions asked when implementing DFA are also applicable when employing DFAM for printing assemblies (Fig. 3.2). Integrating multiple functions into the product also reduces the total part count to create a single component with greater inherent complexity [37]. However, increasing geometrical complexity may increase the build time and production cost of the part, which can become undesirable for large volume production with AM.

3.4.2.3.2 Structural and mechanical optimization

Designs can be optimized and iterated using structural and mechanical optimization techniques via computational methods. Structural optimization works to find the optimal structure by adjusting quantities defining the part's weight, strength, and stiffness properties while satisfying resource and performance constraints such as material, volume, mass, loads, and boundary conditions.

Topology optimization (TO) maximizes a part's performance by finding the optimal location, shape, size, and connectivity of holes within a given design space, subject to a given set of loads, boundary conditions, and constraints. Using mathematical algorithms, the best structure will be computed with an optimized material and stress distribution as excess material unnecessary to the part's functions is removed. Analyzing and modifying the fundamental geometry to improve functional surfaces with necessary geometric and surface features can ultimately reduce the weight of the part and lower its production costs. Fig. 3.15 shows the transformation of a bracket design via TO. More details on TO will be discussed in Chapter 5.

Parts can also be optimized mechanically using other genetic algorithms, stress analysis, distributed material compositions, linear and nonlinear finite element analysis, simulation, and modeling. The goal is to create parts with maximal strength, minimal material usage, and a desirable stress distribution under a set of given circumstances [37]. Results from optimization often yield

Design methodologies for conventional and additive manufacturing **Chapter | 3** **131**

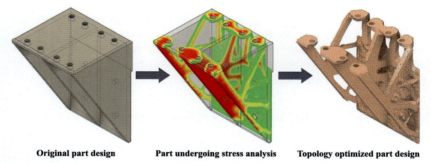

FIGURE 3.15 Topology optimization.

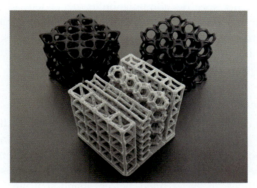

FIGURE 3.16 Different internal cellular structures. *Courtesy HP Smart Manufacturing Application and Research Center (SMARC) Singapore, HP Inc., 2021.*

complex internal geometries that are constructed using internal cellular structures, such as honeycomb, lattice, and foam structures [26]. These geometries usually have channels and structures that are difficult or impossible to fabricate using CM technologies and are only possible via AM processes. Fig. 3.16 shows 3D-printed samples of different internal cellular structures.

The impact of optimization is often more significant in critical safety applications where weight reduction is critical [37]. Progresses in understanding the mechanical properties of cellular structures and advanced optimization techniques have contributed to an increasing number of applications of functionally graded lattice structures in consumer products as well. Still, structural and mechanical optimization may not be necessarily useful, depending on the part's design constraints, functionality, and applications.

3.4.2.4 Detail design

In the detail design stage, a specific process—machine—material combination will be selected and the key dimensions of the part will be determined.

132 Digital Manufacturing

Hence, details of the part's geometrical dimensioning and tolerancing, including its feature size, thickness, tolerances and clearances specific to the selected process, machine precision, and material resolution, need to be defined. A manufacturability analysis should also be carried out prior to sending the part for printing.

3.4.2.4.1 Specific selection of a process−machine−material combination

Following Section 3.4.2.1.3, a specific process−machine−material combination will be selected at this stage based on the design and manufacturing requirements and the availability of AM processes, machines, and materials. The selection of this specific combination is dependent on the desirable properties of the end product and various business considerations, such as build time and process costs. The specific machine capabilities, process characteristics, and intermediate artifacts required to support the production should also be considered to make a wise selection.

3.4.2.4.2 Dimensional specifications of features

It is essential to check through and finalize the geometrical dimensions and tolerances of the part prior to its manufacture. The accuracy and precision of associated physical dimensions and the repeatability of the production are dependent on the part's specifications. Table 3.4 summarizes a list of key feature specifications to be defined before manufacturing. As finer part details such as small features and surface finish may degrade during part removal and postprocessing, extra attention is required to modify the associated dimensions to build a desirable part [26].

Process- and machine-specific characteristics such as the material deposition method and bonding principle will determine the type and nature of raw materials to be used and processed. This in turn affects the support strategies, material properties, and postprocessing procedures of the part. Other factors such as the data file requirement, minimum build resolution, and maximum build dimensions also need to be considered as these characteristics differ according to specific machine parameters, and they can affect the packing for a particular print job. Commercial AM service providers, such as Materialise and Stratasys, have developed several AM design guides for common process and material combinations, which are available for referencing [45,46].

3.4.2.5 Manufacturing and postprocessing

The build parameters of a part, such as its support structures, build orientation, toolpath optimization, and print file specifications, cannot be developed independently from the process, machine, and material parameters. These

TABLE 3.4 A list of key feature specifications to be defined before manufacturing [26,27].

Feature specifications	Remarks
Maximum part size (length, width and, height, aspect ratio, etc.)	The maximum build envelope of the selected machine limits the maximum part size that can be printed. Hence, the part should be scaled for accuracy and precision to prevent warpage and shrinkage
Minimum feature size and spacing	This includes the minimum hole diameter, wall thickness, dimensions of text embossing or debossing, font size and typeface, text finishing, and other geometric structures such as threads, cavities, chains, hinges, and internal springs. The dimensions of these features are determined according to the smallest feature resolution or nozzle diameter of the selected process and machine. The minimum gaps and slot widths between adjacent features should also be considered based on the selected process and machine to ensure that movement within the part or assembly, if intended, is still possible
	For instance, the apex of folded hinges should be rounded to avoid the risk of snapping when the printed part is stretched or pressed. Parts should also be designed to minimize abrupt thickness transitions, which can cause distortions, especially with thermal-driven processes as thicker sections tend to retain heat
Support requirements	Depending on the feature shapes and sizes, internal or external support structures may be required. For supported features, the threshold angle will depend on overhangs, feature sizes, and the overall part aspect ratio. Downward-facing flat surfaces, overhangs, and feature-rich designs should be reduced as much as possible to prevent problems during support removal. Support removal should be planned by designing access holes to prevent unused built material from becoming trapped in hollow or enclosed volumes. Alternatively, the part can be designed or oriented to be self-supporting to reduce scarring on its surface during postprocessing
Recommended assembly fit, tolerances, and clearances	To facilitate the product's assembly, the selection of geometrical tolerances and dimensions of inference, transitional, and clearance fits between mating features should be examined with reference to the specific process-machine capabilities. Interlinked structures should also have sufficient clearances between the individual bodies during fabrication so that any remaining interstitial material is removable after manufacturing. Inclusions, voids, or layer delamination may also happen depending on the process parameters
Edges and corner radii	Sharp inner edges should be avoided, while rounded edges and fillets should be used to prevent the formation of detrimental stress concentrations and difficulty in support removal. Care should be taken when adjusting the corner radii to prevent undercut fillets from forming

Source: Adapted from Refs [26,27].

134 Digital Manufacturing

factors directly affect the quality of the built part and its subsequent postprocessing steps.

3.4.2.5.1 Build parameters

For material extrusion processes, the support structures required, build orientation, and toolpath for production are usually determined by the slicing software. However, most slicing software developments are independent of CAD tools and mostly customizable to a limited list of specific machine settings. Support structures provided by these software programs are often limited to a few types of patterns such as line, grid, and zigzag, while available build positions may also be limited to a 15-degree increment snapping in the three axes, depending on how developed the software is. Fig. 3.17 shows some examples of different support patterns with varying infill patterns. Users are generally unable to control the support structures, build orientation, and toolpath of the print job as each layer is built in the direction of the z-axis. Some CAD tools provide the option to connect to the slicer software via a plug-in so as to create a more seamless workflow when designing for AM. One example is the development of the SolidWorks integration plug-in, where designers can directly export their CAD files from the SolidWorks interface to Ultimaker Cura [47]. The latest design iteration will then directly open in place within the Cura platform.

With greater liberty in more complex AM processes such as powder bed fusion and vat polymerization, manufacturers can customize the support structures, build orientation, and toolpath of a part. Doing so can minimize the build time and costs specific to the selected process—machine—material combination. When customizing internal and external supports, it is critical to check the dimensions of self-supporting overhangs and threshold angles as analyzing where and how to support the part will affect the final product

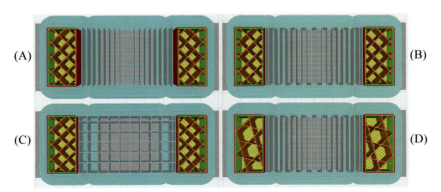

FIGURE 3.17 Line (A), zigzag (B), and grid (C) supports with 15% infills. Part (D) shows a trihexagonal infill variation.

geometry and surface quality. This ensures the part's manufacturability based on the machine's printing accuracy, the part's desired surface roughness, and its anisotropies. Some raw materials serve as natural supports, while some build jobs require a sacrificial build plate or structures to anchor the part to the build plate. Support structures should be minimized by eliminating features and faces needing support and optimizing the part design with self-supporting features. Support structure generation can also be controlled via variables that control support density, placement, touchpoint size, and position [26].

An appropriate build orientation ensures the stability of the part during manufacturing. A poor selection of the build orientation can affect the part's impact strength, surface roughness, tensile, flexural, and compression properties. An undesirable build orientation may also cause residual porosity and induce unwanted anisotropies, resulting in failure initiation and crack propagations at the material deposition interfaces of uneven layers during manufacturing [26]. The lack of support structures can be compensated by orienting the part to enable maximum strength during the printing process [27]. As each machine has a different maximum build envelope, a product may be divided into multiple parts for production and assembled thereafter, although there may be implications on the material and physical properties of the final product. For powder bed fusion, nested packing of the print bed can ensure efficient printing. Inevitably, there will be trade-offs between the build orientation, support material usage, build time, and overall build quality.

The optimization of the toolpath can prevent anisotropies, especially in critical parts for safety or high-yield applications. It presents an opportunity to fabricate complex parts with better precision, surface quality, and mechanical strength within a shorter build time. For instance, Dai et al. successfully customized and printed a bunny model with spatial toolpaths using multidirectional deposition methods [48]. However, a deeper understanding of the benefits and drawbacks of the process—machine—material selections is required to truly optimize the toolpath of a print job. Hence, most AM practitioners still prefer to use the default toolpath recommended by conventional slicing software.

The mesh sizing of the Standard Tessellation Language (STL) or Additive Manufacturing File Format (AMF) part file prepared premanufacture can also affect the surface smoothness and accuracy of the built part. The tessellation quality and slicing scan thickness should be compatible with the machine precision and material resolution of the selected AM process. The layering nature of AM may leave layer marks or small surface transitions known as "stairsteps" along the external surfaces of parts when there is insufficient overlap between successive layers. These "stairsteps" can turn into crack initiation sites and reduce the part's fatigue life [26]. The highest complexity representation of the model should also be retained when importing to and exporting from different software tools. Care should be taken

136 Digital Manufacturing

when preparing the print file for manufacturing based on prior understanding of the specific process, machine, and materials selected.

3.4.2.5.2 Postprocessing

If supports are used during the printing process, they have to be removed to maintain the part's functionality. For metals, the parts are usually anchored to the build platform, and machining is required to remove these parts, followed by cooling. For fine powders, support removal involves unpacking and bead blasting or sandblasting in a processing chamber with personal protective equipment, followed by rinsing to remove the excess loose powder. In special cases, cleaning chains encased in long hollow structures or cleaning brushes modeled with the part's surface folds can be specifically printed to enable easier and faster removal of interstitial powder [44]. Some support structures can also be dissolved and removed in solvents. Manual component handling is often necessary at this stage, resulting in additional labor costs and time.

The support removal process can result in uneven surface finish and low accuracy of the supported surfaces. Other factors such as the layer discretization of the part geometry during printing, build orientation, quality of material feedstock, and other process parameters can also affect surface qualities. Hence, surface finishing processes, such as machining, thermal treatments, coatings, polishing, grinding, dyeing, vibratory tumbling, and curing, are required to achieve the desired surface finish, tactile, and visual qualities of the part. Depending on the part's application, surfaces can adorn glossy, satin, or matte finishes. These processes can also improve material properties and increase mechanical strength by altering grain structures and anisotropies.

3.4.3 Challenges of Design for Additive Manufacturing

The current adoption of DFAM and AM technologies primarily lies within industries where the additional value generated via AM outweighs the higher production costs incurred. With AM, engineering-intensive businesses such as aerospace, medical, and automotive can accelerate prototyping, explore new lightweight structures, and create fully customizable products at minimal extra costs, or none at all [49]. High-value, low-volume production businesses also benefit from the more flexible manufacturing processes with fewer parts, shorter TM, and less material wastage, hence improving production efficiency [49]. Spare parts-intensive or assets-heavy businesses in fields such as maintenance, repair, and overhaul can achieve faster time-to-market, freedom from obsolete parts, and more local or on-demand production opportunities with independence from traditional suppliers [49]. Given the digital nature of designing for AM, companies may also choose to diversify their operations with localized production via home-based manufacturing by individuals (centralized applications) or distribute a global AM production chain with remote teams at different locations (decentralized applications) [27]. These possibilities are empowered by the lowering costs of desktop 3D printers, the prevalence of open-source digital model repositories whereby individuals can instantly upload and download CAD models for 3D-printing, and the ubiquity of cloud storage to enable collaborations regardless of timezones and geographical locations.

To better reap the benefits of manufacturing with AM, establishing standardized information on designing with AM is required. While ISO/ASTM has published AM standards for industrial adoption and practices, there are still several challenges faced by the academic community and industry practitioners when implementing DFAM in the product design process.

3.4.3.1 Lack of universal validation for a common framework and additive manufacturing–related standards

Most DFAM knowledge emerged from experimentation and is specific to a particular process, machine, material, or a combination of these specifications. Furthermore, most DFAM research works are developed independently and do not have interfaces between existing DFAM elements. Majority of the earlier research works introduced in Section 3.4.1 also focus on investigating design opportunities and restrictions due to AM in contrast to the design limitations imposed by CM. Without a well-validated DFAM framework to justify design decisions, the designer's imagination may be constrained to his or her existing knowledge of DFMA, CM, and AM process capabilities. It will be challenging to adapt their thinking and experiences for exploiting AM capabilities [37].

As the capabilities of AM improve and become more consistent, there is a need for a standardized DFAM framework that is well-validated to support the product design process and to guide AM practitioners throughout all design stages and beyond. This is because AM design principles can differ in their applicability and support for different skill levels of AM practitioners at different stages of the product design process. For instance, a lightweight design can be optimized using TO in the embodiment design stage by a design engineer or using infill structures in the manufacturing stage by a manufacturer, depending on the specificity of guidance available to these key actors.

AM processes also have their fundamental differences that require additional design guidance. However, the highly complex interactions between the process parameters and part properties make it more challenging to develop a common database with empirical values specific to each AM process, machine, and material [25]. Sets of generalized design knowledge applicable across different machines utilizing similar AM processes can be proposed in future studies. Extensive validation of the proposed value ranges is also required for practitioners to gain sufficient confidence when relying on such information. Doing so can ensure that procedures and

138 Digital Manufacturing

qualifications are repeatable and standardized across organizational and industrial practices.

To resolve the lack of comprehensive AM-related standards, ISO and ASTM have joint efforts, bringing researchers and industry practitioners together to rapidly define and develop international standards supporting DFAM and various AM-related topics [27,50]. With the increased standardization for AM technologies, many process- and material-specific standards that promote DFAM knowledge and stimulate DFAM research and implementation are in development. These efforts are vital as fields with a high level of safety requirements may warrant certifications of AM-fabricated products and their associated design and manufacturing processes.

3.4.3.2 Limited studies on additive manufacturing process—machine—material selection

Though existing softwares have limited support for the selection between CM and AM technologies, these process selection tools are necessary to help engineers make informed comparisons between CM and AM, especially when AM is evaluated as a competitive alternative to CM. The lack of knowledge on the long-term performance of AM materials also hinders the AM process—machine—material selection process. Decision support systems that include a catalog of AM processes and their characteristics will be helpful in the rapid identification of suitable choices, thus advancing the adoption of AM [37]. The integration of AM material knowledge in these tools will also help designers analyze options over a wide range of criteria and provide appropriate suggestions for process—machine—material selections among CM and AM. As it may be time-consuming to validate local machine capabilities individually, manufacturers can use process and material benchmarks to develop local DFAM rules, compare process parameters, and inform production strategies [27].

3.4.3.3 Need for additive manufacturing—specific software tools

Current solid 3D modeling and parametric/NURBS-based CAD systems are already capable of creating AM-conformal designs via available CAD functions. Comprehensive design feature libraries and user-friendly interfaces will enable easier brainstorming and iterations compared to current systems. Databases of AM artifacts can also serve as association aids and standardized metrics for parts [25]. Incorporating underlying constraint relations in such software tools can also facilitate the information management of manufacturing constraints and help designers prepare more precise geometric models. The development of DFAM-specific software support tools can allow design engineers to exploit AM potentials better within its process and material restrictions based on the intended design objectives and hence, greatly improving the applicability of DFAM frameworks.

Current TO-based design tools run slowly due to the massive memory consumption. Difficulties can build up, especially when manipulating interior volumes and structures, as the number of complex physical features increases with the rise in the hierarchical order of magnitude [39]. These tools often use finite element analysis and optimal control theory to optimize the internal topology according to specified objectives, resulting in unpredictable internal structures [39]. More efforts can be focused on developing cellular structure design software, which can generate predictable multiscale structured parts at comparatively higher speeds and create parts with varied densities within a single component [39]. These software programs should also be developed with the ability to specify material distributions and distributed structural properties within the part.

The current STL format only defines the surface mesh information of a CAD model while AMF provides additional representations for color, texture, material, and other properties. Hence, there is also a growing need for standardized interchangeable data file formats to handle analytical information on the shape and material distribution, the granularity of the build, and more [27].

3.4.3.4 Need to update design and engineering education

As AM technologies progress, there is a need to update the current education curriculum with AM-specific knowledge for both design and engineering disciplines. Educational materials related to DFAM should be accessible at all levels for all engineering professions [27]. Recognizing and defining the knowledge required at the different design stages can also lead to more customized guidance and effective design education [37]. This can be achieved by ensuring that there are support modules of varying abstraction levels adjustable to different types of designers and engineers based on their skills, experiences, and product knowledge [25]. Future research can provide an understanding of the inherent characteristics, material properties, and process-machine specifications of different AM processes that designers need to be informed of. Declaring the context validity of research findings and incorporating relevant practitioners' inputs can also better foster adoption and understanding of DFAM knowledge in practical settings [37].

Beyond the rise of AM, other advances in digital manufacturing also call for an evolution in traditional engineering education. To augment the knowledge and capabilities of engineers, design thinking skills are increasingly integrated into the transdisciplinary engineering curriculum to promote a human-centered innovation mindset and design process, leading to better products and services [51]. Hence, new programs and curriculum developments with an emphasis on design for digital manufacturing are required to support design activities and trainings in educational institutions and industries. Most importantly, the dissemination of design knowledge representation, analysis, and

140 Digital Manufacturing

optimization tools from the academia and hobby community to industries, especially within the fields of digital manufacturing, is necessary [27].

3.5 Summary

Design methodologies must continue to adapt to state-of-the-art manufacturing technologies as they advance and evolve. Design methodologies that apply to CM, such as the Boothroyd-Dewhurst DFMA, will need to be updated with information on AM process, machine, and material parameters to stay relevant in today's digital manufacturing landscape. Concurrently, new DFAM knowledge needs to be organized into general frameworks and specific rules to guide different levels of AM practitioners throughout the product design process and to exploit current AM capabilities and their emerging potentials fully.

New research directions and efforts will continue to improve the applications of AM and the adoption of DFAM. With progress driven from top-down management and bottom-up in research and practices, DFAM knowledge will push the boundaries of new design possibilities, product opportunities, and production paradigms. With greater degrees of freedom in material, geometrical, hierarchical, and functional complexities, the advent of AM technologies has dramatically changed the way parts are manufactured, from rapid prototyping to the production of end-use products. As the possibilities enhanced by AM continue to expand, a new era in digital manufacturing is just beginning.

References

[1] F.C. Moon, Franz Reuleaux: contributions to 19th century kinematics and theory of machines, Journal Applied Mechanics Reviews 56 (2) (2003) 261−285.

[2] F. Reuleaux, The Kinematics of Machinery: Outlines of a Theory of Machines, Courier Corporation, 2013.

[3] C.M. Eastman, Design for manufacture and assembly: the Boothroyd-Dewhurst experience, Design for X: Concurrent Engineering Imperatives, Springer Science & Business Media, 2012, pp. 19−40.

[4] K.N. Otto, Product Design: Techniques in Reverse Engineering and New Product Development, 清华大学出版社有限公司, 2003.

[5] G. Pahl, W. Beitz, Engineering Design: A Systematic Approach, Springer Science & Business Media, 2013.

[6] C.-H. Chen, L. Occena, S.C. Fok, CONDENSE: a concurrent design evaluation system for product design, International Journal of Production Research 39 (3) (2001) 413−433.

[7] S. Ashley, Cutting costs and time with DFMA, Mechanical Engineering 117 (3) (1995) 74−77.

[8] I. Bettles, Design for manufacture & assembly (DFMA)—the Boothroyd & Dewhurst Approach, in: 1992 Third International Conference on Factory 2000, 'Competitive Performance Through Advanced Technology', 1992, IET.

Design methodologies for conventional and additive manufacturing Chapter | 3 **141**

[9] C.M. Eastman, Case experiences with design for assembly methods, Design for X: Concurrent Engineering Imperatives, Springer Science & Business Media, 2012, pp. 41−71.

[10] S. Gao, S.P. Low, K. Nair, Design for manufacturing and assembly (DfMA): a preliminary study of factors influencing its adoption in Singapore, Architectural Engineering and Design Management 14 (6) (2018) 440−456.

[11] T. Ohashi, et al., Extended assemblability evaluation method (AEM), JSME International Journal Series C Mechanical Systems, Machine Elements and Manufacturing 45 (2) (2002) 567−574.

[12] C. Barnes, et al., Assembly sequence structures in design for assembly, in: Proceedings of the 1997 IEEE International Symposium on Assembly and Task Planning (ISATP'97)-Towards Flexible and Agile Assembly and Manufacturing, 1997, IEEE.

[13] M. Solutions, A metric for product improvement. Available from: <https://www.dfma.com/resources/mglobal.asp>, 2020 (cited 02.12.20).

[14] V. Frey, The one million dollar story—our way to success with DFMA. Available from: <https://www.dfma.com/resources/endress.asp>, 2020 (cited 02.12.20).

[15] W. Ni, Re-design on vehicle front support structure using DFMA. Available from: <https://www.dfma.com/pdf/wanpaper.pdf>, 2020 (cited 02.12.20).

[16] Building and Construction Authority, Design for manufacturing and assembly (DfMA). Available from: <https://www1.bca.gov.sg/buildsg/productivity/design-for-manufacturing-and-assembly-dfma>, 2020 (cited 02.12.20).

[17] V.W. Tam, J.J. Hao, Prefabrication as a mean of minimizing construction waste on site, International Journal of Construction Management 14 (2) (2014) 113−121.

[18] G. Sohlenius, Concurrent engineering, CIRP Annals 41 (2) (1992) 645−655.

[19] T. Tomiyama, et al., Design methodologies: industrial and educational applications, CIRP Annals 58 (2) (2009) 543−565.

[20] J.G. Bralla, Design for Excellence, McGraw-Hill Professional Publishing, 1996.

[21] F. Laverne, et al., DFAM in the design process: a proposal of classification to foster early design stages, in: CONFERE, Sibenik, Croatia, 2014, p. 116.

[22] F. Laverne, et al., Assembly based methods to support product innovation in design for additive manufacturing: an exploratory case study, Journal of Mechanical Design 137 (12) (2015).

[23] D.L. Bourell, M.C. Leu, D.W. Rosen, Roadmap for Additive Manufacturing: Identifying the Future of Freeform Processing, The University of Texas at Austin, Austin, TX, 2009, pp. 11−15.

[24] I. Gibson, et al., Design for additive manufacturing, Additive Manufacturing Technologies, Springer, 2021, pp. 555−607.

[25] M. Kumke, H. Watschke, T. Vietor, A new methodological framework for design for additive manufacturing, Virtual and Physical Prototyping 11 (1) (2016) 3−19.

[26] International Organization for Standardization, ISO/ASTM 52910: Additive Manufacturing Design—Requirements, Guidelines and Recommendations, 2018.

[27] M.K. Thompson, et al., Design for additive manufacturing: trends, opportunities, considerations, and constraints, CIRP Annals 65 (2) (2016) 737−760.

[28] U.M. Dilberoglu, et al., Simulator of an additive and subtractive type of hybrid manufacturing system, Procedia Manufacturing 38 (2019) 792−799.

[29] A. Jacob, et al., Integrating product function design, production technology optimization and process equipment planning on the example of hybrid additive manufacturing, Procedia CIRP 86 (2019) 222−227.

142 Digital Manufacturing

[30] Z. Zhu, et al., A review of hybrid manufacturing processes—state of the art and future perspectives, International Journal of Computer Integrated Manufacturing 26 (7) (2013) 596–615.

[31] T. Häfele, et al., Hybrid additive manufacturing—process chain correlations and impacts, Procedia CIRP 84 (2019) 328–334.

[32] D. Strong, et al., Hybrid manufacturing—integrating traditional manufacturers with additive manufacturing (AM) supply chain, Additive Manufacturing 21 (2018) 159–173.

[33] G.A. Adam, D. Zimmer, Design for additive manufacturing—element transitions and aggregated structures, CIRP Journal of Manufacturing Science and Technology 7 (1) (2014) 20–28.

[34] H. Bikas, A. Lianos, P. Stavropoulos, A design framework for additive manufacturing, The International Journal of Advanced Manufacturing Technology 103 (9–12) (2019) 3769–3783.

[35] N. Boyard, et al., A design methodology for parts using additive manufacturing, in: 6th International Conference on Advanced Research in Virtual and Rapid Prototyping, 2013.

[36] S.B. Maidin, I. Campbell, E. Pei, Development of a design feature database to support design for additive manufacturing, Assembly Automation 32 (2012).

[37] P. Pradel, et al., A framework for mapping design for additive manufacturing knowledge for industrial and product design, Journal of Engineering Design 29 (6) (2018) 291–326.

[38] K. Salonitis, Design for additive manufacturing based on the axiomatic design method, The International Journal of Advanced Manufacturing Technology 87 (1–4) (2016) 989–996.

[39] S. Yang, Y.F. Zhao, Additive manufacturing-enabled design theory and methodology: a critical review, The International Journal of Advanced Manufacturing Technology 80 (1–4) (2015) 327–342.

[40] K. Ishii, Life-cycle engineering design, Journal of Vibration and Acoustics 117 (B) (1995) 42–47.

[41] International Organization for Standardization, American Society for Testing and Materials (ISO/ASTM), 52900:2021, Additive Manufacturing - General Principles - Fundamentals and Vocabulary, ISO/ASTM, 2021.

[42] R.H. Kazi, et al., DreamSketch: early stage 3D design explorations with sketching and generative design, in: UIST, 2017.

[43] Y. Li, et al., SweepCanvas: sketch-based 3D prototyping on an RGB-D image, in: Proceedings of the 30th Annual ACM Symposium on User Interface Software and Technology, 2017.

[44] A. Ju, M. Baker, J. Jun, The Fundamentals of Making With MJF, HP Immersive Experiences Lab, HP Inc, 2018.

[45] Materialise, Design guidelines. Available from: <https://www.materialise.com/en/academy-manufacturing/resources/design-guidelines>, 2021 (cited 08.02.21).

[46] Stratasys. Design guidelines. Available from: <https://www.stratasysdirect.com/resources/design-guidelines?resources = 43022f6bc24f4beca37b7205c8bc6c27&sortIndex = 0>, 2021 (cited 08.02.21).

[47] M. Jani, Ultimaker Cura 3D printing software—SolidWorks integration for a better workflow. Available from: <https://ultimaker.com/learn/ultimaker-cura-3d-printing-software-solidworks-integration-for-a-better>, 2018 (cited 31.05.21).

[48] C. Dai, et al., Support-free volume printing by multi-axis motion, ACM Transactions on Graphics (TOG) 37 (4) (2018) 1–14.

[49] J. Bromberger, R. Kelly, Additive manufacturing: a long-term game changer for manufacturers, in: E. Backwell, T. Gambell, V. Marya, C. Schmitz (Eds.), The Great Re-Make: Manufacturing for Modern Times, McKinsey & Company, 2017, pp. 59–66.

[50] G. Moroni, S. Petrò, H. Shao, On standardization efforts for additive manufacturing, in: Proceedings of 5th International Conference on the Industry 4.0 Model for Advanced Manufacturing, 2020, Springer.

[51] J.-Y. Kuo, et al., Fostering design thinking in transdisciplinary engineering education, in: The 28th ISTE International Conference on Transdisciplinary Engineering, 2021.

Chapter 4

Additive manufacturing for digital transformation

Yu Ying Clarrisa Choong
LRQA Limited, Singapore

List of abbreviation

3D	three dimensional
ASTM	American Society for Testing and Materials
ALM	additive layer manufacturing
AMF	Additive Manufacturing Format
BEAM	electron-beam additive manufacturing
CAD	computer-aided design
BJ	binder jetting
CLIP	Continuous Liquid Interface Production
DDM	direct digital manufacturing
DED	directed energy deposition
DLP	digital light processing
DMD	direct metal deposition
DMLS	direct metal laser sintering
EBM	electron-beam melting
FDM	fused deposition modeling
FFF	free-form fabrication
FFF	fused filament fabrication
LENS	laser engineered lens shaping
LOM	laminated object manufacturing
MJF	multijet fusion
MJM	Multijet Modeling
PPE	personal protective equipment
PBF	powder bed fusion
SFF	Solid Freeform Fabrication
SLA	stereolithography apparatus
SLM	selective laser melting
SLS	selective laser sintering
STL	Surface Tessellation Language
UAM	ultrasonic additive manufacturing

Digital Manufacturing. DOI: https://doi.org/10.1016/B978-0-323-95062-6.00002-4
© 2022 Elsevier Inc. All rights reserved.

UC	ultrasound consolidation
UV	ultraviolet
VPP	vat photopolymerization
WAAM	wire arc additive manufacturing

4.1 Introduction to additive manufacturing

4.1.1 Definitions and terminologies

Additive manufacturing (AM), more commonly referred to as three-dimensional (3D) printing, is defined by the American Society for Testing and Materials [1] (ASTM) as *the process of joining materials to make parts or objects from 3D model data, usually layer upon layer, as opposed to subtractive manufacturing and formative manufacturing.* Fig. 4.1 illustrates the process of subtractive manufacturing which involves continuously removing material (via laser cutting, milling, drilling, or lathe) from a solid block of material. In contrast, the AM process involves laying down successive layers of materials to form a 3D object. AM technology enables digital files to be transformed into physical objects and vice versa, transforming the digital–physical interface for product design, development, and manufacturing. Objects with complex geometrical shapes can be created using computer-aided design (CAD) or 3D scanners.

The term *additive manufacturing* is the official industry terminology defined by the ASTM Committee F42 and used for all applications of the technology. Currently, AM is frequently referred to as *3D printing* (3DP) [3],

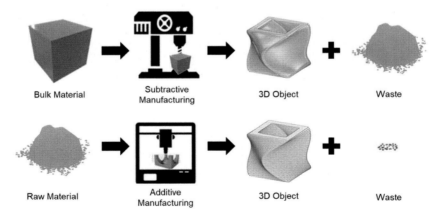

FIGURE 4.1 Difference between the subtractive and additive manufacturing processes [2]. *Adapted from C. Chen, X. Wang, Y. Wang, D. Yang, F. Yao, W. Zhang, et al., Additive manufacturing of piezoelectric materials. Advanced Functional Materials, 30 (52) (2020) 2005141.*

Additive manufacturing for digital transformation Chapter | 4 **147**

TABLE 4.1 Alternative terminologies for additive manufacturing in different contexts.

Alternative terminologies	Context
3DP	Global
Rapid prototyping/manufacturing	Global
Free-form fabrication (FFF)	USA
Solid Freeform Fabrication (SFF)	USA
DDM	USA
Generative manufacturing	Germany
eManufacturing	Germany
Constructive manufacturing	Germany
3D ALM	EADS

3DP, 3D printing; *ALM*, Additive layer manufacturing; *DDM*, direct digital manufacturing; *EADS*, European Aeronautic Defense and Space; *USA*, United States.

where these two terms can be used interchangeably. 3D modeling, fit and function prototyping, and direct part manufacturing are the key applications of AM. Depending on the industrial scope and applications, there are various commonly used terminologies over the last three decades which encompass terms such as rapid prototyping, rapid manufacturing, free-form fabrication, and others as summarized in Table 4.1.

4.1.2 Overview of the additive manufacturing market

4.1.2.1 Additive manufacturing—a building block of digital manufacturing

Transformation is currently taking place in the industry and signaling a change as we move away from the traditional manufacturing landscape. Traditional manufacturing can no longer meet the stringent challenges and requirements of the rapidly changing industry world. In the new digital manufacturing era, connectivity, intelligence, and flexible automation are the key technology trends that drive this transformation. As we embrace ourselves and advance toward the new paradigm shift in the manufacturing workflow, new technological breakthroughs will help to blend the physical and digital worlds together. Fig. 4.2 depicts the building blocks that are powering the digital transformation in the manufacturing industry today. AM forms one of the key technologies that transform the manufacturing operations and customer experience.

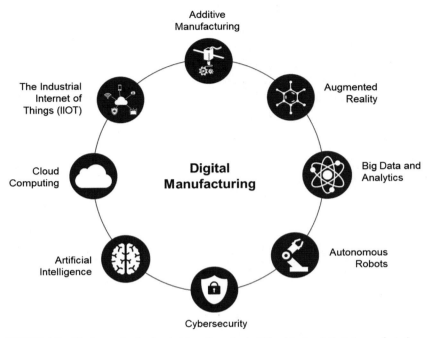

FIGURE 4.2 The key technologies that transform the building blocks of digital manufacturing.

4.1.2.2 The rise of additive manufacturing

The AM technology has come a long way since its launch in the 1980s when Charles W. Hull from 3D Systems introduced and filed a patent for the first stereolithography apparatus (SLA) in 1984 [4]. AM has been in existence for more than 30 years, but this technology has only recently increased in popularity, drawing the attention of both technology experts and the public. AM is currently at a technological and economic inflection point with the ability to reinvent and transform the $\sim$ \$12 trillion global manufacturing market [5]. It can disrupt conventional cost structures and supply chains, give emerging economies equal opportunities to close their competitiveness gap, and bring manufacturing closer to end consumers. The AM market is rising at more than 25% per year over the last 30 years and accounting for about \$14 billion invested in 2019 [6].

4.1.2.3 The global additive manufacturing market

According to the International Data Corporation, the United States and Western Europe account for nearly two-thirds of the worldwide AM market as seen from Fig. 4.3. Despite its rapid growth, AM only contributes to $\sim$ 0.1% of the global manufacturing output. Sales from 3D printers constitute

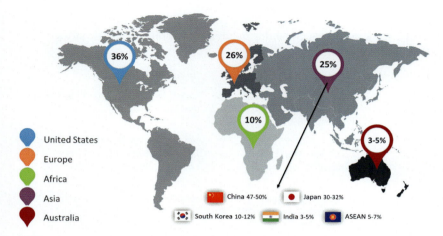

FIGURE 4.3 The overview of the global AM market [7]. *AM*, Additive manufacturing. *Statistics obtained from Thyssenkrupp analysis (J. Wöhrmann, Additive Manufacturing Adding up Growth Opportunities for ASEAN, Thyssenkrupp, Singapore, 2019).*

the largest AM market share of 40%, while other AM services account for 30%, material suppliers take up 15%, and software providers make up the remaining share [7].

4.1.2.4 Additive manufacturing systems market

Due to the technological advancements made in metal AM systems in recent years, the technology has become increasingly appealing to the industry. Even though metal AM systems incur a higher production cost, the interest in AM metal parts for the industry is higher than that of polymer parts. This may be attributed to why AM metal parts have higher quality functionality compared to its polymer counterparts. Nevertheless, the sales of industrial polymer printers have risen marginally over the last 3 years. Potential reasons lie in polymer systems are becoming more affordable and falling below the cap for industrial printers. Wohlers Report 2019 has revealed that 3D home printer sales rose dramatically from 162,199 units in 2014 to 591,079 units in 2018 [8].

4.1.2.5 Expiration of key patents and industrial trends

Key patents, as well as technological barriers, are relevant determinants that may affect the AM systems market. After the expiration of the last major patent for fused deposition modeling (FDM) in 2009, it was possible to manufacture printers without the infringement of intellectual property, which, in turn, generated new interest and investment in AM technologies. Within a limited period since the expiration of FDM patents, a large number of firms

150 Digital Manufacturing

TABLE 4.2 Key patents and industrial trends in additive manufacturing (AM) systems market [9].

AM system	Patent info	Industrial trends
FDM	Patent: US5121329 A Published: October 30, 1989 Expired: 2009	Large number of firms selling low-cost FDM machines
SLS	Patent: US5597589 Published: May 31, 1994 Expired: January 28, 2014	No comparable development as FDM as SLS machines are more complex and requires professional competency
SLM	Patent: DE19649865 Published: December 2, 1996 Expired: 2016	Analysts anticipate a substantial rise in market participants with 983 to 1768 units sold from 2016 to 2017

AM, Additive manufacturing; *FDM*, fused deposition modeling; *SLM*, selective laser melting; *LS*, selective laser sintering.
Source: Adapted from P. Minetola, L. Iuliano, G. Marchiandi, Benchmarking of FDM machines through part quality using IT grades, Procedia CIRP 41 (2016) 1027–1032.

have entered the AM market selling low-cost FDM machines [9] (Table 4.2). However, the expiration of the patent for selective laser sintering (SLS) machines in 2014 did not experience comparable development as with FDM. SLS machines are more complex and require professional competency to build new systems. On the other hand, analysts anticipated a substantial rise in market participants when the patents for selective laser melting (SLM) machines ended in 2016 due to the high market attractiveness. The number of SLM units sold from 2016 to 2017 increased from 983 to 1768 units.

4.1.3 Industry drivers for additive manufacturing adoption

Several headlines have postulated how AM will change the market environment, how it will reshore manufacturing to high-wage markets, disrupt existing industries, and bring consumers and manufacturers closer together. AM provides a suite of compelling market advantages that companies in almost every segment of the economy can leverage in numerous ways across the supply chain. Technological developments, advancements in application areas, rapid low-cost product development, and strong financial support from governments around the world are key factors driving the global AM industry.

The breadth and impact of AM continue to expand as the technology gains acceptance and functionality, making it a viable manufacturing process in various industries. Four main sectors—aerospace, automotive,

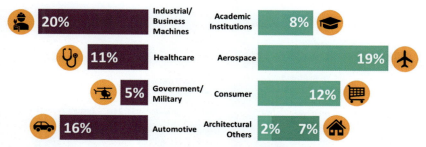

FIGURE 4.4 The key industrial AM drivers and distribution of AM revenue [10]. *AM, Additive manufacturing. Statistics obtained from the Wohlers Report 2018 (I. Campbell, et al., Wohlers Report 2018: 3D Printing and Additive Manufacturing State of the Industry: Annual Worldwide Progress Report, Wohlers Associates, 2018).*

industrial/business machines, and healthcare—account for nearly 65% of the global AM market (as seen from Fig. 4.4) and are the key sectors driving the AM market. Among these sectors the aerospace industry has been the trendsetter in AM, given its low-volume, high-mix parts, and lightweight structural requirements, leading to fuel efficiency and cost savings, which directly harness the main attributes of AM.

4.1.3.1 Additive manufacturing versus conventional manufacturing

Traditional business models are changing, which has also radically changed demand patterns for products and services across sectors. Globalization and interconnectivity are one of the most noticeable megatrends that impact business today. The truly globalized flow of data and increasingly fragmented and distributed value chains have enabled decentralized manufacturing through facilities situated in different parts of the world. Within AM, many designers and companies all over the world can freely share their 3D printable designs through various online platforms. The manufacturing of 3D parts can be produced quickly and on-demand in short timeframes, even during severe supply chain disruptions.

The AM technology, as compared to traditional manufacturing such as injection molding or casting, could potentially reduce the need for logistics through digital inventories, data sharing, and decentralized manufacturing. The use of digital tools such as 3D modeling also allows designers to make infinite design iterations quickly and easily at no extra cost. The digitally enabled system adds more agility to the product development cycle, enabling rapid adaptation and accelerated productions. The ability to shorten prototyping time and deliver new and enhanced products within a short lead time has created a competitive advantage for AM, as illustrated in Fig. 4.5.

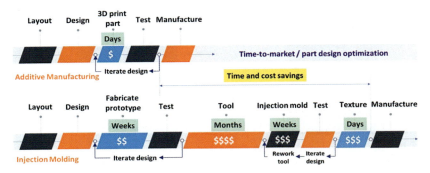

FIGURE 4.5 Comparisons of the product development cycles between additive manufacturing and injection molding.

4.2 Additive manufacturing process chain

Every product development goes through different stages of the process. The AM process chain as shown in Fig. 4.6 involves a combination of steps for successful implementation.

4.2.1 Level of additive manufacturing implementation

Before initiating any AM adoption, it is essential to identify the value proposition of using the technology. AM implementation can be divided into four levels: (1) rapid prototypes and tooling; (2) direct part replacement; (3) part consolidation; and (4) design for AM optimized. The higher the level of AM implementation, the more the number of benefits that can be harnessed from using AM as seen from Fig. 4.7.

4.2.1.1 Level 1: Using additive manufacturing for rapid prototyping and tooling

- Prototypes help to derisk system designs at low cost.
- Costly iterations can be avoided before committing to tooling and manufacturing.
- There is no intention to use AM for production parts.
- Main purpose: low volume parts are directly manufactured from CAD.

4.2.1.2 Level 2: Using additive manufacturing for direct part replacement

- Part must be reproduced as closely as possible to the original part (no change to the part is allowed).
- This level of implementation is mainly for low-risk applications and on-demand production of spare parts with a shorter lead time.
- Main purpose: reproduction of parts that avoid complex manufacturing.

Additive manufacturing for digital transformation **Chapter | 4 153**

FIGURE 4.6 AM process chain. *AM*, Additive manufacturing.

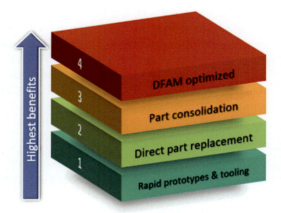

FIGURE 4.7 AM implementation divided into four levels, where the highest benefits can be derived from the highest level. *AM*, Additive manufacturing.

4.2.1.3 Level 3: Using additive manufacturing for parts consolidation

- External shapes might be changed.
- Changes are made to the form of the part (often internal sections).
- The use and function of the part remain unchanged.
- This helps to reduce material wastage.
- Main purpose: complex parts that adapt for AM to simplify assembly and enhance reliability.

4.2.1.4 Level 4: Using additive manufacturing for new product designs

- The entire part is redesigned to maximize AM benefits.
- The use of AM will have an impact on the product structure and part configuration.
- Main purpose: new product designs that add value and offer mass customization.

154 Digital Manufacturing

4.2.2 Design, optimization, and simulation

This step involves the software technology role in the AM process chain. The usage of design software allows scalability and flexibility in accessing digital data and making iterations and adjustments fast.

4.2.2.1 Design generation

CAD modeling: The creation of digital workpiece can be performed using CAD software such as CATIA, SOLIDWORKS, Autodesk, Onshape, Siemens NX, Rhinoceros 3D, AutoCAD, and Fusion 360.

3D scanning: The advent of AM has also inspired users to explore its potential in automated repair and reconstruction using 3D scanning. The data acquisition from the 3D scanners generates "point clouds" retrieved from the part's surface and can be digitally constructed into OBJ or Surface Tessellation Language (STL) file which can be saved, modified, or even printed directly using a 3D printer.

4.2.2.2 Optimization and simulation

Optimization: Creating new designs with AM involves optimizing the parts for advance structures such as free-form designs (lattice and structures), conformal cooling channels, and complex geometries. Structural and topology optimization can be performed in an iterative process for optimal distribution of elements, but the complexity of the end part is commonly restricted by high level of manufacturing technicality. Design guidelines and constraints (such as print orientation, support removal or powder removal, and angle dependency) have to be taken into consideration.

Simulation: Physics-based process models have been identified as being foundational to the qualification of additively manufactured parts. Simulation can be carried out at different length and time scales: (1) part scale—finite element analysis, (2) powder scale—effects of material attributes and process conditions using discrete element method (e.g., LIGGGHTS, LAMMPS) and computational fluid dynamics (e.g., OpenFOAM and ALE3D), and (3) microstructural scale—process conditions and crystallization behavior using phase-field simulation, for example, OpenFOAM, CALPHAD method (Thermo-Calc software). Realistic modeling and simulation can accurately predict and capture the structural, physical, and material features of the processes.

4.2.2.3 Conversion to printable file format (STL, AMF format)

STL format: The term "STL" was created by 3D Systems with several backronyms such as "Standard Triangle Language" and "Standard Tessellation Language." STL files describe only the surface geometry of a 3D object with a series of triangular facets, without the representation of any color and texture.

AMF format: The International Standards Organization (ISO) teamed up with ASTM to publish the AM File Format (AMF) to address the expanding needs of AM in a wider range of industries, including medical device manufacturing. Unlike its predecessor STL format, AMF has native support for multiple colors, textures, lattices, and constellations. AMF is the only international standard for 3D files used in AM, which is endorsed by the National Institute of Standards and technology, America Makes and is actively maintained by the ISO/ASTM joint technical committee TC261.

4.2.3   Material selection

The type of material that is intended to be fabricated plays a significant role in the process of determining which type of AM process to be used. The material properties and its availability are determining factors in identifying the right material type.

The availability and development of AM materials are largely tied to the market demand and the economic drivers according to the application value. Table 4.3 lists the material range available in the AM market.

4.2.4   Manufacturing

The AM or 3DP process involves the fabrication of 3D objects layer upon layer from 3D model data, through material deposition using a print head or nozzle.

AM builds the objects through successive addition of materials layer by layer. Every layer is derived from the virtual cross section of the part from the slice data of the 3D CAD model and transferred to the AM machines. According to ISO/ASTM 52900:2021, there are seven AM process classifications, or one of the ways to classify the processes broadly is by the initial form of its materials as shown in Table 4.4.

The adjustment of processing parameters (such as printing speed, layer thickness, laser power, build orientation, and build environment temperature) by an experienced or competent AM machine operator is critical in this step to determine the success or failure of the print job.

4.2.5   Postprocessing

Postprocessing methods are often required to either remove the printed parts from the build platform or to achieve certain level of finishing and final mechanical strength in the part.

156 Digital Manufacturing

TABLE 4.3 Materials available in the additive manufacturing market.

Organic	Ceramic	Polymer	Metal
Waxes	Alumina	ABS	Aluminum
Living cells	Mullite	PA (nylon)	Tool steel
Wood/paper	Zirconia	PLA	Titanium
Food stuffs	Silicon carbide	PEEK and EKK	Inconel
Beta-tricalcium phosphate		Thermosetting Epoxies	Cobalt-chrome
	Silica (sand)	Hydrogels	Copper
	Plaster	PMMA	Gold/platinum
	Graphite	PC	Hastelloy
	Concrete	Polyphenyl-sulfone	Tungsten
	Glass	ULTEM (PEI)	Shape memory alloys
	Ceramic (nano)-loaded epoxies		Stainless steel
	Filled polymers		
		Polymer-coated metallic powders	
Pharmaceuticals			
Multimaterial—multifunctional systems			

ABS, Acrylonitrile butadiene styrene; *PC*, polycarbonate; *PEEK*, Polyaryletherketone; *PLA*, polylactic acid.

- Ultraviolet (UV) curing: Parts that are fabricated using SLA or digital light processing (DLP) have to be placed in a UV oven for postcuring to ensure that the parts are fully polymerized.
- Machining: Unwanted support structures are to be removed or sliced away.
- Heat treatment: Parts that are subjected to heating and cooling cycles, such as directed energy deposition (DED), SLS, or SLM parts, tend to accumulate high level of residual stress. Heat treatment methods, such as hot isostatic press (HIP) or annealing can reduce pores, relieve residual stress, or modify microstructure.
- Shot peening: Parts with rough finishing such as DED or electron-beam melting (EBM) parts can be blasted with high-speed metal powder to smoothen the surfaces.

TABLE 4.4 Classifications of additive manufacturing processes.

ASTM classification (ISO/ASTM 52900:2021)	Related acronyms	Raw material form
Vat photopolymerization	• SLA • DLP	Liquid
Material jetting	• Polyjet	
Powder bed fusion	• SLS • SLM • EBM	Powder
Directed energy deposition	• Laser engineered lens shaping (LENS)	
Binder jetting	• Binder jetting (BJ)	
Material extrusion	• FDM • Fused filament fabrication (FFF)	Solid
Sheet lamination	• LOM	

AM, Additive manufacturing; *ASTM*, American Society for Testing and Materials; *DLP*, digital light processing; *EBM*, electron-beam melting; *FDM*, fused deposition modeling; *LOM*, laminated object manufacturing; *SLA*, stereolithography apparatus; *SLM*, selective laser melting; *SLS*, selective laser sintering.

4.2.6 Process monitoring and validation

The last step involves ascertaining whether the printed parts are of consistent quality, strength, and reliability.

- Process monitoring is often carried out during the manufacturing process to enable real-time detection of the process abnormalities or indications that can be correlated with defects in the built part. The data acquired during process monitoring helps to facilitate and support documentation and traceability of the manufacturing process by storing recorded data. Some of the key techniques and sensors that are used to conduct studies on monitoring and control for metal-based AM processes are using infrared thermography or photodiodes to monitor melt pool, reflective topography or displacement sensors for deposition/layer height control, and neutron diffraction for residual stress detection.
- Validation of the printed parts can be in terms of metrology or material properties. Compliance with technical specifications will require geometrical measurements such as dimensional accuracy, thin walls, pin diameter, and shrinkage or distortion measurements, while material properties can be benchmarked against ASTM or ISO standards.

158 Digital Manufacturing

4.3 Additive manufacturing technologies and processes

The ASTM Committee F42 publishes the standard terminology for AM technologies in an effort to standardize the terminology used by AM users, producers, researchers, educators, press, or media. According to ISO/ASTM 52900:2021, the AM technologies are generically defined into seven process classifications, specifically vat photopolymerization (VPP), material extrusion (MEX), material jetting (MJT), sheet lamination, powder bed fusion (PBF), binder jetting (BJ), and DED [1]. A summary table of the seven ASTM classification of AM technologies can be found in Table 4.5.

4.3.1 Vat photopolymerization

This section introduces the VPP technology, listing its standard definition as defined by ISO/ASTM 52900:2021 benefits, drawbacks, range of materials available in the current AM market, and applications associated with the technology.

4.3.1.1 Introduction and definition

VPP is defined as

An additive manufacturing process in which liquid photopolymer in a vat is selectively cured by light-activated polymerization. [1]

In the beginning, VPP processes utilized only UV laser and x-y scanning mirrors on computer-controlled galvanometers to direct the laser beam to scan and cure the top surface of the photopolymer resin in a vat. SLA is the oldest scanning type of VPP process, which was patented in 1984 by Charles (Chuck) W. Hull. It is the most widely used rapid prototyping technology. Subsequently, DLP was introduced as the projection-type of VPP process which uses a lamp or light-emitting diodes (LEDs) as the source of UV energy to project an image onto the surface of a vat and cure an entire cross-sectional area at once.

4.3.1.2 Working principles

The VPP process is based fundamentally on the following principles:

1. The conventional scanning type SLA process uses a UV laser beam to scan and cure the surface of the photocurable resin layer by layer, while the projection-type DLP process utilizes an LED light to project a whole cross-sectional area of mask projection onto the resin surface.
2. The photosensitive resins are mainly acrylic or epoxy-based monomers and a polymerization chain reaction that solidifies the resin layer is initiated when exposed to UV light.

TABLE 4.5 A summary of the seven American Society for Testing and Materials (ASTM) classifications of additive manufacturing (AM) technologies [11].

ASTM classification	Working principle	AM technology	Benefits	Drawbacks	Materials	Major equipment manufacturers
Vat photopolymerization	Photopolymer in a vat is selectively cured by light-activated polymerization	• SLA • DLP • CLIP	• High accuracy • Excellent surface finishes • Easily accessible resin vat • Highly isotropic parts	• Requires postcuring • Tedious manual removal of support structures • Limited availability of photopolymer	• Thermoset photopolymers • Ceramics	• 3D Systems, USA • EnvisionTEC, Germany • Asiga, USA • Formlabs, USA • Carbon3D Inc, USA
Material extrusion	Material is selectively dispensed through a nozzle or orifice	• FDM • Fused filament fabrication (FFF) • MJM	• Low cost • High speed • Office friendly • Water-soluble support structures	• Low accuracy • Rough surface • Weak mechanical strength	• Thermoplastic • Composites	• Stratasys, USA • Markforged, USA • Ultimaker, the Netherlands • Makerbot, USA
Material jetting	Droplets of build material are selectively deposited	• 3D inkjet technology • Direct Ink Writing	• High accuracy and resolution • Multimaterial and multicolor capabilities • Freedom of composition	• Relatively costly due to closed system that permits only its own proprietary material	• Thermoset photopolymers • Ceramics • Composites • Biologicals	• Stratasys, USA • 3D Systems, USA • WASP, Italy • Nano Dimension, USA

(Continued)

TABLE 4.5 (Continued)

ASTM classification	Working principle	AM technology	Benefits	Drawbacks	Materials	Major equipment manufacturers
			• Soluble supports	• Degradation of material—vulnerable to heat and sunlight		
Sheet lamination	Sheets of material are bonded to form an object	• LOM • Ultrasound consolidation/ ultrasonic additive manufacturing (UC/UAM)	• Low cost • High speed • Colored objects	• Inferior surface quality • Low dimensional accuracy • Limitation in manufacturing complex shapes	• Polymers • Metal sheets • Ceramics • Paper	• 3D Systems, USA • MCor, Ireland
Powder bed fusion	Thermal energy selectively fuses regions of a powder bed	• SLS/SLM • EBM • DMLS • MJF	• High level of complexity • Powder bed acts as integrated support structure • Reusable powder	• Relatively slow • High cost • Rough surface finish • Requires postprocessing to reduce porous surface	• Thermoplastic • Metals • Ceramics • Composites	• EOS, Germany • ACAM, Sweden • Concept Laser, Germany • Renishaw, the United Kingdom • HP Inc, USA • SLM Solutions, Germany

Binder jetting (BJ)	Liquid bonding agent is selectively deposited to join powder materials	• 3D Inkjet Technology	• Relatively low cost • Very large parts • Complex geometries	• Brittle due to high porosity • Rough details • Limited material selection	• Polymers • Ceramics • Composites • Metals	• HP Metal Jet, USA • ExOne, USA • PolyPico, Ireland • ZCorp (purchased by 3D Systems, USA)
Directed energy deposition	Focused thermal energy is used to fuse materials by melting as they are being deposited	• DMD • LENS • BEAM • WAAM	• Fast build • Fully dense parts • Suitable for fabricating large workpieces • Reduce material waste • Excellent for repair and retrofitting	• Low resolution and accuracy • Rough surface finish • Expensive • Limitation in manufacturing complex shapes with fine details • Requires postprocessing	• Metal powder • Hybrid	• Trumpf, Germany • DMG Mori, Germany • Optomec, USA • Sciaky, USA

3D, Three-dimensional; *AM*, Additive manufacturing; *BEAM*, Electron-Beam AM; *CLIP*, Continuous Liquid Interface Production; *DLP*, Digital Light Processing; *DMD*, direct metal deposition; *DMLS*, direct metal laser sintering; *EBM*, electron-beam melting; *FDM*, fused deposition modeling; *LENS*, Laser Engineered Lens Shaping; *LOM*, laminated object manufacturing; *MJF*, multijet fusion; *MJM*, multijet modeling; *SLA*, stereolithography apparatus; *SLS/SLM*, selective laser sintering/melting; *USA*, United States; *WAAM*, Wire Arc AM.

Source: Modified from S.A. Tofail, et al., Additive manufacturing: scientific and technological challenges, market uptake and opportunities. Materials Today, 21 (1) (2018) 22−37.

162 Digital Manufacturing

3. The VPP processes can be operated based on two mechanisms: (1)a top-down configuration in which the laser scans from above, and the build platform is lowered after curing each layer; or (2) a bottom-up configuration where the resin is cured from the transparent base of the vat, and the build platform is raised after each layer is cured.
4. The resin flows in after each adjustment of the build platform and the process repeats until all layers have been built.
5. Upon completion the printed part is elevated from the vat, and the unreacted resin can be washed away using isopropyl alcohol.
6. A postcuring process using a UV oven is required to ensure that the printed parts are fully cured to achieve the desired mechanical performance.

Note for users: Laser scanning speed, laser or LED power, exposure time, layer thickness, and hatch spacing are key processing parameters that affect the properties of the SLA/DLP printed parts [12]. To ensure good adherence between layers yet avoiding overcuring, the curing depth or penetration depth must be slightly larger than the layer thickness [13,14]. The curing depth determines the minimum layer thickness and laser scanning speed/exposure time per layer.

4.3.1.3 Benefits

The VPP process has the following benefits:

- VPP processes are recognized to provide the highest accuracy, excellent surface finishing, and fine details. The SLA technology is also notable for large object sizes.
- SLA/DLP parts are highly isotropic due to a continuous polymer network within and between the layers, which enables better resistance to directional forces.
- The open system of an accessible resin vat allows for the ease of material development such as ceramic—polymer composites [15,16].
- The bottom-up configuration process uses lesser amounts of resin, which makes it more economically efficient.

4.3.1.4 Drawbacks

The VPP process has the following drawbacks:

- The manual removal of support structures for overhanging features can be tedious and time-consuming.
- Postcuring process for complete curing and stability is required, especially for functional resins for engineering, and mandatory for dentistry and jewelry applications.

Additive manufacturing for digital transformation **Chapter | 4** **163**

- The range of photopolymer materials for VPP processes is very limited.
- The strong adhesion between the cured layer and the build platform can cause delamination or detachment of the printed parts for bottom-up configuration processes.

4.3.1.5 Range of materials available

Table 4.6 lists the range of materials available for VPP process:

4.3.1.6 Applications

The applications of VPP process can be divided into different areas:

- functional prototyping,
- molds and casting patterns for jewelry and audiology,
- dental applications such as dentures and aligners, and
- healthcare and medical applications such as prototypes for visual aids or surgical training purposes due to availability of transparent materials.

TABLE 4.6 Range of materials available for vat photopolymerization processes [3].

Rigid opaque materials	• 3D Systems (SLA)—Accura Xtreme, VisiJet FTX Silver, and VisiJet SL Tough
	• EnvisionTEC (DLP)—ABS TRU 3SP, LS600, and Photosilver
Transparent materials	• 3D Systems (SLA)—Accura Phoenix, VisiJet SL Clear, and Accura 60
	• EnvisionTEC (DLP)—E-Glass 3SP, E-Shell 450, and E-Clear 3SP
Rubber-like materials	• EnvisionTEC (DLP)—E-Gum
High-temperature materials	• EnvisionTEC (DLP)—HTM140 (heat deflection 140°C)
Biocompatible materials	• 3D Systems (SLA)—VisiJet SL Clear
	• EnvisionTEC (DLP)—E-Shell 200
Dental materials	• 3D Systems (SLA)—Accura e-Stone
	• EnvisionTEC (DLP)—E-Dent 100 and E-Guard
Castable materials	• 3D Systems (SLA)—VisiJet FTX Cast and VisiJet SL Jewel
	• EnvisionTEC (DLP)—EC500, EPIC, and PIC100

3D, Three-dimensional; *ABS*, acrylonitrile butadiene styrene; *DLP*, digital light processing; *SLA*, stereolithography apparatus.

164 Digital Manufacturing

4.3.2 Material extrusion

This section introduces the MEX technology, listing its standard definition as defined by ISO/ASTM 52900:2021, benefits, drawbacks, range of materials available in the current AM market, and applications associated with the technology.

4.3.2.1 Introduction and definition

MEX is defined as

> *An additive manufacturing process in which the material is selectively dispensed through a nozzle or orifice. [1]*

MEX processes are commonly referred to as FDM or fused filament fabrication (FFF). In the late 1980s the FDM system was developed by S. Scott Crump and commercialized by Stratasys in 1990 [17]. The expiration of FDM patent has led to a vast proliferation of commercial and self-made FDM widely available as low-cost 3D printers in the wide open-source market. FDM is the second most widely used rapid prototyping technology.

4.3.2.2 Working principles

The MEX process is based fundamentally on the following principles:

1. Solid thermoplastic filament materials are wound on a coil and fed into an extrusion nozzle that can activate and disable the flow of materials.
2. A numerically operated mechanism using a stepper motor or servo motor, which is directly controlled by a computer-aided manufacturing system, enables the nozzle head to be moved in both horizontal and vertical directions.
3. The filament is heated into a semiliquid state to be extruded through the nozzle and deposited onto the build platform layer by layer.
4. The thin layer cools and solidifies while binding to the layer beneath it until the built part is completed.

Note for users: Layer thickness, width, the orientation of filaments, and air gap (present in the same layer or between layers) are key processing parameters that influence the mechanical properties of the printed FDM parts [18]. To prevent warping, large flat areas should be avoided, and fillets can be added in sharp corners.

4.3.2.3 Benefits

The MEX process has the following benefits:

- Low cost, high speed, and simplicity of the process are the main benefits of FDM.
- FDM printers are office friendly, quiet, and easy to set up. It practically requires only electricity, raw materials, and a computer.

Additive manufacturing for digital transformation Chapter | 4 **165**

- FDM is reasonably quick in production for small pieces in the order of a few cubic meters.
- FDM printers are widely available with the ability to build fully functional parts.

4.3.2.4 Drawbacks

The MEX process has the following drawbacks:

- Parts built by FDM are generally low in accuracy and not amenable to fine details.
- Due to the discretized layer-by-layer printing process, FDM parts typically have stair-stepping surfaces, which lead to relatively rough surface finishing.
- FDM parts are usually weaker in mechanical properties due to vertical anisotropy, which may be attributed to distortion within or between layers [19].
- Further postprocessing is required after printing to remove support structures. Support structures that are fabricated for overhanging geometries can be removed by manually breaking them away or simply washed away for water-soluble support materials.

4.3.2.5 Range of materials available

Table 4.7 lists the range of materials available for MEX process.

4.3.2.6 Applications

The following lists the range of materials available for MEX process:

- concept modeling and rapid prototyping,
- architectural models,
- marketing and graphic design, and
- food printing.

4.3.3 Material jetting

This section introduces the MJT technology, listing its standard definition as defined by ISO/ASTM 52900:2021, benefits, drawbacks, range of materials available in the current AM market, and applications associated with the technology.

4.3.3.1 Introduction and definition

MJT is defined as

> *An additive manufacturing process in which droplets of build material are selectively deposited. [1]*

166 Digital Manufacturing

TABLE 4.7 Range of materials available for material extrusion processes [3].

Tough material	• ABS—general purpose plastic
	• ASA—UV stable material for outdoor applications
	• PLA—low shrinkage and warping, popular for desktop 3D printer
	• PET—translucent and elastic for bottle manufacturing
	• PC—high tensile and flexural strength
Flexible material	• Nylon/PA—high fatigue and break resistant for snap fits
	• PP—flexible yet tough
	• TPU—flexible and rubber-like
Flame-retardant material	• ULTEM—high heat resistant and thermal stability for aerospace applications
Particle-filled material	• Carbon fiber—stiff and lightweight for drone parts
	• Graphite particles—electrically conductive
	• Phosphor particles—glow in the dark

3D, Three-dimensional; *ABS*, acrylonitrile butadiene styrene; *ASA*, acrylonitrile styrene acrylate; *PA*, polyamide; *PC*, polycarbonate; *PET*, polyethylene terephthalate; *PLA*, polylactic acid; *PP*, polypropylene; *TPU*, thermoplastic polyurethane; *UV*, ultraviolet.

The MJT process was first invented in the late 1980s by a Japanese company Brother Industries, Ltd., with its patent being issued in 1991. Similar patents emerged later from Solidscape Inc. in 2000 and Objet Geometries Ltd. in 2001. The merger between Objet and Stratasys in 2012 brought about the Polyjet 3D inkjet printing technology, which has now become well known.

4.3.3.2 Working principles

The MJT process is based fundamentally on the following principles:

1. Droplets of photocurable resin for build material and support material are selectively deposited through one or more inkjet printing heads.
2. UV lamps are typically attached to the sides of the print heads to solidify or cure the droplets instantly. Multiple print heads are used to speed up the build process.
3. Multiple materials (such as rigid or flexible materials) of varying colors can be deposited through different print heads simultaneously for added functionality or esthetics.

4. After each layer is cured the build platform lowers and the process is repeated till completion.
5. Postprocessing is usually required to remove support materials using a waterjet.

Note for users: MJT systems are usually closed systems with restrictions to tuning their process parameters, but users can choose between different printing modes such as digital or single material, high quality, or high speed. The density of the support materials can be of standard, heavy, or lite grid, while the surface finish of the printed parts can be either matte or glossy.

4.3.3.3 Benefits

The MJT process has the following benefits:

- The main benefit of MJT is the ability to distribute and mix multimaterials and colors selectively.
- The multiple print heads also allow for the freedom of compositions that can blend the photocurable resins to create composites called "digital materials."
- Soluble support materials are available that ease the process of support removal and allows complex parts with undercuts to be fabricated.

4.3.3.4 Drawbacks

The MJT process has the following drawbacks:

- MJT is generally slower than other processes and the materials are relatively more costly due to its closed system that permits only its own proprietary material [20].
- The printed parts are vulnerable to heat and sunlight, causing them to become brittle and degrade over time.

4.3.3.5 Range of materials available

Table 4.8 lists the range of materials available for MJT process.

4.3.3.6 Applications

The following lists the range of materials available for MJT process:

- sophisticated prototypes,
- presentation models such as biomedical models,
- prototypes for investment casting,
- electronics components [21,22],
- 4D printing (which uses the same techniques of 3DP through computer-programmed deposition of material in successive layers to create a 3D object. 4D printing adds the dimension of transformation over time).

168 Digital Manufacturing

TABLE 4.8 Range of materials available for material jetting processes [3].

Rigid opaque materials	• Stratasys (Polyjet)—Digital ABS, Rigur, and Vero family
	• 3D Systems (MultiJet)—VisiJet M2R-BK, VisiJet M3-X, and VisiJet CR-WT
Transparent materials	• Stratasys (Polyjet)—RGD720 and VeroClear
	• EnvisionTEC (DLP)—E-Glass 3SP, E-Shell 450, and E-Clear 3SP
Rubber-like materials	• Stratasys (Polyjet)—TangoPlus FLX 930 and Agilus30 FLX935
	• 3D Systems (MultiJet)—VisiJet M2 ENT, VisiJet M2 EBK, and VisiJet CE-NT
High temperature materials	• Stratasys (Polyjet)—RGD525 (heat deflection 67°C, by postprocessing can be increased to 80°C)
	• 3D Systems (MultiJet)—VisiJet M5-X (heat deflection 65°C)
Biocompatible materials	• Stratasys (Polyjet)—MED610
	• 3D Systems (MultiJet)—VisiJet M3 Crystal
Castable materials	• 3D Systems (MultiJet)—VisiJet M3 Procast

3D, Three-dimensional; *ABS*, acrylonitrile butadiene styrene; *DLP*, digital light processing.

4.3.4 Sheet lamination

This section introduces the sheet lamination technology, listing its standard definition as defined by ISO/ASTM 52900:2021, benefits, drawbacks, range of materials available in the current AM market, and applications associated with the technology.

4.3.4.1 Introduction and definition

Sheet lamination is defined as

> *An additive manufacturing process in which sheets of material are bonded to form an object [1].*

In the 1990s, some companies commercialized printers that can cut cross sections using a carbon dioxide laser from special adhesive coated paper and then laminate them together. The use of paper as the build material has significantly reduced the cost to print. This inspired MCor Technologies Ltd to create a different method in 2005 using a tungsten carbide blade to cut the shape from ordinary sheets of office paper and selectively deposit adhesive and pressure to bond the prototype. The related AM techniques are laminated object manufacturing (LOM), paper lamination technology, and ultrasonic additive manufacturing (UAM), which is a new subclass of LOM that incorporates ultrasonic metal seam welding and CNC milling in the process of lamination.

4.3.4.2 Working principles

The sheet lamination process is based fundamentally on the following principles:

1. Thin-layered materials such as aluminum foil or sheets of paper are coated with adhesive and successively glued together layer by layer.
2. A mechanical cutter or laser is used to precisely cut the successive layers before bonding them together (which can be form-then-bond or bond-then-form).
3. The form-then-bond method allows the removal of excess material prior to bonding, which facilitates effective thermal bonding of ceramic and metallic materials and the fabrication of internal features.
4. Excess materials leftover after cutting can be used as support and subsequently removed or recycled.

Note for users: The quality of the fabricated part relies largely on the thickness of the layered materials used.

4.3.4.3 Benefits

The sheet lamination process has the following benefits:

- The ease of material handling results in a reduction of tooling cost and manufacturing time for the sheet lamination process, which is also suitable for manufacturing large structures.
- Sheet lamination is used primarily to create colored objects with a high detailed resolution.
- The 3D objects in the paper are resistant and can be fully colored.
- UAM is the only AM method that can construct metal structures at low temperatures [23].

4.3.4.4 Drawbacks

The sheet lamination process has the following drawbacks:

- Postprocessing may be required as the surface quality and dimensional accuracy are lower as compared to other powder bed processes.
- The removal of excess material after cutting and laminating is a time-consuming process that may not be recommended for complicated shapes.
- The strength and integrity of the fabricated parts depend on the adhesive used.

170 Digital Manufacturing

4.3.4.5 Range of materials available

The following lists the range of materials available for sheet lamination process:

- paper-based materials are most popular,
- thin-layered materials like aluminum foil or metal-filled tapes,
- polymer composites, and
- ceramics.

4.3.4.6 Applications

The following lists the range of materials available for sheet lamination process:

- paper manufacturing,
- foundry industries, and
- embedded electronic devices, sensors, pipes, and other features.

4.3.5 Powder bed fusion

This section introduces the PBF technology, listing its standard definition as defined by ISO/ASTM 52900:2021, benefits, drawbacks, range of materials available in the current AM market, and applications associated with the technology.

4.3.5.1 Introduction and definition

PBF is defined as

> *An additive manufacturing process in which thermal energy selectively fuses regions of a powder bed [1].*

In the mid-1980s, Dr. Carl Deckard and Dr. Joseph Beaman at the University of Texas at Austin invented and patented selective laser sintering (SLS). Scanning laser sintering coalesces powder into a solid mass by heating. In contrast, selective laser melting (SLM) does not use sintering for powder granule fusion, but it uses a high-energy laser to completely melt the powder to produce fully dense materials with similar mechanical properties to traditional processed metals in a layer-wise process. EBM is another PBF process for metal parts that produces components in a high vacuum by melting metal powder layer by layer with an electron beam. EBM parts are fully dense, free of porosity, and very strong since they are melted above melting point. Multijet fusion (MJF) from HP Inc. is similar to the SLS process but uses an infrared energy source to sinter cross-sectional areas that are deposited with a fusing agent to promote the absorption of infrared light.

4.3.5.2 Working principles

The PBF process is based fundamentally on the following principles:

1. A thin layer of powder is spread by a roller or blade recoater over the surface of the build platform.
2. The powder is selectively sintered or melted and fused together using a heat source, such as Yb-fiber or CO_2 laser or electron beam. In the case of HP MJF, fusion heaters traverse across the powder bed fusing the regions where the fusing agents have been deposited.
3. The build platform is lowered after each layer for the next layer of powder to be recoated and fused together until the build part is completed.
4. Upon completion the build platform is elevated and excess powder can be brushed off or vacuumed.

Note for users: Laser power and scanning speed are the main parameters affecting the printing process [24,25]. Other key parameters include layer thickness and hatch spacing, while energy density is a function of these parameters. Powder size distribution and packing density are determining factors on the density of the printed part and the efficacy of this technology [26]. The cost of MJF and SLS is comparable, but the lead time for MJF is on average 3 days faster.

4.3.5.3 Benefits

The PBF process has the following benefits:

- The key advantages of PBF processes are fine resolution and high-quality printing, making it ideal for printing complex structures.
- In a laser sintering process, unfused powder surrounding the part serves as supports for overhangs and undercuts, thereby eliminating the need for the removal of supports.
- Leftover unfused powders can be collected, sieved, and reused. HP MJF technology can attain consistent performance while achieving up to 80% surplus powder reusability [27].

4.3.5.4 Drawbacks

The PBF process has the following drawbacks:

- PBF processes are generally slow and involve high costs, while parts usually have rough surface finishing.
- High thermal gradients in the build chamber of the SLM process may result in high residual stress and lead to warpage. Support structures and anchors are required to maintain the part.
- Laser sintered parts are not fully melted, which results in high porosity due to fusion of the powders at a molecular level. Hence, postprocessing such as sand-blasting or hot isostatic pressing is required to reduce porous surfaces.

172 Digital Manufacturing

4.3.5.5 Range of materials available

Table 4.9 lists the range of materials available for PBF process.

4.3.5.6 Applications

The following lists the range of materials available for PBF process:

- form/fit and functional testing,
- rapid tooling patterns,
- lightweight structures using lattices,
- biomedical applications such as customized orthopedic components (e.g., titanium alloy cranial or acetabular implants), and
- aerospace applications such as fuel nozzle on General Electric's GE9X engine.

TABLE 4.9 Range of materials available for powder bed fusion processes [9].

Thermoplastic powders for laser sintering	• Nylon/glass-filled nylon
	• PS
	• Polyamide (PA11/PA12)
	• PU
	• PEEK—a high-performance material with flame retardant
Metal powders for laser sintering/melting	• Bronze alloys
	• Copper alloys
	• Aluminum (AlSi10Mg)
	• Cobalt-chrome alloys
	• Titanium alloys (Ti6AL4V)
	• Stainless steel (316 L, PH15−5, PH17−4, and maraging steel)
	• Nickel-based alloys (Inconel 718, Inconel 625)
	• Precious metal alloys: silver, yellow gold, rose gold, red gold, and platinum
Metal powders for electron-beam melting	• Cobalt-chrome alloy
	• Titanium alloys
	• Inconel 718

PEEK, Polyaryletherketone; *PS*, polystyrene; *PU*, polyurethane.

4.3.6 Binder jetting

This section introduces the BJ technology, listing its standard definition as defined by ISO/ASTM 52900:2021, benefits, drawbacks, range of materials available in the current AM market, and applications associated with the technology.

4.3.6.1 Introduction and definition

BJ is defined as

> An additive manufacturing process in which a liquid bonding agent is selectively deposited to join powder materials [1].

In the late 1980s, MIT professors Dr. Emanuel Sachs and Dr. Michael Cima invented the BJ technology, with their patent being released in 1993. MIT licensed this technology to a few companies, including Z Corporation. 3D Systems acquired Z Corporation later in 2012. BJ is also known as inkjet powder printing and 3D printing. BJ is equivalent to material jetting in the use of inkjet printing to dispense the material. The droplet material dispensed by the inkjet nozzles in BJ systems is not the build material but rather a liquid that acts as an adhesive and is selectively deposited to bind powder particles on a powder bed.

4.3.6.2 Working principles

The BJ process is based fundamentally on the following principles:

1. A layer of powder is spread across the build platform with a recoating blade.
2. Droplets of a binding agent (glue) is selectively deposited through a carriage with inkjet nozzles to bond the powder particles together. For full-color BJ the colored inks/binders are also deposited during this step.
3. After each layer is completed the build platform lowers for the next layer to be recoated.
4. Upon completion the part is excavated and excess powder can be removed using compressed air.

Note for users: The BJ process can be considered a cheaper alternative for metal or colored parts if the applications do not require very high performance or only for visual purposes.

4.3.6.3 Benefits

The BJ process has the following benefits:

- BJ can manufacture very large parts without dimensional distortions associated with thermal effects such as warping as bonding occurs at room temperature.

174 Digital Manufacturing

- BJ requires no support structures similar to SLS processes, allowing for freedom in geometrical designs.
- Full-color prototypes can be produced at a much lower cost as compared to material jetting.

4.3.6.4 Drawbacks

The BJ process has the following drawbacks:

- Parts produced directly from BJ process without postprocessing usually have poor mechanical properties (very brittle) due to high porosity.
- BJ parts are generally low in accuracy, and only rough details can be printed.
- BJ offers a limited material selection.

4.3.6.5 Range of materials available

Table 4.10 lists the range of materials available for BJ process.

4.3.6.6 Applications

The following lists the range of materials available for BJ process:

- full-color prototypes for visual purposes.
- widely used for sand molds and casts patterns.

4.3.7 Directed energy deposition

This section introduces the DED technology, listing its standard definition as defined by ISO/ASTM 52900:2021, benefits, drawbacks, range of materials

TABLE 4.10 Range of materials available for BJ processes [3].

Metal powders	• Stainless steel−bronze-infused (60% stainless steel + 40% bronze infiltrant)
	• Iron−bronze-infused (60% stainless steel + 40% bronze infiltrant)
	• Nickel−chrome alloy
	• Chromite
Other materials	• Zircon, silica sand, and ceramic beads—molds used for metal casting
	• PMMA—"burnable" models used in investment casting
	• Plaster (VisiJet PXL)—plaster-like material for full-color 3D printing

available in the current AM market, and applications associated with the technology.

4.3.7.1 Introduction and definition

DED is defined as

> An additive manufacturing process in which focused thermal energy is used to fuse materials by melting as they are being deposited [1].

While cladding and welding have a long history of use in adding materials, David Keicher of Sandia National Labs pioneered the automated use of these techniques to build up materials into 3D near net shape parts. He invented Laser Engineered Lens Shaping (LENS), which was later licensed to Optomec. Common AM processes that use the DED technology include direct metal deposition (DMD), laser solid forming, electron-beam additive manufacturing (BEAM), and wire arc additive manufacturing (WAAM). The energy sources most typically used include arc, laser, or electron beam. Recently, hybrid approaches combining with CNC milling are leading to more extensive adoption of this technology.

4.3.7.2 Working principles

The DED process is based fundamentally on the following principles:

1. Metal powders or wires are continuously blown with a gas (typically argon) through nozzles directed at the focal point of a laser or electron beam.
2. The melted molten material is deposited and fused into the substrate layer by layer with the movement of the nozzle head.
3. The DED system can operate in multiple axes, hence support structures are not needed.

Note for users: The difference between DED and PBF is that powders are fed in DED, while PBF uses powder bed. The DED process is similar to FDM where the feedstock material is melted before deposited layer by layer but using an extremely high amount of energy to melt metals [28].

4.3.7.3 Benefits

The DED process has the following benefits:

- DED process is commonly used for large components with low complexity.
- With its rapid deposition rates, DED can reduce manufacturing time and replacement costs.
- The feedstock materials are melted within an extremely small heat-affected zone, which produces parts that are fully dense, with controlled microstructure and mechanical properties similar to that of wrought material.
- Reduce material waste due to near net shapes fabrication.

176 Digital Manufacturing

4.3.7.4 Drawbacks

The DED process has the following drawbacks:

- DED parts have low accuracy, rougher surface quality, and can only manufacture less complex parts.
- DED systems are typically very expensive.
- Most DED parts require significant postprocessing in the form of secondary machining.

4.3.7.5 Range of materials available

The following lists the range of materials available for DED process:

- aluminum
- copper
- inconel
- magnesium
- nickel
- stainless steel
- titanium alloys
- zirconium.

4.3.7.6 Applications

- Retrofitting and repairing manufactured parts.
- Feature addition for multimaterial application.

4.4 Case studies of additive manufacturing during the COVID-19 pandemic

The prevalence of the highly infectious coronavirus disease—COVID-19 has caused massive health and socioeconomic upheavals worldwide [29]. Major slumps in industrial production due to stringent lockdown measures and export restrictions have led to severe logistical challenges and drastic disruptions to the global supply chains. The devastating impacts of the COVID-19 pandemic were highly evident in the immensely overloaded medical infrastructures and health-care systems, where medical supplies and personal protective equipment (PPE) faced critical shortages in many countries. Rising to the challenges and unprecedented demands, the AM technology has demonstrated operational resilience with timely and innovative responses to help in the global supply efforts. This section reveals the critical roles of AM in navigating the impacts of the pandemic and features various applications that use the AM technology as viable solutions.

4.4.1 Providing rapid emergency responses

This pandemic has culminated in a significant increase of patients in need of respiratory support and thereby resulted in critical shortages of mechanical ventilators and their components in many health-care systems [30]. A hospital in Chiari, Lombardy, was overrun by COVID-19 patients and faced critical shortages of ventilator valves to keep its patients breathing [31−35]. The unavailability of the critical parts has inspired two Italian companies, Isinnova and Lonati SpA, to reverse engineer and replicate these ventilator valves with the help of AM technology [35], in view of resolving this shortage problem quickly [31−35]. An alliance between the Consorci de la Zona Franca (CZFB), HP Inc., Leitat (Technological Center) has also fabricated its first industrialized field 3D-printed emergency respiration device (LEITAT 1·3) to support emergency ventilation of COVID-19 patients [36]. Using AM technology, LEITAT 1·3 (as shown in Fig. 4.8) can be prototyped and fabricated in slightly more than a week [37].

These episodes have clearly demonstrated the potential of AM technology to adapt quickly and provide rapid emergency responses due to its simpler logistic requirements and its ability to change production capacities according to the hospitals' demand in real time with minimal impact [38]. Decentralized 3D printing facilities can also be situated closer to hospitals, reducing the need for long-distance transportation and saving precious time through shorter shipping time. AM technology, thus, reduces logistical and supply chain challenges through digital inventories, data sharing, and decentralized manufacturing.

FIGURE 4.8 3D-printed emergency respiration device, LEITAT 1.3 [37]. *3D*, Three-dimensional. *Data courtesy Leitat and HP Inc.*

178 Digital Manufacturing

4.4.2 Mass customizations

AM technology also offers mass customization possibilities. The World's Advanced Saving Project (WASP) company pivots its production to 3D print customizable face masks using 3D scanning technology [39] to fit the wearer's face ergonomically. WASP has developed an open-source tool, called "My Face Mask," that can construct a personalized mask using photogrammetry done with any smartphone device. Similarly, Bellus3D Inc. also combines its 3D facial scanning technology with 3D printing to develop personalized mask fitters for each individual [40]. The mask fitter is a plastic frame that conforms to the contours of each individual's unique face profile, which can be fitted over surgical or face masks to improve the seal substantially.

4.4.3 Agile operations and accelerated productions

The use of digital tools such as 3D modeling allows designers to make infinite design iterations quickly and easily at no extra cost. The digitally enabled system adds more agility to the product development cycle, enabling rapid adaptation, and catalyzes the production of critical medical devices to address urgent global needs. The ability to shorten prototyping time and deliver new and enhanced products within a short lead time has created a competitive advantage for the AM technology. It entails the utmost use of technologies among other realms of digital technologies in the battle against the pandemic.

Face shields are the simplest to fabricate among all PPE items. A basic face shield typically consists of two main parts: a transparent plastic sheet to serve as a visor to cover the wearer's face and a face shield frame to hold the visor. Only the face shield frame needs to be fabricated using the AM process. Many 3D companies (such as HP Inc. [41], Avid Product Development [42], Budmen Industries [43], and Stratasys Ltd. [44]) and designers have made their 3D printable face shields designs file freely available online. Hence, allowing these face shields to be fabricated immediately in any region around the world with AM machines, eliminating the need for a distant manufacturer. The 3D printable face shield (see Fig. 4.9) is optimized for AM with HP MJF technology using the HP 3D High Reusability PA12 material. It is designed to provide maximum wearing comfort and help hold the face shield in place. Using HP Jet Fusion 4200 and 5200 machines, 78 face shield frames can be printed in a full build and they are EN166 certified according to the COVID-19 regulations [41].

4.4.4 Preserving sustainability and continuity

Supplies of single-use and disposable PPE items may not be sufficient to cope with the surging demands during an extended outbreak or pandemic [45]. Furthermore, used PPE items have also accumulated tremendously as medical waste since the start of the pandemic [46]. In view of alleviating the

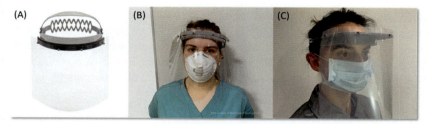

FIGURE 4.9 (A) 3D printable face shield using HP Jet Fusion 5200 and 4200 with certification of EN166 according to COVID-19 regulations; (B) Avid 3D printable face shield; and (C) Budmen face shield with frame that prints in a compact geometry for maximum efficiency. *Courtesy HP Inc.*

FIGURE 4.10 CIIRC RP95−3D half-mask with P3 replaceable external filter. *Courtesy CIIRC CTU 2020 and HP Inc.*

pressure of hospitals to deal with the mass amount of medical waste and prevent potential secondary transmissions, AM technology has offered some promising solutions to conserve precious resources by advocating reusable items and recyclable materials.

A research team in the Czech Institute of Informatics, Robotics and Cybernetics with the Czech Technical University (CIIRC CTU) in Prague had developed 3D printable half-mask respirators (CIIRC RP95−3D) that are intended to maximize reusability after proper disinfection and filter material replacement [47,48] (see Fig. 4.10) to conserve precious resources and

180 Digital Manufacturing

handle the influx of COVID-19 patients. When fitted together with a P3 replaceable external filter, the CIIRC RP95−3D becomes a PPE device that can offer the same or higher degree of protection as the FFP3 respirator. Hence, allowing the CIIRC R95−3D to be used in health-care settings to protect frontline health-care workers dealing with COVID-19 cases [49].

4.5 Summary

Additive manufacturing is yet another technological advancement that provides manufacturing operations with digital versatility and productivity. Without dedicated equipment or tooling, AM allows on-demand manufacturing, unlocks digital resources, and provides groundbreaking performance and unprecedented versatility across industries. Knowledge remains one of the key hurdles to AM's broader acceptance. This chapter provides the essential knowledge from the rise of AM to the working principles and mechanistic insights of the AM processes. The following chapters will provide a more in-depth dissection of each key technology.

References

[1] International Organization for Standardization, American Society for Testing and Materials, ISO/ASTM, 52900:2021, 2021, Additive manufacturing - General principles - Fundamentals and vocabulary, Standard (2021).

[2] C. Chen, X. Wang, Y. Wang, D. Yang, F. Yao, W. Zhang, et al., Additive manufacturing of piezoelectric materials, Advanced Functional Materials 30 (52) (2020) 2005141.

[3] C.K. Chua, K.F. Leong, 3D Printing and Additive Manufacturing: Principles and Applications, World Scientific, Singapore, 2015.

[4] C.W. Hull, C.W. Lewis, Methods and Apparatus for Production of Three-Dimensional Objects by Stereolithography, Google Patents, 1991.

[5] D. Thomas, Costs, benefits, and adoption of additive manufacturing: a supply chain perspective, The International Journal of Advanced Manufacturing Technology 85 (5−8) (2016) 1857−1876.

[6] J.R.M. Co, A.B. Culaba, 3D printing: challenges and opportunities of an emerging disruptive technology, in: 2019 IEEE 11th International Conference on Humanoid, Nanotechnology, Information Technology, Communication and Control, Environment, and Management (HNICEM), 2019, IEEE.

[7] J. Wöhrmann, Additive Manufacturing Adding up Growth Opportunities for ASEAN, Thyssenkrupp, Singapore, 2019.

[8] T. Wohlers, et al., Wohlers Report 2019: 3D Printing and Additive Manufacturing State of the Industry, Wohlers Associates, 2019.

[9] P. Minetola, L. Iuliano, G. Marchiandi, Benchmarking of FDM machines through part quality using IT grades, Procedia CIRP 41 (2016) 1027−1032.

[10] I. Campbell, et al., Wohlers Report 2018: 3D Printing and Additive Manufacturing State of the Industry: Annual Worldwide Progress Report, Wohlers Associates, 2018.

[11] S.A. Tofail, et al., Additive manufacturing: scientific and technological challenges, market uptake and opportunities, Materials Today 21 (1) (2018) 22−37.

[12] Y.Y.C. Choong, et al., High speed 4D printing of shape memory polymers with nanosilica, Applied Materials Today 18 (2020) 100515.

[13] Y.Y.C. Choong, et al., 4D printing of high performance shape memory polymer using stereolithography, Materials & Design 126 (2017) 219−225.

[14] Y.Y.C. Choong, et al., Curing characteristics of shape memory polymers in 3D projection and laser stereolithography, Virtual and Physical Prototyping 12 (2016) 1−8.

[15] H. Eng, et al., Development of CNTs-filled photopolymer for projection stereolithography, Rapid Prototyping Journal 23 (1) (2017) 129−136.

[16] H. Eng, et al., 3D stereolithography of polymer composites reinforced with orientated nanoclay, Procedia Engineering 216 (2017) 1−7.

[17] S.S. Crump, Apparatus and Method for Creating Three-Dimensional Objects, Google Patents, 1992.

[18] O.A. Mohamed, S.H. Masood, J.L. Bhowmik, Optimization of fused deposition modeling process parameters: a review of current research and future prospects, Advances in Manufacturing 3 (1) (2015) 42−53.

[19] A.K. Sood, R.K. Ohdar, S.S. Mahapatra, Parametric appraisal of mechanical property of fused deposition modelling processed parts, Materials & Design 31 (1) (2010) 287−295.

[20] H.W. Tan, et al., Metallic nanoparticle inks for 3D printing of electronics, Advanced Electronic Materials 5 (5) (2019) 1800831.

[21] H. Tan, T. Tran, C. Chua, A review of printed passive electronic components through fully additive manufacturing methods, Virtual and Physical Prototyping 11 (4) (2016) 271−288.

[22] H.W. Tan, et al., Induction sintering of silver nanoparticle inks on polyimide substrates, Advanced Materials Technologies 5 (1) (2020) 1900897.

[23] A. Hehr, M.J. Dapino, Interfacial shear strength estimates of NiTi−Al matrix composites fabricated via ultrasonic additive manufacturing, Composites Part B: Engineering 77 (2015) 199−208.

[24] C.Y.Y. Choong, G.K.H. Chua, C.H. Wong, Investigation on the Integral Effects of Process Parameters on Properties of Selective Laser Melted Stainless Steel Parts, 2018.

[25] G.K.H. Chua, C.Y.Y. Choong, C.H. Wong, Investigation of the Effects on the Print Location During Selective Laser Melting Process, 2018.

[26] B. Utela, et al., A review of process development steps for new material systems in three dimensional printing (3DP), Journal of Manufacturing Processes 10 (2) (2008) 96−104.

[27] F. Sillani, et al., Selective laser sintering and multi jet fusion: process-induced modification of the raw materials and analyses of parts performance, Additive Manufacturing 27 (2019) 32−41.

[28] Y. Choong, K. Chua, C. Wong, Control of process parameters for directed energy deposition of PH15-5 stainless steel parts, in: Industry 4.0−Shaping the Future of the Digital World: Proceedings of the 2nd International Conference on Sustainable Smart Manufacturing (S2M 2019), 9−11 April 2019, Manchester, UK, 2020, CRC Press.

[29] Y.Y.C. Choong, et al., The global rise of 3D printing during the COVID-19 pandemic, Nature Reviews Materials 5 (9) (2020) 637−639.

[30] K. Iyengar, et al., Challenges and solutions in meeting up the urgent requirement of ventilators for COVID-19 patients, Diabetes & Metabolic Syndrome: Clinical Research & Reviews 14 (4) (2020) 499−501.

[31] P. Sophie, Volunteers use 3D-printing to support Italian COVID-19 effort. Available from: <https://www.healthcareitnews.com/news/europe/volunteers-use-3d-printing-support-italian-covid-19-effort>, 2020 (15.05.20).

182 Digital Manufacturing

[32] H. Jennifer, 3D printers fabricate emergency valves for ventilators to keep coronavirus patients breathing. Available from: <https://www.dezeen.com/2020/03/19/3d-printers-valve-ventilators-hospital-coronavirus/>, 2020 (15.05.20).

[33] K. Zoe, Coronavirus: 3D printers save hospital with valves. Available from: <https://www.bbc.com/news/technology-51911070>, 2020 (15.05.20).

[34] F. Amy, Meet the Italian engineers 3D-printing respirator parts for free to help keep coronavirus patients alive. Available from: <https://www.forbes.com/sites/amyfeldman/2020/03/19/talking-with-the-italian-engineers-who-3d-printed-respirator-parts-for-hospitals-with-coronavirus-patients-for-free/#6a85f3e78f1a>, 2020 (15.05.20).

[35] E. Anas, Hospital in Italy turns to 3D printing to save lives of coronavirus patients. Available from: <https://3dprintingindustry.com/news/hospital-in-italy-turns-to-3d-printing-to-save-lives-of-coronavirus-patients-169136/>, 2020 (15.05.20).

[36] Leitat, The first medically validated and industrially producible 3D emergency ventilator. Available from: <https://covid-leitat.org/en/ventilator/>, 2020 (18.05.20).

[37] Covid-Leitat, LEITAT 1 reaches ICU patients as an accredited field ventilator. Available from: <https://covid-leitat.org/en/leitat-1-reaches-icu-patients-as-an-accredited-field-ventilator/>, 2020 (18.05.20).

[38] C.K. Chua, K.F. Leong, 3D Printing and Additive Manufacturing — Principles and Applications, fifth ed., World Scientific Publishing Co. Pte Ltd, 2017.

[39] F. Moretti, Custom 3D printed mask. Available from: <https://www.3dwasp.com/en/3d-printed-mask-from-3d-scanning/>, 2020 (15.05.20).

[40] Bellus3D Inc., How to make Bellus3D Face mask fitter for COVID-19. Available from: <https://www.bellus3d.com/solutions/facemask.html>, 2020 (09.05.20).

[41] HP Inc., 3D printing in support of COVID-19 containment efforts. Available from: <https://enable.hp.com/us-en-3dprint-COVID-19-containment-applications/>, 2020 (15.05.20).

[42] Avid Product Development, 3D printing PPE in support of COVID-19 containment efforts. Available from: <https://avidpd.com/knowledge-base/3d-printing-ppe-in-support-of-covid-19-containment-efforts/>, 2020 (05.05.20).

[43] Budmen Industries, Flatten the curve. Available from: <https://budmen.com/>, 2020 (05.05.20).

[44] Stratasys Ltd., Face shield initiative adds capacity and partners. Available from: <https://www.stratasys.com/explore/blog/2020/face-shield-initiative-adds-capacity-and-partners>, 2020 (05.05.20).

[45] A.A. Chughtai, et al., Policies on the use of respiratory protection for hospital health workers to protect from coronavirus disease (COVID-19), International Journal of Nursing Studies 105 (2020) 103567.

[46] H. Yu, et al., Reverse logistics network design for effective management of medical waste in epidemic outbreaks: insights from the coronavirus disease 2019 (COVID-19) outbreak in Wuhan (China), International Journal of Environmental Research and Public Health 17 (5) (2020) 1770.

[47] TRIX Connections, CIIRC RP95-3D. Available from: <https://www.rp95.cz/en/>, 2020 (05.05.20).

[48] CIIRC CTU, CIIRC CTU develops own prototype of CIIRC RP95 respirator/half mask. Available from: <https://www.ciirc.cvut.cz/covid-2/>, 2020 (05.05.20).

[49] TRIX Connections, Together against coronavirus. Available from: <https://trixconnections.com/>, 2020 (05.05.20).

Chapter 5

Simulation and optimization for additive manufacturing

How Wei Benjamin Teo[1], Kim Quy Le[1], Kok Hong Gregory Chua[2] and Hejun Du[1,2,3]

[1]*HP-NTU Digital Manufacturing Corporate Lab, Nanyang Technological University, Singapore,* [2]*Singapore Centre for 3D Printing, Nanyang Technological University, Singapore,* [3]*School of Mechanical and Aerospace Engineering, Nanyang Technological University, Singapore*

Abbreviations

ALE	Arbitrary Lagrange Eulerian
AM	additive manufacturing
BESO	bidirectional evolutionary structural optimization
CAD	computer-aided design
CFD	computational fluid dynamics
DED	directed energy deposition
DEM	discrete element method
EBM	electron-beam melting
ESO	evolutionary structural optimization
FEA	finite element analysis
FEM	finite element method
iPS	isostatic polystyrene
MD	molecular dynamics
ML	machine learning
PBF	powder-bed fusion
SIMP	solid isotropic material with penalization
SLM	selective laser melting
TO	topology optimization
VOF	volume of fluid

Symbols

Alphabet symbols

A	Area
c_k	small constant
C_p	specific heat

Digital Manufacturing. DOI: https://doi.org/10.1016/B978-0-323-95062-6.00010-3
© 2022 Elsevier Inc. All rights reserved.

184 Digital Manufacturing

d	contact distance
D, D_f	diffusion coefficient
e	coefficient of restitution
F_i^{fric}	frictional force on particle i
F_i^{gra}	gravitational force on particle i
F_i^{total}	total force
$\mathbf{F}_{st}$	surface tension force
$\mathbf{F}_M$	Marangoni force
F	total free energy of solidification
f_{local}	local energy density
f_{grad}	nonlocal energy density
f_s	volume fraction of solid metal
f_{sol}	free energy density got solid phase
f_l	volume fraction of liquid metal
f_{liq}	free energy density got liquid phase
g	gravitational acceleration
g	monotonically increasing function
G	free energy density of misorientation
H	latent heat of fusion
h	specific enthalpy
j	mode number
K	crystal temperature difference
$k, k_{\mathbf{T}}$	thermal conductivity
K_c	permeability coefficient
k_n	elastic constant at normal contact
k_t	elastic constant at tangential contact
M_ψ	mobility of phase-field
M_ϑ	rotational diffusion
m_i	mass of particle i
p	pressure
p	energy interpolating function
$\mathbf{P}_r$	recoil pressure
q_{abs}	absorbed heat from the laser beam
q	double-well function
r	crystal position
r_i	position of particle i
r_d	distance between two particles
R_i	radius of particle i
R_j	radius of particle j
s	excess energy due to misorientation
S	shear modulus
t	time
T	temperature field of the computational domain
T_c	crystallization temperature
T_m	melting temperature
T_m^o	equilibrium melting temperature
T_s	solidus temperature

Simulation and optimization for additive manufacturing Chapter | 5   **185**

T_l	liquidus temperature
$\mathbf{u}$	flow velocity
W	barrier height
x	horizontal crystal growth direction
y	vertical crystal growth direction
Y	Young's modulus

Greek symbol

α	thermal diffusivity
$\overline{\alpha}$	energy distribution
α_1	volume fraction of metallic phase
β	anisotropic growth of crystal
γ	solid/liquid state of metal cells
γ_n	viscoelastic damping constant at normal contact
γ_t	viscoelastic damping constant at tangential contact
Γ	mobility of crystal
$\delta_{t_{ij}}$	tangential displacement vector between two spherical particles
ε	strength of anisotropy
ζ	unstable energy barrier
ζ_o	stable solidification potential
θ	orientation angle
ϑ	normalized crystallographic orientation
ϑ_m	missing orientation angle between solid and liquid phase
κ_o	coefficient of interface gradient
μ	dynamic viscosity
ν	Poisson's ratio
ϕ	phase field
ρ	volume average density
φ	material properties of each calculated cell
φ_s	material property in solid phase
φ_l	material property in liquid phase
φ_g	material property in gaseous phase
ψ	crystal order parameter
ψ_x	crystal order in x direction
ψ_y	crystal order in y direction

5.1 Introduction

Physics-based process models have been recognized as being foundational to the qualification of additively manufactured parts. This section will introduce the theoretical and modeling aspects applied to additive manufacturing (AM) processes.

Simulation modeling is a necessity in the modern world of science and engineering. It provides an analysis of multiple applications in the real world that are complicated to understand. By creating a digital block mesh similar to the actual prototype, the models can generate analytical data that cause

186 Digital Manufacturing

the phenomenon associated with the physical process in the real world, based on mathematical equations and physical parameters. Hence, with the validation of experimental results, the simulation model provides details, such as material behavior and properties, during complicated manufacturing processes and defect analysis.

Besides solving complicated manufacturing processes, a phenomenon that may happen on processes that are yet to be performed can be predicted by simulation models. Simulation models are able to prevent manufacturing [1], optimize the manufacturing and fabrication process [2], and identify failure leading factors [3]. Thus the efficiency of simulation modeling results reduces the cost and time spent in optimizing parameters for manufacturing. The direction where the simulation and modeling are heading to the simulated results can be served as the concurrent feedback to 3D printer program, which allows to adjust to printing parameters. This can be done when the model is able to calculate in large domain and in short period of time using high-performance computing system. Advanced computing systems such as machine learning (ML), artificial intelligent, and cloud computing are necessary to develop.

In the current technology the models using in AM simulation process can be classified from a physics point of view or from the simulation scale. It can be classified by physics behaviors, including heat transfer, melting/solidification process, powder deposition, and phase-field model. In addition, the models can also be classified by scale such as macroscale, mesoscale, and microscale. Depending on the objectives of the study, different simulation types and simulation scales should be chosen. For instance, to investigate on the warpage behavior of the printed part, heat transfer model at macroscale should be considered. Besides, the melting and solidification process can be observed by employing the conservation of mass, momentum, energy, and volume of fluid (VOF) model at the mesoscale simulation. However, some certain behaviors are unable to be captured by macroscale or mesoscale simulation such as the interaction of molecules during printing. In this case, microscale simulation needs to be performed. The following sections explain in detailed various simulation scales and simulation models typically applied in AM simulation.

5.1.1 Macroscale modeling

In AM, macroscale modeling is often employed to simulate the heat transfer, shrinkage, warpage, or mechanical properties of the printing/printed part. One of the popular methods that have been used for macroscale modeling in AM is the finite element method (FEM). This method has been widely applied in different AM processes, including the powder-bed fusion (PBF) process [4] and the material extrusion process [5]. Stender et al. adopted FEM to simulate temperature history, displacement, and stresses in directed energy deposition (DED) [6]. Processes in PBF, such as powder spreading and powder interaction, can be simulated by discrete element method (DEM) to determine powder distribution and the packing density of powders in powder bed [7]. These simulation models clearly display the effects of the print job from the variation of printing parameters, such as temperature, packing density, and feed rate. The method will be further elaborated in Section 5.2.1.

5.1.2 Mesoscale modeling

Mesoscale simulation involves processes that change with time. Such processes allow observing in detail the heat transfer, fusing process, phase change at powder-scale during manufacturing. As the simulation is in such details, different types of defects can be captured, including the formation of pores, hump, and discontinuity of the printed melt track [8−10]. Other simulation methods can simulate the AM process at mesoscale, including the Lattice-Boltzmann method, VOF method, computational fluid dynamics (CFD) method, and Arbitrary Lagrange Eulerian method [11]. Through the time-dependent modeling, the changes within the printed material can be observed during processes of AM. These results provide further investigations, such as pore formation, melt pool flow, powder coalescence, and crystal growth. The simulation models using the CFD model will be elaborated in Sections 5.2.2 and 5.2.3.

5.1.3 Microscale modeling

Microscale modeling in AM often refers to the phase-field model, employed to simulate the dendritic formation [12]. The phase-field model in microscale modeling provides the study of crystal growth when materials undergo the cooling process. The simulation model shows the crystallization process of the material, which shows the nucleation of crystals and the growth of dendrites with respect to temperature and time. The simulation is achieved through Lattice-Boltzmann or Ginzburg−Landau models where material viscosity is required. The crystallization behavior may have close relation toward mechanical and physical properties of printed materials. Hence, through the simulation of crystal morphology, crystallization behavior of printed materials can be studied. The other type of microscale simulation in AM uses molecular dynamics (MD) simulation to investigate the interactions among atoms [11]. MD model requires extremely high computational expense. However, it is meaningful in investigating new alloys or new materials. Looking to further understand microscale modeling via phase-field modeling, the Ginzburg−Landau and Granasy models will be elaborated in Section 5.2.4.

188 Digital Manufacturing

5.1.4 Parameters optimization

In AM, multiple input parameters, including laser power, scanning speed, layer thickness, powder size and shape, and heating/cooling conditions, can directly affect the mechanical properties, geometry accuracy, pores, or surface roughness formation. Thus to produce printed parts with high mechanical properties and minimized defects, it requires the optimization process of input parameters. This process can be chosen by trial−error experimental testing. It is evident that using the experimental method, a direct result can be observed. However, applying experimental method in all the optimization processes is time- and cost-consuming. Parallelly, using optimization methods such as topology optimization (TO) or using ML to predict material properties is a promising method for choosing input parameters. In the scope of this chapter, TO will be discussed in detail in Section 5.3.

5.1.5 Objectives

Hence, to achieve an effective and efficient print job, it is essential to employ simulation and optimization modeling in assisting the AM processes. In Section 5.2, various simulation models will be introduced and how they are used to analyze complex problems from the AM process. In Section 5.3, TO will be presented and discussed on the usefulness of these models to AM. The objectives of this chapter are:

- To understand the fundamental physics of AM processes which contains multiple interacting physical phenomena.
- To propose theoretical models at different length and time scales (e.g., part scale, powder scale, microstructural scale) applicable to capture the structural, physical, and material features and reduce experimental testing.

5.2 A review of models employing in additive manufacturing simulations

To produce quality standards, AM functional parts, trial and error experiments are typically chosen. However, it could be time- and cost-consuming. Therefore various simulation methods have been employed to investigate the AM process. The current models can range from macroscale to mesoscale and even to microscale simulation of the AM process. For example, FEM is developed to predict the temperature distribution of complex geometrical AM part at the macroscale [13]. The powder interaction and melt pool dynamics have also been captured using modeling in mesoscale [8]. In addition, microscale simulation includes that MD model is employed to investigate the interactions among atoms in AM process [11]. Furthermore, the microstructure of the AM printed part has also been predicted close to the experiment [14].

In this chapter, mesoscale simulations of the PBF process, including powder interaction and melt pool dynamics, are discussed in detail. Additionally, the crystallization behavior in microscale simulation is also presented by adopting the phase-field model. The contents of the chapter offer a broader method apart from conventional continuous solid mechanics modeling. Some of the models, such as the heat source model and the heat transfer model, are applicable to other simulation methods, such as the simulation of multitrack in PBF process using the FEM approach.

5.2.1 Powder interaction

Powder interaction has been investigated in different PBF processes such as selective laser sintering, selective laser melting (SLM), electron-beam melting (EBM), Multi Jet Fusion (MJF), and DED. The DEM is one of many simulation models to investigate the interactions of powder with other powders, powder-bed surfaces, or printed surface layers. For better simulation studies within powder bed, the heat transfer equation is also required in DEM. Thus, DEM is a valuable channel to further investigate powder behaviors during the 3D printing process.

Powder deposition simulation can be performed through DEM modeling. In the model, Newton's second law is used to determine the position of the powder particles in the powder bed, and the movement of powder particle i can be determined by [15],

$$m_i \ddot{r}_i = F_i^{total} \tag{5.1}$$

where m_i and r_i are mass and position of particle i, respectively. F_i^{total} is the total force applying on the particle i.

In this model the total force is the combination of frictional and gravitational force, and it is calculated as,

$$F_i^{total} = F_i^{fric} + F_i^{gra} \tag{5.2}$$

where F_i^{fric} and F_i^{gra} are the frictional and gravitational forces on the particle i, respectively.

When the particles do not overlap, the only active force is the gravitational force, which is described as [15],

$$F_i^{gra} = -m_i g \hat{z} \tag{5.3}$$

In the overlapping case the distance r_d between two particles of radius R_i and R_j is less than their contact distance $d = R_i + R_j$ as shown in Fig. 5.1. In this case, both gravitational and friction forces are exerted on the particles.

The friction force between two granular particles is determined as [16],

$$F_i^{fric} = (k_n \delta_{n_{ij}} - \gamma_n \nu_{n_{ij}}) - (k_t \delta_{t_{ij}} - \gamma_t \nu_{t_{ij}}) \tag{5.4}$$

190 Digital Manufacturing

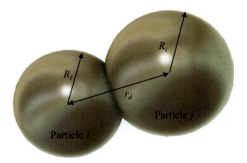

FIGURE 5.1 Schematic representation of the overlapping particles.

where $\delta_{n_{ij}} = d - r_d$ is the overlap distance between two particles, k is the elastic constant, γ denotes the viscoelastic damping constant, and ν represent the relative velocity of the two particles, respectively. The subscripts n and t denote normal and tangential contact, respectively, $\delta_{t_{ij}}$ indicates the tangential displacement vector between two spherical particles. In the current model, sliding friction was employed to simulate the sliding interaction of particles. However, rolling friction has not been considered yet. The rolling force should be included to improve the accuracy of the model in future studies.

From the material properties the calculation of k_n, γ_n, k_t, and γ_t can be found in Han's study [7], where Y represents Young's modulus, ν indicates Poisson's ratio, G denotes shear modulus, and e describes the coefficient of restitution.

5.2.1.1 Software

DEM modeling can be simulated through software packages such as EDEM, Yade, or LIGGGHTS.

EDEM is used to simulate and analyze the behavior of individual particles, integrating particles, fluid, and machine dynamics. EDEM is employed in various industrial problems such as agriculture, chemical processing, and pharmaceuticals [17]. By adding discrete particles to the continuum model, EDEM CFD coupling can solve particle–fluid flows and the solid–liquid phase.

Yade is another flexible open-source package written in C++, which allows simulating of various discrete element types, shapes, and sizes.

LIGGGHTS (LAMMPS Improved General Granular-Granular Heat Transfer Simulation) is an open-source DEM particle simulation software based on LAMMPS, which is an MD code for different materials. The DEM model built-in LIGGGHTS can also be easily modified as an open-source code software.

In powder deposition simulation, stainless steel 316 L (SS316L) was employed, and the simulation is performed on the LIGGGHTS platform. The simulation parameters and the material properties applied in the DEM model are tabulated in Table 5.1. The powders were assumed as solid spheres with

Simulation and optimization for additive manufacturing **Chapter | 5** **191**

TABLE 5.1 Simulation parameters and material properties of SS316L using in the DEM model.[a]

Particle's properties	Values	Refs.
Density (kg/m^3)	7800	[18]
Young's modulus (GPa)	1.1	—
Poisson's ratio	0.3	[18]
Coefficient of friction	1.0	[19]
Coefficient of restitution	0.3	—

[a]K. Q. Le, "Computational modeling of selective laser melting process," 2020.

their radii following a Gaussian distribution, and the powder particles have the size of 27 ± 10 μm.

Fig. 5.2A—E illustrates the snapshots of the powder depositing onto the substrate from $t = 0$ to $t = 1.6$ s. The 3D model has the domain of $x \times y \times z = 2100 \times 800 \times 600 \ \mu m^3$, where the gravitational force is along the y-axis. A blanket of 1800 powder particles was inserted at the area above $y = 300 \ \mu m$. Due to the gravitational force, all the particles settled down and reached equilibrium at $t = 0.036$ s. After that, a blade having a size of $x \times y \times z = 10 \times 750 \times 600 \ \mu m^3$ was adopted such that the top of the blade was located at $y = 800 \ \mu m$, and it was moving from left to right with the velocity of 0.001 m/s. Due to the recoating of the blade, a uniform powder layer thickness of 50 μm was formed. To validate the DEM model a contact image sensor was attached to the powder recoater module to capture powder distribution [20]. Unfortunately, the DEM method cannot model the melting process, which is a critical phenomenon in PBF printing.

5.2.2 Heat transfer and melt pool dynamics

In many AM processes, the material is heated up until it becomes partially melted or fully melted. Then, the melted material coalesces with the surrounding material and the material from the previous layer. Therefore it is important to investigate the heat transfer and the flow dynamics of the molten material during the AM process. The FEM has been employed to investigate the heat transfer in both mesoscopic and macroscopic models [13]. The CFD model is employed to simulate both the heat transfer and melt pool dynamics. This model provides a detailed understanding of the melting and solidification process of AM. Further development of the CFD model predicts the microstructure of the printed material.

The CFD model is employed to simulate laser-powder-bed interaction, heat transfer, and fluid flow of the melt track. In this model, SS316L was

192 Digital Manufacturing

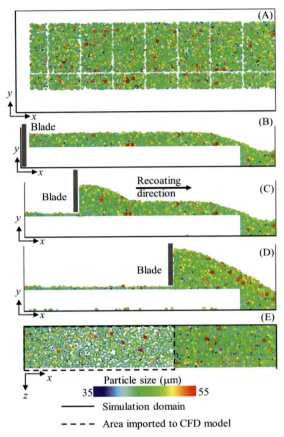

FIGURE 5.2 Powder bed deposition model viewed from (A) the front at $t = 0$ s, (B) the front at $t = 0.036$ s, (C) the front at $t = 1.12$ s, (D) the front at $t = 1.6$ s and (E) the top at $t = 1.6$ s.

chosen as the material of the powder and substrate [21]. The metallic and gaseous phases were determined by α_1 and $(1 - \alpha_1)$, respectively. $\alpha_1 = 1$ refers to a cell completely occupied by the metallic phase and $\alpha_1 = 0$ represents a fully gaseous cell. Any value ranges from 0 to 1 describes that the material is in the melting process. A temperature-dependent variable γ is used to indicate the solid or liquid state of the metal cells and is calculated as [22],

$$\gamma = \begin{cases} 0 & , \quad T \leq T_s \\ \dfrac{T - T_s}{T_l - T_s} & , \quad T_s < T < T_l \\ 1 & , \quad T \geq T_l \end{cases} \quad (5.5)$$

Simulation and optimization for additive manufacturing **Chapter | 5** **193**

where T_s and T_l represents solidus and liquidus temperature of the metal powder, respectively. T indicates the temperature field of the computational domain. Thus the volume fractions of solid metal f_s and liquid metal f_l are $\alpha_1(1 - \gamma)$ and $\alpha_1\gamma$, respectively. Consequently, the material properties $\overline{\varphi}$ of each cell are calculated based on the material properties in each state and its volume fraction as represents [23],

$$\overline{\varphi} = \alpha_1(1 - \gamma)\varphi_s + \alpha_1\gamma\varphi_l + (1 - \alpha_1)\varphi_g \tag{5.6}$$

where s, l, and g indicate the solid, liquid, and gaseous state, respectively. In addition, the volume fraction of the metallic phase satisfies the following condition,

$$\frac{\partial(\overline{\rho}\alpha_1)}{\partial t} + \nabla\cdot(\overline{\rho}\mathbf{u}\alpha_1) = 0 \tag{5.7}$$

where $\mathbf{u}$ represents the flow velocity, t describes time, and $\overline{\rho}$ denotes the volume average density.

The mass conservation is computed as,

$$\frac{\partial\overline{\rho}}{\partial t} + \nabla\cdot(\overline{\rho}\mathbf{u}) = 0 \tag{5.8}$$

while the momentum equation is determined as [23,24],

$$\frac{\partial\overline{\rho}\mathbf{u}}{\partial t} + \nabla\cdot(\overline{\rho}\mathbf{u}\otimes\mathbf{u}) = -\nabla p + \nabla\cdot(\mu\nabla\mathbf{u}) - K_c\left[\frac{(1-f_l)^2}{f_l^3 + c_K}\right]\mathbf{u} + \overline{\rho}\mathbf{g} + \mathbf{F}_{st} + \mathbf{F}_M + \mathbf{P}_r \tag{5.9}$$

where p denotes the pressure, μ indicates the dynamic viscosity, and $\mathbf{g}$ signifies gravitational acceleration. The Darcy's term explains the contribution of the damping term in the mushy zone, and it is calculated as $-K_c\left[\frac{(1-f_l)^2}{f_l^3 + c_k}\right]\mathbf{u}$. The last three forces on the right-hand side of Eq. (5.12) describe the surface tension force, $\mathbf{F}_{st}$, Marangoni force, $\mathbf{F}_M$, and the recoil pressure, $\mathbf{P}_r$.

The energy conservation is expressed as [25,26],

$$\frac{\partial\overline{\rho}h}{\partial t} + \nabla\cdot(\overline{\rho}\mathbf{u}h) = \nabla\cdot\left(D_f\nabla h\right) + q_{abs} - q_{loss}|\nabla\alpha_1|\frac{2\overline{\rho}}{\rho_s(1 - \gamma) + \rho_l\gamma + \rho_g} \tag{5.10}$$

where h indicates the specific enthalpy, q_{abs} is the absorbed heat from the laser beam. The diffusion coefficient, D_f, is given as,

$$D_f = (1 - \alpha_1)\frac{k_g}{C_p} + \alpha_1\frac{k_s\alpha_1(1 - \gamma_l) + k_l\alpha_1\gamma}{C_p} \tag{5.11}$$

where k represents thermal conductivity, and C_p describes specific heat. The heat loss at the metal/gas interface due to convection, radiation, and emission is described by $q_{loss} = q_{conv} + q_{rad} + q_{evap}$ [27]. The details of heat loss terms can be found in Le's work [11].

5.2.2.1 Software

Various CFD software packages using the VOF method are used to simulate the SLM process, including ANSYS FLUENT, FLOW-3D, and OpenFOAM. ANSYS FLUENT offers high-performance computing to solve complex and large models efficiently and cost-effectively. It was adopted to simulate different AM processes, including the modeling of temperature field and grain growth of SLM of AlN/AlSi10Mg composite [28].

FLOW-3D is a commercial software that has been widely adopted to simulate the PBF process, such as SLM and EBM. OpenFOAM, on the other hand, is an open-source software package employed in many different fields of engineering and science. OpenFOAM provides solutions to simulate various problems such as water flow, heat transfer, multiphase interaction, and phase change [8].

One of the case studies on single-track simulation was performed using our own developed CFD model running on the OpenFOAM platform. The simulated temperature and velocity distribution of the melt track at the cross section along the scanning direction is depicted in Fig. 5.3. This model allows capturing the complex molten flow inside the melt pool, which gives information about melt pool geometry and formation of pores and defects of the printed part. Thus input parameters such as laser power and scanning speed can be optimized by choosing the input parameters to produce the

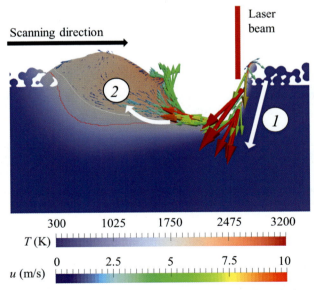

FIGURE 5.3 Simulated temperature and velocity distribution of single-track along the scanning direction. K.Q. Le, C. Tang, C.H. Wong, On the study of keyhole-mode melting in selective laser melting process, *International Journal of Thermal Sciences*, 145 (2019) 105992.

most suitable melt pool geometry and lessen the defects. To validate the sintering or melting process in PBF processes, experiments employing high-speed camera or in situ X-ray imaging are the useful technologies to capture the molten flow and powder spattering.

5.2.3 Light source simulation

Different light sources have been adopted in the AM processes, including UV light, laser, electron beam, and infrared light. The light source can be modeled as a volume heat source applied on top of the material, called the simplified heat source, as depicted in Fig. 5.4A. The simplified heat source is suitable for simulating the sintering process or melting process with shallow penetration of the melt pool. The advantage of using a simplified heat source is that light source model is simple yet can still predict the melting process close to the experiment. However, for the deep penetration, the simplified heat source is limited in producing high accuracy in the melt pool geometry. Thus a more accurate light source model called the ray-tracing model is developed to mimic the multiple reflections of the light rays (see Fig. 5.4B). This model can simulate the reflection of the light source on powder particles and gas–molten material interface. This model is generally more expensive in terms of computational expense. It can, however, produce higher accuracy, especially for deep penetration. Furthermore, complex physics like humping and pore formation can be captured more precisely by applying ray-tracing model. The detail of these two models can be found in Le's work [11].

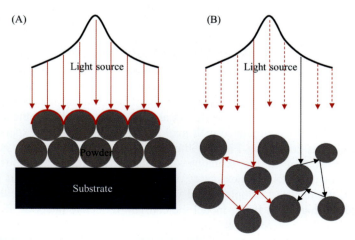

FIGURE 5.4 A schematic diagram of (A) simplified heat source and (B) ray-tracing model.

196 Digital Manufacturing

5.2.3.1 Melt pool simulation using simplified heat source

Both simulation and experiment were performed to validate the CFD model using a simplified heat source. A 3D computational domain has been modeled in the simulation, where the gravitational force is exercised along the y-axis.

In the experiment, single-melt tracks were printed on an SS316L substrate without powder. Different laser powers and scanning speeds as listed in Table 5.2 were employed to print these melt tracks. To simplify the number of variables of the model, the model is validated by using the single-melt track without powder on the substrate. This test is an initial validation work to assure the accuracy of the model.

The melt pool depths and widths of the simulation results were compared with the experimental results to verify the accuracy of the model. Table 5.2 shows in most cases the simulated melt pool depths and widths are closed to the experimental ones where the deviations are smaller than 15%, except for sample *S8*. Furthermore, Fig. 5.5 illustrates the melt pool geometry of different samples where samples *S2*, *S3*, *S4*, and *S5* have the same energy density (see Fig. 5.5A–D). As noticed, the melt pools in the simulation and the ones in the experiment share the geometrical similarity. However, the simulated melt pool depth and width of sample *S8* is far different from the experimental ones as seen in the deviation values listed in Table 5.2. Also, Fig. 5.5E shows that melt pool depth in the experiment is much deeper than the one in the simulation. Therefore it is challenging to achieve a deep-keyhole melt pool by employing the simplified heat source model due to the omission of multiple reflections of the laser rays. To simulate deep keyhole, it is necessary to consider the multiple reflections of laser rays. Therefore the simplified heat source model is suitable for simulating the melt pool with shallow penetration.

5.2.3.2 Melt pool simulation using ray-tracing model

The single-tracks were modeled by employing our own developed CFD model where the light source was simulated using a ray-tracing model. The simulated melt pool depths and widths were measured and compared with the experimental ones done by Bertoli et al. [29] as tabulated in Table 5.3.

Table 5.3 lists the simulated melt pool depths and widths, the experimental ones, and its deviations. Fig. 5.6 illustrates the melt pool geometry at vertical cross section to the scanning direction at the center of different simulated melt tracks. In most cases, the deviations are smaller than 15%. However, the depth of sample A and width of sample K express the deviations greater than 15%. In the model the constant reflective index of SS316L was employed. However, in the real printing, the reflective index is varied depending on the temperature of the material. Furthermore, some material properties like the specific heat of liquid metal are applied constantly at

TABLE 5.2 The melt pool depths and widths of simulation and experiment of the single-melt track.[a]

$P(W)/S$ (mm/s)[b]	Depth			Width		
	Simulation (μm)	Experiment (μm)	Deviation, ε (%)	Simulation (μm)	Experiment (μm)	Deviation, ε (%)
(*S1*) 200/1600	40 ± 2	42.42 ± 5	5.70	104 ± 0	98.76 ± 7	5.31
(*S2*) 100/400	60 ± 0	51.51 ± 4	16.48	114 ± 2	111.84 ± 10	1.93
(*S3*) 200/800	80 ± 0	83.04 ± 5	3.66	118 ± 2	120.40 ± 5	1.99
(*S4*) 300/1200	88 ± 0	91.13 ± 4	3.43	134 ± 6	122.27 ± 7	9.59
(*S5*) 400/1600	90 ± 2	95.29 ± 7	5.55	132 ± 4	125.11 ± 12	5.51
(*S6*) 300/1000	102 ± 2	115.84 ± 6	11.95	126 ± 2	123.46 ± 8	2.06
(*S7*) 400/100	140 ± 0	157.71 ± 8	11.23	138 ± 2	136.69 ± 11	0.96
(*S8*) 200/400	132 ± 4	212.25 ± 11	37.81	140 ± 4	179.68 ± 9	22.08

[a]K. Q. Le, "Computational modeling of selective laser melting process," 2020.
[b]P is laser power (W), S is laser scanning speed (mm/s).

198 Digital Manufacturing

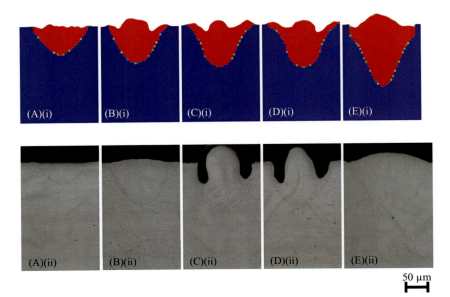

FIGURE 5.5 Melt pool geometry of sample (A) *S2*, (B) *S3*, (C) *S4*, (D) *S5*, and (E) *S8* from the (i) simulation at vertical cross section to the scanning direction and (ii) experiment. *K.Q. Le, Computational Modeling of Selective Laser Melting Process, 2020.*

775 J/kg.K. In reality, it should be temperature-dependent values. Thus, the model cannot simulate the melt pool width as near to the experiment when the laser penetrates concentrated into the substrate.

In summary, different light source models can be employed depending on the printing parameters and accuracy requirements of the model. The simplified heat source can simulation a narrow melt pool, and it saves the computational expense. However, to consider the reflection of the light source, it is necessary to adopt the ray-tracing model. Even though the calculation time is longer, it offers high accuracy of the simulation results. It is possible to validate the melt track using the melt pool geometry of the printed melt track.

5.2.4 Crystallization/microstructure simulation

In the polymer printing process, coalescences of powders allow the radical molecules from various polymer chains to form longer polymer chains. After the coalescence of the powders, the melt polymer will be cooled to room temperature. During this process, long polymer chains will start to bend and fold into crystal lamellar [30,31]. The lamellae will form at the crystallized temperature in an almost perfectly round shape known as spherulite. To date, the understanding of the relationship between the formation of spherulite and the mechanical properties of printed parts is very limited [32].

TABLE 5.3 Melt pool depth and width in simulation and experiment.[a]

$P(W)/S$ (mm/s)[b]	Depth			Width		
	Simulation (μm)	Experiment [29] (μm)	Deviation, ε (%)	Simulation (μm)	Experiment [29] (μm)	Deviation, ε (%)
(A) 300/1500	60 $\pm$ 0	50	20.00	86 $\pm$ 2	85 $\pm$ 11	1.18
(B) 400/2000	57.3 $\pm$ 2.7	64	10.47	92 $\pm$ 4	105 $\pm$ 18	12.38
(C) 300/900	120 $\pm$ 0	128	6.25	100 $\pm$ 8	109 $\pm$ 21	8.26
(D) 500/1500	112 $\pm$ 0	123	8.94	100 $\pm$ 4	99 $\pm$ 33	1.01
(E) 300/750	152 $\pm$ 0	146	4.12	112 $\pm$ 4	115 $\pm$ 25	2.61
(F) 300/600	198 $\pm$ 0	198	0.00	114 $\pm$ 2	130 $\pm$ 29	12.31
(G) 400/800	194 $\pm$ 2	220	11.82	108 $\pm$ 4	120 $\pm$ 22	10.00
(H) 200/300	276 $\pm$ 4	266	3.76	110 $\pm$ 2	130 $\pm$ 10	15.38
(K) 300/450	278 $\pm$ 6	302	7.95	110 $\pm$ 2	140 $\pm$ 25	21.43

[a]K. Q. Le, C. Tang, and C. H. Wong, "On the study of keyhole-mode melting in selective laser melting process," International Journal of Thermal Sciences, vol. 145, p. 105992, 2019.
[b]P is laser power (W), S is laser scanning speed (mm/s).

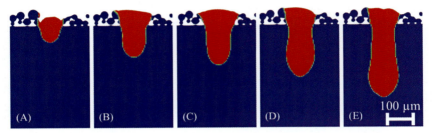

FIGURE 5.6 Melt pool geometry of sample (A) *A*, (B) *C*, (C) *E*, (D) *F*, and (E) *K* at vertical cross section to the scanning direction at the center of simulated melt track. K.Q. Le, C. Tang, C.H. Wong, On the study of keyhole-mode melting in selective laser melting process, *International Journal of Thermal Sciences, 145 (2019) 105992.*

Metal printing, on the other hand, presents a different concept in crystallization. Unlike polymer, metals are commonly presented in the form of alloys. Hence, the melting temperatures of each metal element are considered during the process. Through the model, the grain growth of the alloy is observed based on the different crystallographic orientations [33]. It is also noted that the microstructure of alloys, such as steel, undergo phase transformation at a given thermal condition [34]. It is known that the microstructure of alloys varies with a different temperature gradient.

Through the concept of AM, the formation of crystals in printed materials occurs during the cooling period. The spherulite morphology consists of a nucleated radial nucleus with an arrangement of thin lamellae coming off the nucleus. The importance of investigating the morphology of spherulites provides the key factors related to the deformation of materials. The morphology of the spherulites differs at different supercooling range [35], and it can be categorized given the temperature at which crystallization occurs.

Phase-field modeling can be performed with various numerical software, such as OpenFOAM and MATLAB®. As mentioned, OpenFOAM is an open-source software package that provides solutions to simulate problems related to fluid dynamics, heat transfer, multiphase interaction, and phase change [8]. MATLAB is a commercialized software that can create models and perform numerical solutions, which widely are used in the engineering field.

In this section, some of the common phase-field models will be elaborated to better grasp the concepts behind the formation of the spherulites. The morphology of the crystal will be observed from the models, and the properties of the different spherulite growth are discussed. This section will be focusing on the modeling of polymers as case studies.

5.2.4.1 Time-dependent Ginzburg–Landau model

The time-dependent Ginzburg–Landau model performs crystal growth through the transition of one phase to another. In polymer materials the phase is differentiated between the amorphous (metastable) and the solid crystal (stable) phases. The evolution of crystal growth is equipped by a nonconserved phase-field equation associated with physical quantities, which is important to crystallization, including latent heat and concentration of the material. Unlike alloys, polymers are assumed melt homogeneously. Hence, the change in concentration by phase segregation is neglected. The thermal conduction of latent heat is another important aspect of crystallization. Thus the phase-field model for polymers consists of a phase-field equation and a heat conduction equation [36].

The time-dependent model is written based on the theory developed by Ginzburg and Landau [36]. From this theory the phase-field equation is written as,

$$\frac{\partial \psi(r, t)}{\partial t} = -\Gamma \frac{\delta F(\psi, T)}{\delta \psi(r, t)} \tag{5.12}$$

where $\psi(x, t)$ is the crystal order parameter with respect to time t and position r. The mobility of crystal, Γ, is known to be inversely proportional to the viscosity of melt [36]. The total free energy of the solidification system, $F(\psi, T)$, consists of a local free energy density and a nonlocal gradient term,

$$F(\psi, T) = \int [f_{local}(\psi, T) + f_{grad}(\psi)] dV \tag{5.13}$$

It is known that a completed crystallization is unachievable due to imperfect polymer crystals, and it should be included in the phase-field model. Thus a temperature-dependent local free energy density is developed using the Harrowell−Oxtoby relation,

$$f_{local}(\psi, T) = W \int_0^{\psi} \varnothing(\varnothing - \zeta)(\varnothing - \zeta_0) d\varnothing \tag{5.14}$$

where $\zeta = \zeta(T)$ represents the unstable energy barrier, and ζ_o represents stable solidification potential. A dimensionless coefficient, W, describes the energy barrier height. ψ is ranged between 0 and ζ_o, and ζ_o represents the degree of crystal perfection. ζ_o can be determined as fraction between the melting temperature T_m and equilibrium melting temperature T_m^o of the polymer.

Fig. 5.7 shows that the local energy density (f_{local}) is lower as the temperature is lower than the melting temperature (T_m). From the graph, it is observed that $\psi = \zeta_o$ is achieved at the T_m of the polymer, where f_{local} is found to be 0 representing the perfect stable crystal phase. As the temperature gets lower, f_{local} tends to be 0 before attaining a stable crystal phase, suggesting defective crystal growth at $\psi = \zeta$.

The gradient of the crystal order parameter that determines the growth process is represented by the nonlocal free energy density, which is written as,

$$f_{grad}(\psi) = \frac{1}{2} \kappa_o^2 \beta^2(\theta)(\nabla \psi)^2 \tag{5.15}$$

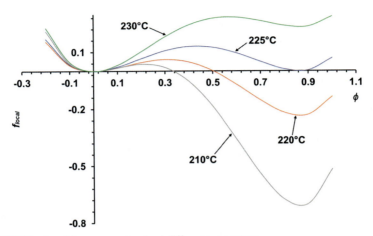

FIGURE 5.7 Local free energy density at different temperature.

where κ_o defines as the coefficient of interface gradient, $\beta(\theta)$ describes a function of orientation angle θ which represents the anisotropic growth of the crystal, and it is written as,

$$\beta(\theta) = 1 + \varepsilon \cos(j\theta) \tag{5.16}$$

where j indicates the mode number, and ε denotes the strength of anisotropy. Followingly, the orientation angle can be calculated as,

$$\theta = \tan^{-1}\left(\frac{\partial \psi / \partial y}{\partial \psi / \partial x}\right) \tag{5.17}$$

By substituting all the equations to Eq. (5.13), the phase-field equation is written as,

$$\frac{\partial \psi(x,t)}{\partial t} = -\Gamma\left[\frac{\partial f_{local}(\psi, T)}{\partial \psi} - \frac{\kappa_0^2}{2}\nabla \cdot \left(\frac{\partial}{\partial \psi_x}[\beta^2(\theta)(\nabla \psi)^2] + \frac{\partial}{\partial \psi_y}[\beta^2(\theta)(\nabla \psi)^2]\right)\right] \tag{5.18}$$

The heat conduction equation is another main factor in polymer isothermal crystallization due to the release of latent heat. It is noted that latent heat can affect the growth of the crystal interface significantly. Thus the heat conduction equation can be modeled based on the conservation law of enthalpy,

$$\frac{\partial T}{\partial t} = \alpha \nabla^2 T + K \frac{\partial \psi}{\partial t} \tag{5.19}$$

where α is the thermal diffusivity of the material, which is defined as $\alpha = k_T/(\rho C_p)$, while $K = \Delta H/C_p$. ρ, C_p, and k_T represent the density, specific heat capacity, and thermal conductivity of the material, while ΔH is the latent heat of fusion of the material.

During crystallization, heat diffusion is much quicker than solute diffusion, which provides different time scales between the phase-field equation and the heat conduction equation. Hence, to generate the calculation of this model with lower computational cost, a higher diffusion coefficient, D, or a smaller thermal diffusivity is applied. Hence, the interface thickness is widen during this process and provides testable predictions to crystal growth [37].

In this case study, OpenFOAM open-source software is used to generate crystal growth of isostatic polystyrene (iPS) during the isothermal crystallization process. The equilibrium melting temperature is determined at 242°C. The melting and crystallization (T_c) temperature are related based on the Hoffman–Week approach. The model is solved based on the finite difference method with a mesh size of 512 μm × 512 μm. The temporal step size is fixed at $\Delta t = 1 \times 10^{-6}$ s and the spatial step size is given at $\Delta x = \Delta y = 1 \times 10^{-7}$ m.

It is known that crystallization morphology and crystallization kinetics are greatly affected by the supercooling ($T_m^o - T_c$) of the material. The crystal morphology, however, is also affected by many other factors other than supercooling. Thus the morphology can be different under the similar supercooling condition in the same polymer material. These factors include thermal history, chain conformation, and the thickness of the polymer film. [37]. Therefore the anisotropic parameters ε and j are assumed based on the conditions of the material.

Fig. 5.8 shows that the crystallization morphology is shown by the phase-field model with anisotropy mode $j = 6$ and anisotropy strength $\varepsilon = 0.06$. With T_m given at 233°C and T_c set at 210°C, the crystal interface is presented in a hexagonal shape. It is witnessed that the crystal size increases with time, and the crystal is nucleated when $t = 0$ s. As supercooling increases, the facets of the crystal tend to concave with the interface structures transformed into dendritic snowflakes. A further increment reduces the size of the crystal core and increases secondary branches from the six major branches.

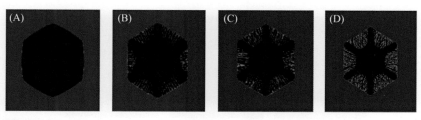

FIGURE 5.8 Crystal morphology at different crystallized temperature at (A) 210°C, (B) 200°C, (C) 195°C, and (D) 180°C with $j = 6$ and $\varepsilon = 0.06$.

204 Digital Manufacturing

Fig. 5.9 shows the crystal morphology at $T_m = 229°C$ and $T_c = 195°C$, with an anisotropy mode $j = 6$ and a series of anisotropy strength 0.01, 0.03, and 0.06. With low anisotropy strength, the crystal morphology appears to be less hexagonal, with numerous secondary branches spread externally. As anisotropy strength increases, the center core is shaped into a precise hexagonal shape with the vertices form thick dendrite branches. It is also observed that the number of secondary branches increases with higher anisotropy strength. It is noted that the crystal morphology differs as crystallization temperature changes. As mentioned, the anisotropy strength of the crystal may be related to other factors.

In Fig. 5.10 the crystal morphology shows the isotropic growth of spherulite. In this case the anisotropy mode $j = 0$. From here, any changes to the anisotropy strength will not have an effect to the crystal morphology as β is constant. It is observed that the crystal grows into a circular shape during this process. As supercooling increases, dendrite-like branches start to form with a smaller crystal core. More secondary crystal branches are found with higher supercooling, suggesting a close relation to the amount of latent heat release when crystallization occurs at a lower temperature [37].

5.2.4.2 Granasy model

Granasy et al. developed another phase-field model to simulate another polymer crystal growth from nucleation of a needle-like fiber, creating new

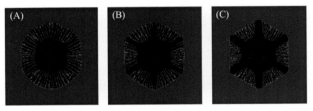

FIGURE 5.9 Crystal morphology at different anisotropy strength, ε, at (A) 0.01, (B) 0.03, and (C) 0.06 with crystallized temperature 195°C and $j = 6$.

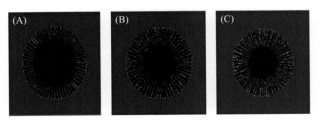

FIGURE 5.10 Crystal morphology at a different crystallized temperature at (A) 200°C, (B) 195°C, and (C) 180°C with $j = 0$.

"branches" at the fibers' ends. This crystallization process forms into a crystal "sheaf" that spreads apart as it grows. Eventually, the sheaves will develop two uncrystallized region on each nucleation area and form into a spherical crystal [38]. It is noted that this crystal growth is witnessed in polymeric materials. Takaki et al. modified the model to investigate the crystal growth of the polymer [39]. Hence, this proposed model provides another perspective to evaluate the growth of spherulites. An illustration of this crystal growth is found in Fig. 5.11.

In this model the free energy function is derived as,

$$F(\psi, T) = \int \left[\frac{\beta^2(\theta)}{2} (\nabla \psi)^2 + f(\psi, T) + \mathbf{G}(\psi, |\nabla \psi|) \right] dA \qquad (5.20)$$

which is relatively similar to Eq. (5.13), with the free local density, $f(\psi, T)$, is now written as,

$$f(\psi, T) = \mathbf{p}(\psi) f_{sol}(T) + (1 - \mathbf{p}(\psi)) f_{liq}(T) + Wq(\psi) \qquad (5.21)$$

where W indicates the height of energy barrier, $f_{sol}(T)$ and $f_{liq}(T)$ represent free energy densities for the solid and the liquid phases, respectively. The energy difference $f_{sol}(T) - f_{liq}(T) = \Delta H(T - T_m)/T_m$ is the driving force for this model. The energy interpolating function, $p(\psi)$, is given as $p(\psi) = \psi^3(10 - 15\psi + 6\psi^2)$ and the double-well function, $q(\psi)$, is written as $q(\psi) = \psi^2(1 - \psi)^2$. The new term, $\mathbf{G}(\psi, |\nabla \vartheta|)$, is an excess free energy density due to misorientation at the solid−liquid interface and the grain boundary. It is noted that the branching effect of fibrous ribbon or secondary nucleation is simulated by the new term, which is defined as,

$$\mathbf{G}(\psi, |\nabla \vartheta|) = \mathbf{g}(\psi) s |\nabla \vartheta| \qquad (5.22)$$

From here, s refers to a positive constant that represents the excess energy due to misorientation. $g(\psi)$ denotes a monotonically increasing function that complies $g(1) = 1$ and $g(0) = 0$ to exclude misorientation effects in liquid phase. Hence, $g(\psi) = p(\psi)$ to exclude the orientation of the liquid phase.

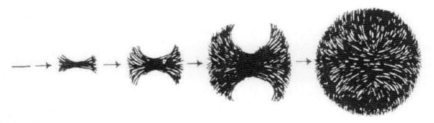

FIGURE 5.11 Crystal growth through Granasy model [38].

206 Digital Manufacturing

In this model, $\beta(\theta)$ is written in two formats, which are,

$$\beta_1(\theta) = \frac{\overline{\alpha}}{1+\varepsilon}\left[1 + \varepsilon\cos\left\{j\left(\theta - \frac{2\pi s}{j}\right)\right\}\right] \tag{5.23}$$

$$\beta_2(\theta) = \frac{\beta_1(\vartheta_m + \vartheta)}{\cos(\vartheta_m)}\cos(\theta - \vartheta) \tag{5.24}$$

$\beta_1(\theta)$ is a modified version from Eq. (5.17). ϑ is the normalized crystallographic orientation between the solid and liquid phases. $\overline{\alpha}$ is a constant which may represent an energy distribution per length. ϑ_m goes as the missing orientation angle between the solid and the liquid phase. Given that $\varepsilon < (j^2 - 1)$, then $\beta(\theta) = \beta_1(\theta)$ for all angle of θ.

This model consists of a phase-field equation, ψ, a heat conduction equation, and an orientation equation, ϑ. These equations are written as,

$$\frac{\partial\psi}{\partial t} = M_\psi\left[\nabla \cdot (\beta^2\nabla\psi) - \frac{\partial}{\partial x}\left(\beta\frac{\partial\beta}{\partial\theta}\frac{\partial\psi}{\partial y}\right) + \frac{\partial}{\partial y}\left(\beta\frac{\partial\beta}{\partial\theta}\frac{\partial\psi}{\partial x}\right) - \frac{\partial f(\psi,T)}{\partial\psi} - \frac{\partial g(\psi)}{\partial\psi}s|\nabla\psi|\right] \tag{5.25}$$

$$\frac{\partial\vartheta}{\partial t} = M_\vartheta\nabla \cdot \left[g(\psi)s\frac{\nabla\vartheta}{|\nabla\vartheta|}\right] \tag{5.26}$$

$$\frac{\partial T}{\partial t} = \alpha\nabla^2 T + K\frac{\partial p(\psi)}{\partial\psi}\frac{\partial\psi}{\partial t} \tag{5.27}$$

M_ψ and M_ϑ drive the phase-field and orientation equations. M_ψ represents the mobility of phase-field, which is established by the crystallization rate and developed from kinetic theory [38]. M_ϑ is associated with rotational diffusion, and it is a function of viscosity. The mobilities are temperature-dependent, and the ratio, M_ϑ/M_ψ, is related to the degree of the second nucleation.

5.2.5 Summary

This section shows various modeling examples used to investigate material behavior when undergoing AM processes. It should be noted that the simulation models listed in this chapter are some of the many models that can be used in modeling AM processes. Besides, other models including CFD–DEM coupling model, FEM model in heat transfer and phase-field simulation, and MD model, are available in showing outcome of AM processes in multiscale simulation are not discussed here. The presented models adopt governing equations to define the physics behind every process for AM, including the spreading of materials, thermal expansion of materials when heated, melt flow of melted materials and crystallization process when materials are cooled. The models provide the information required to simulate the processes of materials undergoing AM at a multiscale level. By understanding the material's behavior during the AM process, the optimization of the print job can be adjusted by setting the correct parameters on the printer. Thus TO can be adopted to achieve the best parameters for the required print job.

5.3 Topology optimization

TO is a numerical method first introduced in 1988 to solve structural problems using the homogenization method [40]. The methodology computes the optimal distribution of the elements with variable holes for the structure to satisfy the given design requirements.

A topology-optimized part requires the designer to specify the design domain, boundary conditions, loads, and constraints to generate an optimal design. Designers commonly applied this method in many areas of the engineering field to develop a primitive design in the early design phase. Although an optimal part will be generated after applying TO, the complexity of the part is commonly restricted by the high level of manufacturing technicality.

Recently, many software companies introduced a new module called the generative design. It has the capability to generate many variations of designs within the given requirements. Despite no clear definition of what generative design is, it can be seen as a big umbrella that encompasses TO. Software companies such as Siemen NX, nTopology, ANSYS offer generative design as part of the software package [41]. However, the focus of Section 5.3 will be on TO.

Both finite element analysis (FEA) and TO use FEM to solve design problems based on the given loads, boundary conditions, and constraints to achieve the optimal design. FEA helps to validate the computer-aided design (CAD) model whether it satisfies the given boundary conditions, loads, and constraints through simulation so that the model meets the overall performance. TO, on the other hand, generates a shape that satisfies the loads, with the boundary condition and constraints introduced at the start of the simulation. During the process, elements subjected to negligible stress and strain energy under the loading conditions were removed by the TO solver. This methodology of obtaining the optimal design does neither esthetic nor manufacturing capability into consideration. As such, the shape generated by the software is usually a complex natural shape with nonsmooth structural geometry, which requires fine-tuning for manufacturability. Lastly, FEA tools validate the fine-tuned design to satisfy the design requirements. Further discussion about the workflow for TO is discussed in Section 5.3.3.

In the past, TO is often employed in the early stages of the designing process. The results surfaced were often difficult to manufacture due to its

208 Digital Manufacturing

naturally complex shapes. The dawn of AM is introduced as a material joining process, layer by layer, whereby it can fabricate parts with complex shapes. Over the decades, AM technologies have gradually proven their capability to fabricate parts using different materials, ranging from polymers to metal parts. The combination of TO and AM offers excellent benefits and opportunities for design in industrial manufacturing.

Section 5.3 consists of four subsections. Section 5.3.1 discussed the different types of structural optimization for AM, such as TO and infilled latticing. These were some of the strategies used to reduce the weight of the parts. Section 5.3.2 focuses on the different types of methodologies used for TO. In this section, we will learn to differentiate their advantages and disadvantages. Section 5.3.3 provides a generic framework on TO workflow for AM. This section elaborates on the key factors, such as creating design and nondesign space, meshing the design space, defining loads and boundary conditions, defining design responses. It also defines the objective and objective functions, defines constraints, postprocessing of optimization results, and the necessary function that requires to run TO. Section 5.3.4 will introduce some of the available commercial software for TO in the market.

5.3.1 Structural optimization

AM has become more and more popular as many companies have turned to them to fabricate application parts. The role of structural optimization in AM is rising across industries as they offer lightweight strategies while finding an optimum relationship between weight and stiffness for mechanical structures. Structural optimization techniques are commonly employed to the design of automotive and aerospace structures where weight savings are essential. Structural optimization is classified into three distinct problem categories: shape, sizing, and TO [42].

- Shape optimization helps find the optimal shape for a model while fulfilling the desired design response and the given constraint. The optimization algorithm will alter the part such as changing fillets, radius, and material thickness; however, it will not include additional or remove holes to the existing model.
- Sizing optimization differs from shape optimization. It is commonly used for perfecting the geometry of truss-like structures, such as bridges, support bars, and frames. The optimization algorithm will remove the nonsupporting trusses if it is not required. The design variables are geometrical parameters such as length, width, and thickness.
- TO is a combination of both sizing and shape optimization. TO will provide the most efficient design based on the desired objective, design response, loads, and constraints in a given design space. No material will be added to the model during the process.

5.3.1.1 Topology optimized lattice-based structure

Lattice structures have become an important structure design feature in AM. It is usually introduced as a lightweight strategy. Although lattice structure is not a part of TO optimization, it has been increasingly seen as an integration function for AM designs. Also, latticing modules are provided by most of the CAD software and are frequently used to replace low-stress areas for weight reduction. The advantages of lattice structures are high in strength-to-weight ratio, high specific energy absorption, high heat removal rate, time and material saving, and good control over vibration and noise dampening [43]. Some studies also demonstrate that close packing lattice support structures can reduce process-induced residual stress, thus mitigating warpages during the print. Proper selection of lattice structure types must be taken into consideration for powder removal [44]. Currently, most work involving lattice and TO use an unpenalized SIMP density-based solution. This prediction method allows to optimize material distribution when certain inputs are given, including design space, load cases, boundary conditions, manufacturing constraints, and performance requirements. A combination of lattice and TO can be classified into three categories [45]:

- Intersected lattice replaces the density-based TO solution with a uniform material density lattice structure and fills the entire design.
- Graded lattice maps the lattice with varying material density onto the grayscale density-based TO solution. Only density values between "0" and "1" are considered where "0" refers to no material and "1" refers to a full solid.
- Scaled lattice, also known as variable density lattice, maps the lattice with varying material density onto a rescaled grayscale density-based TO solution where the density values for the solution are bounded between "0" and "1."

5.3.1.2 Intersected lattice

Improving the manufacturability of the lattice structure via AM has attracted significant interest from many researchers. In one of the studies, Gorguluarslan et al. introduced intersected lattice to the design and performed a two-phase optimization on a beam that is subjected to bending [46]. The two-phase optimization process aims to first perform optimization to remove unnecessary trusses, and second to provide manufacturing constraint by setting the minimum cross section value of the trusses that can be fabricated by AM machine. In their study, they successfully fabricated three different parts using their two-phase optimization process. They also highlighted the importance of the identification of the minimum cross section values for the truss.

5.3.1.3 Graded lattice

Researchers have recently produced a graded lattice structure by mapping with density-based TO results in the lattice structure [47]. Zhang et al. performed a density-based TO on a beam to obtain the optimized topology on a structural level [48]. Later, the lattice structure is mapped according to the material density distribution previously obtained from the TO results. The results show that TO graded lattice beam performed better under a three-point bending test compared to a homogenous lattice beam with an average flexural stiffness of 275 and 200 N/mm, respectively. Robbins et al. conducted a density-based TO on a structure and mapped the result to a lattice structure with properties determined from the homogenization method [49]. The meshing was carried out using Sandia's Sculpt tool, which allows adaption and smoothening of the mesh at the boundaries, thus creating a ready-to-print STL file. The part was printed successfully using the PBF process, demonstrating the feasibility of producing a ready-to-print file.

5.3.1.4 Scaled lattice

Cheng et al. performed a density-based TO and used a mapping technique to convert the results into a scaled lattice [50]. Asymptotic homogenization is used to compute the mechanical properties of the lattice structure as a function of its relative densities. Both simulation and experimental results achieved good agreement with the homogenized model results, indicating good accuracy and efficiency of the method they proposed. In a recent paper, Wu et al. presented a method of distributing lattice structures to adhere to both the principal stress direction and the boundary of the optimized shape obtained from shape optimization [51]. They fabricated the optimized lattice structure and compared it with an optimized solid structure. Experimentally, the optimized lattice structure with varying distribution of lattice structure can support twice the compression load compared to the optimized solid structure.

5.3.2 Types of topology optimization methodologies

The three commonly used methods for TO, solid isotropic material with penalization method (SIMP), level set method, and evolutionary structural optimization (ESO), will be briefly covered in this section.

5.3.2.1 Solid isotropic material with penalization method

The SIMP method is the most popular method among TOs. The method is a gradient-based model that predicts an optimal material with the given requirement and design space [52]. It started with discretizing a model into finite elements. The elements are filled with either material required or voids (no material is needed). The density assigned to the material is given a binary value between "0" and "1," where "0" refers to material removed, and "1" refers to material required. A "penalization" value is introduced in the SIMP equation is to mitigate the formation of the intermediate densities in the structure. The arbitrary number is usually between "1" and "4" with no physical meaning in the optimization. The advantage of using the SIMP method is its fast and stable convergence. However, it tends to result in poor structural boundaries.

5.3.2.2 Level set method

The level set method, a subset of boundary-based structural optimization, provides a smooth structural boundary through the optimization process. It is preferred over the SIMP method as the boundaries for SIMP are very much tessellated. The model produced using the level set method can be easily manufactured as the maximum local curvature can be limited in producing a smooth surface. It is carried out by mapping the level set field to the FEM model through the approximate Heaviside project, where the accuracy of analysis and the intermediate densities around boundary areas are reduced. The extended FEM is introduced under the level set framework to fix the boundary loads, and other boundary phenomena should be included in the simulation [53].

5.3.2.3 Evolutionary method

The ESO is also another popular method for TO. Similar to the SIMP method, it iteratively removes material with low stresses from the model [54]. It does not have the intermediate density elements as in the case with SIMP. Thus it is easy to implement complex structural problems. On the downside, the ESO method has a lack of algorithmic convergence and poor rejection criteria. Furthermore, the model obtained from the optimization has a stepped contour, which is not ideal for fabrication. An enhanced extension pack, bidirectional evolutionary structural optimization (BESO), is later introduced by Yang et al. to improve the structural integrity further [55]. The advantage of implementing BESO is that it adds elements to the structure in the high-stressed area. Huang and Xie also investigated modifying the BESO method to solve nonconvergence and mesh dependency problems [56].

5.3.3 Topology optimization workflow for additive manufacturing

This section describes a typical TO workflow for commercial software and addresses the necessary practices for new users. Fig. 5.12 illustrates a general workflow for TO that is shared across different software. The section is separated into seven categories. Each will cover the fundamental approach that the user should perform while carrying out the TO simulation.

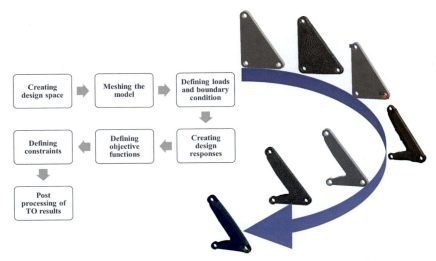

FIGURE 5.12 TO workflow for additive manufacturing [57]. *TO*, Topology optimization.

5.3.3.1 Creating the design space

In the initial design stage the user is required to prepare a CAD model before carrying out TO. The CAD model is further divided into two parts, a design space region and a nondesign space region. Although TO is not performing on the nondesign space region, it is still connected to the other parts of the model for analysis.

- Design space is a space where the TO is performed on the defined geometry.
- Nondesign space is a space where the defined geometry will remain unchanged.

5.3.3.2 Meshing the model

Both the design space and nondesign space are meshed using the same tool into discrete elements. A good rule of thumb is to perform initial simulation using relative coarse mesh, as it will greatly help reduce computational time. However, to get a more accurate result, a smaller mesh size is required, and the computational time will increase exponentially. Loads and boundary conditions are applied to the model after meshing. The TO software calculates the stress and displacement distribution of the model and removes elements that do not contribute significantly to support the loads.

5.3.3.3 Defining loads and boundary condition

Like any finite element simulations, TO is required to provide the necessary loading and constraints for the model. All the loads and constraints are applied to the nondesign space. It is critical to provide proper loads and boundary conditions as the results are load direction sensitive. Yet, missing loads or excessively simplified loads will influence the TO results. Therefore it is necessary to provide all the possible loads and boundary condition variations to ensure the robustness of the design outcome.

5.3.3.4 Creating design responses

When performing TO, design responses are specified before the simulation objective and constraints. They are the parameters that are evaluated and controlled. For example, a typical TO goal is to minimize the displacement at a node so that it can maximize the stiffness or a constraint that forces the weight to reduce by 50%. Weight, volume, displacement, rotation, and strain energy are some of the typical design responses found in all the TO software.

5.3.3.5 Defining objective functions

The objective function is defined to either minimize or maximize the design responses. For example, a structural TO problem usually require minimizing its compliance. Therefore displacement and weight are specified as constraints. Weight is included in the simulation to remove material from the object. The frequency of loads may be included as the constraint to make the simulated object more rigid or flexible. Successful optimization is when the results converge with the increasing number of iterations while fulfilling all constraints.

5.3.3.6 Defining constraints

There are two types of constraints in TO, geometric constraints and performance constraints. Before the simulation starts, the geometric constraints are applied to the model. The user can have the option to:

1. Freeze the area or the elements, ensuring that there will not be any changes to the selected geometry.
2. Propose symmetry conditions so that the results will be symmetrical.
3. Limiting the upper and lower limits of the member size, ensuring no cross section is either too big or small.

However, increasing the number of geometric constraints introduced will reduce the design area significantly. Therefore the user should only implement minimum geometric constraints to obtain the ideal solution.

Performance constraints are constraints that control the software to ensure that the simulation objective is met. They are introduced at the start of the simulation to ensure that the simulation meets its objective. Some of the performance constraints are volume, weight, moment of inertia, reaction forces and moments, displacement, rotation, center of gravity, etc. Unlike geometric

214 Digital Manufacturing

constraints, it is preferred to include all the necessary performance constraints to obtain a robust solution.

5.3.3.7 Postprocessing of optimization results

After TO the optimized geometry usually presents itself as a rough geometry, and it does not look esthetically pleasing. Remodeling of the optimized geometry is necessary to smoothen out all the sharp-pointed surfaces. Finite element simulation is recommended to perform on the remodeled design to ensure it meets the given requirements.

5.3.4 Available commercial software for topology optimization

As TO has been heavily promoted in the AM industry, a significant increase of software providers has risen. They provided software packages that offer users the capability to produce the files for immediate fabrication. This allows the software providers to create competition among themselves, while, at the same time, developing a more robust algorithm and approaches for structurally optimized AM part. This section will cover some of the commercial software that is available to perform TO and generative optimization. Table 5.4 shows some of the commercial software that is available in the market [41].

5.3.4.1 Altair HyperWorks OptiStruct

Altair Engineering Inc., the creator of HyperWorks suite, provides a structural analysis solver for the linear and nonlinear problems with static and dynamic load. Besides the TO module, Altair also offers topography optimization for thin wall structures, free size optimization to identify the optimal

TABLE 5.4 Commercially available software.

Company	Software product
Altair	Hyperworks OptiStruc
ANSYS	ANSYS Discovery AIM
Autodesk	Autodesk Generative Design
COMSOL	Optimization Module
Dassault Systèmes SE	Tosca Structure
nTopology	Element free/Pro
PTC	Generative Topology Optimization extension
Siemens	Solid Edge Generative Design

thickness distribution in machined metallic structures, etc. The capabilities of the software are demonstrated with a feasible study from big companies such as RUAG, EOS, EADS, and Renishaw.

5.3.4.2 ANSYS Discovery AIM

ANSYS Inc. presents Discovery AIM, a platform to perform TO and design exploration. Notably, ANSYS offers a highly intuitive user interface to guide the user through the simulation process. This process helps the user to focus on the engineering problem and spend lesser time figuring out how to use the software. Besides TO, ANSYS Discovery AIM also provides a wide range of simulation packages such as fluid, thermal, electromagnetic, and multiphysics simulations.

5.3.4.3 Autodesk Generative Design

Autodesk Inc. recently developed a generative design module for the user of Fusion 360. Although it is similar to TO, the generative design offers more design choices for the user. It also provides an intuitive interface for the user so that the user will not have trouble setting up the simulation process.

5.3.4.4 COMSOL Optimization module

COMSOL Inc. introduced an add-on optimization module to COMSOL's Multiphysics product to allow the user to perform TO. Their TO uses a gradient-based method called method of moving asymptotes written by Prof. K Svanberg. Although COMSOL does not have a highly intuitive interface for the user, it is one of the preferred choices by many experienced users.

5.3.4.5 Dassault Systèmes SE Tosca Structure

Dassault Systèmes SE, a French software company, developed Tosca Structure that provides structure optimization such as topology, shape, sizing, and bead optimization. One of the software's strength is that the user can perform all three processes (designing, TO, remodeling) in one platform. It is used as a stand-alone module that is compatible with most high-end software, such as Abaqus and ANSYS.

5.3.4.6 nTopology Element Pro

nTopology Element Pro provides a one-stop solution from geometry construction to TO and verification process. One of the capabilities of the software is the automatic function of converting TO results into editable geometry. In addition, the software also focuses on infusing lattice into the design, allowing the user to create lightweight structures.

216 Digital Manufacturing

5.3.4.7 PTC Generative Topology Optimization extension

PTC Inc. developed both generative and TO extension that automatically creates innovative product designs based on the given requirement. The extension can be found in CREO 7.0. Similar to other software mentioned earlier, CREO 7.0 can also have FEA capabilities.

5.3.4.8 Siemens Solid Edge Generative Design

Siemens Solid Edge 2020 offers both generative and TO, providing the users with an end-to-end AM manufacturing solution. As the topology module is powered by Frustum, the advantage of using the Siemen platform is that it is made ready for manufacturing, and it does not require the rebuilding of the model. Solid Edge 2020 also further allows the user to add manufacturing constraints to generate the part that is suitable for its application without remodeling.

5.4 Summary

Simulation modeling has provided analytical solutions to complicated processes within AM techniques. Governed by the mathematical formulations and physical parameters, the models can simulate the actual manufacturing process digitally. Thus these models represent the manufacturing process and material behavior during the printing process and provide analytical data to understand the phenomenon of such techniques. Powder interaction models simulate the forces acting on between particles when powders are spread across the powder bed. Heat transfer and melt pool dynamic show the coalescence of materials when powders are melted from the heat introduced to the model. The crystallization model, such as the phase-field model, provides information on how the crystals are formed during the cooling process and how morphology differs due to a change in properties.

On the other hand, TO provides an optimal solution in design for a given space, boundary conditions, loads, and constraints to maximize the performance of the printed part. These models have proven their effectiveness in reducing material wastage and achieving a better strength-to-weight ratio for the part designed. This can be very useful to the aerospace and automotive industry in creating a lightweight structure with high mechanical performance.

References

[1] H. Mokhtarian, A. Hamedi, H.P.N. Nagarajan, S. Panicker, E. Coatanea, K. Haapala, Probabilistic modelling of defects in additive manufacturing: a case study in powder bed fusion technology, Procedia CIRP 81 (2019) 956−961.

[2] T.A. Krol, C. Seidel, M.F. Zaeh, Prioritization of process parameters for an efficient optimisation of additive manufacturing by means of a finite element method, Procedia CIRP 12 (2013) 169−174.

[3] A. Zargarian, M. Esfahanian, J. Kadkhodapour, S. Ziaei-Rad, Numerical simulation of the fatigue behavior of additive manufactured titanium porous lattice structures, Materials Science and Engineering C 60 (2016) 339−347.

[4] W. Zhang, M. Tong, N.M. Harrison, Resolution, energy and time dependency on layer scaling in finite element modelling of laser beam powder bed fusion additive manufacturing, Additive Manufacturing 28 (2019) 610−620.

[5] S.-I. Park, D.W. Rosen, S.-K. Choi, C.E. Duty, Effective mechanical properties of lattice material fabricated by material extrusion additive manufacturing, Additive Manufacturing 1 (2014) 12−23.

[6] M.E. Stender, et al., A thermal-mechanical finite element workflow for directed energy deposition additive manufacturing process modeling, Additive Manufacturing 21 (2018) 556−566.

[7] Q. Han, H. Gu, R. Setchi, Discrete element simulation of powder layer thickness in laser additive manufacturing, Powder Technology 352 (2019) 91−102.

[8] K.Q. Le, C. Tang, C.H. Wong, On the study of keyhole-mode melting in selective laser melting process, International Journal of Thermal Sciences 145 (2019) 105992.

[9] C. Tang, K.Q. Le, C.H. Wong, Physics of humping formation in laser powder bed fusion, International Journal of Heat and Mass Transfer 149 (2020) 119172.

[10] K. Le, C.H. Wong, K.H.G. Chua, C. Tang, H. Du, Discontinuity of overhanging melt track in selective laser melting process, International Journal of Heat and Mass Transfer 162 (2020) 120284.

[11] K.Q. Le, Computational Modeling of Selective Laser Melting Process, 2020.

[12] J.H.K. Tan, S.L. Sing, W.Y. Yeong, Microstructure modelling for metallic additive manufacturing: a review, Virtual and Physical Prototyping 15 (2020) 87−105.

[13] W. Yan, et al., An integrated process−structure−property modeling framework for additive manufacturing, Computer Methods in Applied Mechanics and Engineering 339 (2018) 184−204.

[14] J.A. Koepf, M.R. Gotterbarm, M. Markl, C. Körner, 3D multi-layer grain structure simulation of powder bed fusion additive manufacturing, Acta Materialia 152, (2018) 119−126.

[15] R. Ganeriwala, T.I. Zohdi, Multiphysics modeling and simulation of selective laser sintering manufacturing processes, Procedia CIRP 14 (2014) 299−304.

[16] S. Haeri, Y. Wang, O. Ghita, J. Sun, Discrete element simulation and experimental study of powder spreading process in additive manufacturing, Powder Technology 306 (2017) 45−54.

[17] Z. Ma, Y. Li, L. Xu, Discrete-element method simulation of agricultural particles' motion in variable-amplitude screen box, Computers and Electronics in Agriculture 118 (2015) 92−99.

[18] H. Wei, Y. Zhao, J. Zhang, H. Saxén, Y. Yu, LIGGGHTS and EDEM application on charging system of ironmaking blast furnace, Advanced Powder Technology 28 (2017) 2482−2487.

[19] H. Chen, Q. Wei, S. Wen, Z. Li, Y. Shi, Flow behavior of powder particles in layering process of selective laser melting: numerical modeling and experimental verification based on discrete element method, International Journal of Machine Tools and Manufacture 123 (2017) 146−159.

[20] L. Tan Phuc, M. Seita, A high-resolution and large field-of-view scanner for in-line characterization of powder bed defects during additive manufacturing, Materials & Design 164 (2019) 107562.

[21] C.Y. Yap, et al., Review of selective laser melting: materials and applications, Applied Physics Reviews 2 (2015) 041101.

218 Digital Manufacturing

[22] F. Rösler, D. Brüggemann, Shell-and-tube type latent heat thermal energy storage: numerical analysis and comparison with experiments, Heat and Mass Transfer 47 (2011) 1027.

[23] C. Qiu, C. Panwisawas, M. Ward, H.C. Basoalto, J.W. Brooks, M.M. Attallah, On the role of melt flow into the surface structure and porosity development during selective laser melting, Acta Materialia 96 (2015) 72−79.

[24] K.Q. Le, C. Tang, C.H. Wong, A study on the influence of scanning strategies on the levelness of the melt track in selective laser melting process of stainless steel powder, JOM 70 (2018) 2082−2087.

[25] J.L. Tan, C. Tang, C.H. Wong, A computational study on porosity evolution in parts produced by selective laser melting, Metallurgical and Materials Transactions A 49 (2018) 3663−3673.

[26] W. Yan, et al., Multi-physics modeling of single/multiple-track defect mechanisms in electron beam selective melting, Acta Materialia 134 (2017) 324−333.

[27] C. Tang, J.L. Tan, C.H. Wong, A numerical investigation on the physical mechanisms of single track defects in selective laser melting, International Journal of Heat and Mass Transfer 126 (2018) 957−968.

[28] D. Dai, D. Gu, R. Poprawe, M. Xia, Influence of additive multilayer feature on thermodynamics, stress and microstructure development during laser 3D printing of aluminum-based material, Science Bulletin 62 (2017) 779−787.

[29] U. Scipioni Bertoli, A.J. Wolfer, M.J. Matthews, J.-P.R. Delplanque, J.M. Schoenung, On the limitations of volumetric energy density as a design parameter for selective laser melting, Materials & Design 113 (2017) 331−340.

[30] F. Liu, T. Sun, P. Tang, H. Zhang, F. Qiu, Understanding chain folding morphology of semicrystalline polymers based on a rod-coil multiblock model, Soft Matter 13 (2017) 8250−8263.

[31] T.D. Tap, et al., SAXS investigation on morphological change in lamellar structures during propagation steps of graft-type polymer electrolyte membranes for fuel cell applications, Macromolecular Chemistry and Physics 221 (2019).

[32] B. Crist, J.M. Schultz, Polymer spherulites: a critical review, Progress in Polymer Science 56 (2016) 1−63.

[33] L.-Q. Chen, Phase-field models for microstructure evolution, Annual Review of Materials Research 32 (2002) 113−140.

[34] F. Klocke, M. Mohammadnejad, R. Hess, R. Harst, A. Klink, Phase field modeling of the microstructure evolution in a steel workpiece under high temperature gradients, Procedia CIRP 71 (2018) 99−104.

[35] X.-D. Wang, J. Ouyang, J. Su, W. Zhou, A phase-field model for simulating various spherulite morphologies of semi-crystalline polymers, Chinese Physics B 22 (2013).

[36] R. Kobayashi, Modeling and numerical simulations of dendritic crystal growth, Physica D: Nonlinear Phenomena 63 (1992) 410−423.

[37] H. Xu, R. Matkar, T. Kyu, Phase-field modeling on morphological landscape of isotactic polystyrene single crystals, Physical Review E 72 (2005) 011804.

[38] L. Granasy, T. Pusztai, G. Tegze, J.A. Warren, J.F. Douglas, Growth and form of spherulites, Physical Review E 72, (2005) 011605.

[39] T. Takaki, M. Asanishi, A. Yamanaka, Y. Tomita, Phase-field simulation during spherulite formation of polymer, Key Engineering Materials 345−346 (2007) 939−942.

[40] M.P. Bendsoe, N. Kikuchi, Generating Optimal Topologies in Structural Design Using a Homogenization Method, 1988.

[41] S.N. Reddy K, I. Ferguson, M. Frecker, T.W. Simpson, C.J. Dickman, Topology optimization software for additive manufacturing: a review of current capabilities and a real-world

example, in: International Design Engineering Technical Conferences and Computers and Information in Engineering Conference, vol. 50107, American Society of Mechanical Engineers, 2016, p. V02AT03A029.

[42] T.R. Marchesi, et al., Topologically optimized diesel engine support manufactured with additive manufacturing, IFAC-PapersOnLine 48 (2015) 2333−2338.

[43] T. Tancogne-Dejean, A.B. Spierings, D. Mohr, Additively-manufactured metallic micro-lattice materials for high specific energy absorption under static and dynamic loading, Acta Materialia 116 (2016) 14−28.

[44] A. Hussein, L. Hao, C. Yan, R. Everson, P. Young, Advanced lattice support structures for metal additive manufacturing, Journal of Materials Processing Technology 213 (2013) 1019−1026.

[45] A. Panesar, M. Abdi, D. Hickman, I. Ashcroft, Strategies for functionally graded lattice structures derived using topology optimisation for additive manufacturing, Additive Manufacturing 19 (2018) 81−94.

[46] R.M. Gorguluarslan, U.N. Gandhi, Y. Song, S.-K. Choi, An improved lattice structure design optimization framework considering additive manufacturing constraints, Rapid Prototyping Journal 23 (2017).

[47] D. Brackett, I. Ashcroft, R. Hague, Topology optimization for additive manufacturing, in: Proceedings of the Solid Freeform Fabrication Symposium, Austin, TX, vol. 1, 2011, pp. 348−362.

[48] P. Zhang, et al., Efficient design-optimization of variable-density hexagonal cellular structure by additive manufacturing: theory and validation, Journal of Manufacturing Science and Engineering 137 (2015).

[49] J. Robbins, S. Owen, B. Clark, T. Voth, An efficient and scalable approach for generating topologically optimized cellular structures for additive manufacturing, Additive Manufacturing 12 (2016) 296−304.

[50] L. Cheng, et al., Natural frequency optimization of variable-density additive manufactured lattice structure: theory and experimental validation, Journal of Manufacturing Science and Engineering 140 (2018).

[51] J. Wu, W. Wang, X. Gao, Design and optimization of conforming lattice structures, IEEE Transactions on Visualization and Computer Graphics 27 (2019).

[52] W. Zuo, K. Saitou, Multi-material topology optimization using ordered SIMP interpolation,", Structural and Multidisciplinary Optimization 55 (2017) 477−491.

[53] N.P. van Dijk, K. Maute, M. Langelaar, F. Van Keulen, Level-set methods for structural topology optimization: a review, Structural and Multidisciplinary Optimization 48 (2013) 437−472.

[54] E. Cervera, J. Trevelyan, Evolutionary structural optimisation based on boundary representation of NURBS. Part I: 2D algorithms, Computers & Structures 83 (2005) 1902−1916.

[55] X. Yang, Y. Xie, G. Steven, O. Querin, Bidirectional evolutionary method for stiffness optimization, AIAA Journal 37 (1999) 1483−1488.

[56] X. Huang, Y. Xie, Convergent and mesh-independent solutions for the bi-directional evolutionary structural optimization method, Finite Elements in Analysis and Design 43 (2007) 1039−1049.

[57] L. Meng, et al., From topology optimization design to additive manufacturing: today's success and tomorrow's roadmap, Archives of Computational Methods in Engineering 27 (2019) 1−26.

Chapter 6

Polymer materials for additive manufacturing

Jia An
Singapore Centre for 3D Printing, School of Mechanical and Aerospace Engineering, Nanyang Technological University, Singapore

List of abbreviation

AM	additive manufacturing
AOR	angle of repose
DSC	differential scanning calorimetry
FDM	fused deposition modeling
FTIR	Fourier-transform infrared spectroscopy
IR	infrared spectroscopy
MFI	Melt Flow Index

6.1 Introduction

Polymers are essential and ubiquitous in our daily life, and our understanding of polymers advances with new characterization techniques and new synthetic polymer materials. Many classifications of polymers are proposed and accumulated over time. These classifications are classic as each represents a unique angle of view and a unique understanding of different aspects of polymers. At the fundamental level, they are interlinked by the molecular model of polymers. Before moving on to polymers for additive manufacturing (AM), it is necessary to understand the concepts behind each classification. Fig. 6.1 summarizes the typical classifications of polymers based on origin, monomer, synthesis, backbone, structure, crystallinity, intermolecular forces, tacticity, and thermal response [1,2]. These classifications can be broadly grouped into two categories, molecular material−related and molecular structure−related.

Digital Manufacturing. DOI: https://doi.org/10.1016/B978-0-323-95062-6.00011-5
© 2022 Elsevier Inc. All rights reserved.

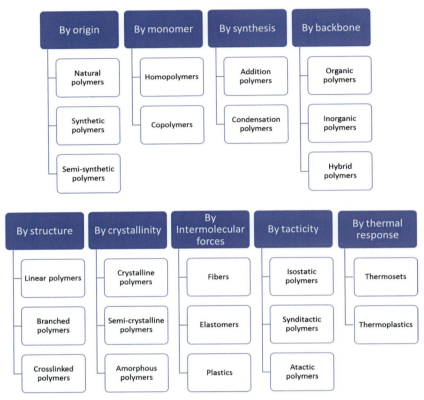

FIGURE 6.1 Classifications of polymers.

6.1.1 Molecular material–related classifications

Natural polymers are materials existing in nature, while synthetic polymers are man-made and nonexistent in nature. Semisynthetic polymers are chemically modified natural polymers. The term "resin" is more of natural origin in analog with plant resins. For synthetic resins, it is interchangeable with plastics; be it thermosetting plastics or thermoplastics, there is no clear differentiation.

Homopolymers are polymers consisting of identical monomers, while copolymers are polymers composed of different monomers. Addition polymers are an integration of monomers. Condensation polymers are also an integration of monomers but by removing small molecules or by-products during the chain elongation, resulting in ester or amide linkages in the chain. The molecular weight in condensation polymers is not a multiple integral of monomers.

Organic polymers are carbon based; the backbone is a string of carbon atoms. Inorganic polymers comprise a string of noncarbon atoms in the

Polymer materials for additive manufacturing Chapter | 6 **223**

backbone. When the backbone is mixed with carbon and noncarbon atoms, it is called hybrid polymers.

6.1.2 Molecular structure—related classifications

Linear polymers have no side chains as a contrast to branched polymers. In general, they have better packing density and regularity as well as a higher melt point. Cross-linked polymers can be physically cross-linked (intermolecular forces) or chemically cross-linked (covalent bonds). Depending on the cross-linking density, they can be hard, rigid, and brittle.

Polymer chains can be regularly folded in high orders (crystalline phase) or irregularly entangled (amorphous phase). These are the two basic phases in almost any polymer. The degree of crystallinity is used for quantification. There is 100% amorphous polymer, but there is almost no 100% crystalline polymer. Most polymers are semicrystalline. Polymers with an extremely high degree of crystallinity are called crystalline polymers.

Polymers are a collection of long chains with functional groups. The intermolecular forces among the functional groups can be high, medium, and low, which gives rise to fibers, plastics, and elastomers. In fibers the high intermolecular forces such as H-bonding often result in high crystallinity or a large percent of regularly packed long chains. In plastics, be it thermoplastics or thermosetting plastics, the intermolecular forces are intermediate. Hence there is a partial crystallinity. Note the intermolecular forces are interactions among functional groups and should not be confused with the covalent cross-links in thermosetting plastics. In elastomers the intermolecular forces are the weakest and hence an amorphous network of polymer chains.

Tacticity refers to the spatial orientation of the functional groups along the main chain. If all the functional groups are on the same side, it is called isostatic polymers. If they appear in alternative patterns, it is called syndiotactic polymers. If the orientations are randomly distributed along the main chain, it is called atactic polymers. Like the interactions among functional groups, the relative positions among functional groups or tacticity can also affect the physical properties of polymers.

Thermosets do not melt when being heated but burn. Therefore they cannot be remolded to another shape. On the other hand, thermoplastics can melt during heating and remold to different shapes following heating and cooling.

6.1.3 Polymer classification for additive manufacturing

The existence of many classifications implies that the current understanding of polymers is advanced and intertwined. Polymers are viewed from different angles to gain a better picture of their ultimate identity and behaviors. It is important to understand the concepts in these classifications because

224 Digital Manufacturing

investigating the potentials of polymers for AM requires the assessment of 3D printability, which sets many restrictions on polymer compositions and behaviors. From the angle of AM the classification by thermal response is more favorable because thermosets and thermoplastics are more manufacturing process-related, and they match the descriptions of feedstock materials in most polymer AM processes. Therefore the discussion in this chapter continues with thermosets and thermoplastics.

6.2 Thermosets

Thermosets or thermosetting polymers are polymers that are irreversibly solidified from prepolymer liquid (e.g., monomer solution) by heat or radiation. They are moldable from prepolymer liquid but generally not remoldable by heat once they are turned into solidified forms. Making a thermosetting part is always one way and irreversible, owing to a chemical process termed curing.

6.2.1 Curing

Curing refers to covalent bond formation. It can be a bond to join monomers or a bond to join polymer chains. Both thermosetting and thermoplastic liquid solutions can be cured via monomer conversion. The fundamental difference between thermosets and thermoplastics is whether there are covalent cross-linkers between polymer chains. A major phenomenon during curing is gelation or solidification of the liquid polymer solution. Other phenomena during curing include the release of heat and shrinkage.

6.2.1.1 Gelation

The gelation phenomenon can be explained by the polymerization and cross-linking at the molecular level [3,4]. Upon excitation by heat or light the initiators in a monomer solution become reactive and trigger the polymerization process, which progressively grows the polymer chains until termination. If the functionality of a monomer is 2, the monomer is bifunctional, and it forms linear chains by addition or condensation polymerization. For example, thermoplastic resins can be cured by heat via linear polymerization without cross-links. If the functionality of a monomer is more than 2, branched chains and cross-links among polymer chains can be formed (Fig. 6.2), and this is how thermosetting resins are cured. Bifunctional, trifunctional, and tetrafunctional monomers and oligomers can be mixed in a solution system as precursors for complex polymer network formation. The concentrations of multifunctional monomers and oligomers determine the extent of branches and cross-links.

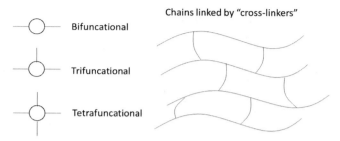

FIGURE 6.2 Gelatin via monomer conversion.

6.2.1.2 Release of heat

In general, polymerization and cross-linking are exothermic processes, and heat is released during the gelation at every point of the curing volume. The greater the conversion from monomers, the greater the heat generation. Therefore materials with low-to-medium conversion are usually preferred. If heat cannot be dissipated immediately, it will accumulate over time and accelerate the reactions, especially if the monomers are thermally cured in a free radical polymerization system. This is known as Trommsdorff effect or autoacceleration [5]. Autoacceleration can cause deterioration of polymer properties, or in a worse scenario, the heated resin may explode and cause a safety hazard. However, the amount of heat released can differ in thermal curing and light curing [6] even though the heat accumulated in vat polymerization is a challenge for developing a capacity to print large parts. Recently, the high area rapid prototyping process solved this challenge by using a circulating fluid to remove excessive heat and, thus, it was able to print a part in more than 1 m in length [7].

6.2.1.3 Shrinkage

Cure shrinkage is another phenomenon in thermosetting resins. The volume of the resin decreases during curing due to the conversion of monomers from diffusion state into compacted physical polymer networks. The higher the monomer conversion, the larger the shrinkage. Since the shrinkage is highly relevant to the accuracy and resolution of a 3D printing process, materials with a low shrinkage rate are preferred.

6.2.2 Curing characteristics

Characterizing the curing process requires a quantitative assessment of polymerization and cross-linking during gelation. The most common indicators are the degree of monomer conversion and cross-linking density.

226 Digital Manufacturing

6.2.2.1 Monomer conversion

Most photopolymers used for 3D printing are acrylate based, which have a carbon—carbon double bond for chain formation. During polymerization the double bonds are converted to single bonds to form long chains and networks. Therefore the extent of double bond conversion can be used to describe the degree of polymerization.

Double bond conversion can be experimentally determined by infrared spectroscopy (IR), Fourier-transform infrared spectroscopy (FTIR), or mid-infrared spectroscopy [8,9]. Upon IR irradiation the atoms of a molecule are energized and vibrate in patterns like stretching, bending, and rotation. Different bonds absorb the IR energy at different frequencies, and hence their presence can be identified by the peak absorption in the absorption spectrum. The peak absorption of the carbon—carbon double bond is reduced in height after polymerization. The ratio between pre- and postpolymerization peak absorptions can be used to calculate the conversion:

$$\alpha = 1 - \frac{A}{A_0} \tag{6.1}$$

where α is extent of conversion, A_0 is absorbance postpolymerization, and A is absorbance before polymerization.

Differential scanning calorimetry (DSC) can also be used to determine the conversion. Converting double bonds to single bonds is exothermic because the bond energy of a double bond is higher than a single bond. The heat released during polymerization can be captured in DSC curves. The heat of polymerization is then obtained from the areas under DSC curves for either cured or uncured monomers depending on the experimental design [6,10,11]. The heat for 100% conversion can be calculated from the difference in bond energies for a given mass. Therefore the ratio between the heat of polymerization and the heat for 100% conversion can be used to estimate the degree of double bond conversion.

$$\alpha = \frac{H}{H_{100\%}} \tag{6.2}$$

where H is the heat released after polymerization, and $H_{100\%}$ is the heat released if at 100% conversion.

The degree of conversion has also been corelated to glass transition temperature [4].

$$\frac{1}{T_g} = \frac{1 - \alpha}{T_{g0}} + \frac{\alpha}{T_{g\infty}} \tag{6.3}$$

where T_g is the corresponding glass transition temperature at the conversion of α, T_{g0} is the resin T_g before polymerization, and $T_{g\infty}$ is resin T_g after polymerization. Therefore the conversion can be estimated by measuring glass transition temperatures.

Volumetric shrinkage is another way to estimate the conversion if the reduction in resin volume is due to polymerization at a constant temperature [12].

$$\alpha = \frac{V_s}{V_{s100\%}} \tag{6.4}$$

where V_s is the observed volumetric shrinkage after polymerization, and $V_{s100\%}$ is the total volumetric shrinkage for 100% cured polymer, which can be calculated based on molar volume contraction (volume per mole of reacted monomers).

6.2.2.2 Cross-link density

When two polymer chains or two regions within a chain are connected via a covalent bond or a sequence of covalent bonds such as chain segments, they are called cross-linked. The functionality of the cross-linkers can be greater than 2, resulting in larger and more complex networks. Cross-link density is defined as the number of the cross-linkers per unit volume, and hence the unit of cross-link density is mole per volume. As the cross-links may have an average molecular weight, the cross-link density can also be expressed as the molecular weight between two cross-link points. Cross-link density is responsible for many bulk properties of polymers. Increasing cross-link density makes polymer networks much denser in a defined volume. It also increases the modulus, strength, hardness, chemical, and heat resistance. Consequently, as a trade-off, the elongation, toughness, fatigue, and thermal expansion will decrease.

Cross-link density cannot be directly measured as no equipment can chemically identify all the cross-linkers. However, cross-link density can be measured indirectly through the effects they have on the polymer networks. The most common method is equilibrium swelling. Polymer networks consist of free volume, which can absorb solvent molecules to expand or swell. Cross-linkers restrict the configurations of polymer chains hence oppose the expansion effect until an equilibrium is reached. Therefore the cross-link density can be related to measurable volumes of polymer and solvent. The Flory–Rehner theory describes this relationship in Eq. (6.5) for a lightly cross-linked polymer system [3,4,13].

$$X_c = \frac{\ln\left(1 - v_p\right) + v_p + \chi v_p^2}{V_s\left[v_p^{1/3} - \frac{v_p}{2}\right]} \tag{6.5}$$

where X_c is cross-link density (mol/cm^3), V_s is molar volume of solvent, v_p is volume fraction of polymers after swelling, and χ is the Flory–Huggins interaction parameter.

228 Digital Manufacturing

For a highly cross-linked polymer system, the molecular weight between two cross-link points can be obtained by the theory of rubber elasticity and determined by thermogravimetric analysis [4].

$$M_c = \frac{3dRT}{E'} \tag{6.6}$$

where d is density, R is gas constant, T is temperature, and E' is the dynamic storage modulus in the rubbery plateau region.

6.2.2.3 Curing rate

Curing rate or polymerization rate is defined as how fast the resin solidifies upon exposure to heat or light. Higher curing rate will result in a higher degree of bond conversion and polymerization within a given time. Hence in AM, curing rate largely determines processing parameters such as power, exposure time, and scan speed.

Curing rate has been correlated to chemical reaction rate constant [4].

$$\frac{d\alpha}{dt} = K(1-\alpha)^n \tag{6.7}$$

where K is reaction rate constant, and n is the order of the reaction. The reaction rate constant is dependent on many factors such as the concentration of initiators and monomers, the presence of a catalyst, presence of inhibitor, temperature, light intensity, and whether the monomers are thermocure, photocure, or a mixture of thermo-photo-cure.

By taking the derivative of Eq. (6.8) for calculating bond conversion, the curing rate can also be expressed in terms of heat rate.

$$\frac{d\alpha}{dt} = \frac{1}{H_{100\%}} \frac{dH}{dt} \tag{6.8}$$

6.2.2.4 Curing depth

In AM, curing involves using a laser beam or a lamp, and each time only one thin layer of the resin is cured. Curing of a 2D (two-dimensional) layer does not necessarily lead to the successful building of a 3D object because a strong interlayer bonding is critical. Usually, the interlayer bonding is achieved by an overlap between two adjacent layers; that is, the layer thickness in 3D (three-dimensional) printing must be smaller than the curing depth. The curing depth measures the extent of a solidified resin in a single exposure and relates to the penetration depth of a laser beam [14,15].

$$D_c = D_p \ln\left(\frac{E}{E_t}\right) \tag{6.9}$$

where D_c is curing depth, D_p is penetration depth, E_t is threshold exposure energy for solidification, and E is exposure energy. The exposure energy is determined by laser power, scanning speed, and exposure time. Curing depth profile can be directly measured on the 3D printed specimens.

6.2.3 Dynamic covalent bonds

The abovementioned understanding of curing is based on permanent covalent cross-links in polymer networks. In the last decade, there is a growing interest in dynamic covalent bonds, which are chemical bonds that can exchange or switch between different molecules [16,17]. Introducing dynamic covalent bonds into a polymer network can significantly change the properties of polymers, making thermoset resins reprocessable, repairable, and self-healing. However, the fundamental of thermosets being not melting nor remoldable remains unchanged.

6.3 Thermoplastics

Unlike thermosets, thermoplastic polymers consist of linear or branched chains. There is no chemical bond between polymer chains. Upon heating, thermoplastics can melt and be molded to other shapes. Thermoplastics can dissolve upon interaction with a solvent, as a contrast to swelling seen in thermosets. Therefore polymer melt and solution are common in thermoplastics but not so in thermosets.

6.3.1 Polymer melt

Intermolecular forces usually determine the state of a material. When the forces are strong, the polymers manifest themselves as solids. Temperature is related to the average kinetic energy of molecules. When polymers are heated to a certain temperature, some heat will be converted to the kinetic energy of polymer chains. If this kinetic energy overcomes the intermolecular forces among polymer chains, the solid state of polymers will be gradually lost and replaced by a fluid state, which is polymer melt. Due to the presence of amorphous and crystalline phases in thermoplastics, the melting kinetics can be complicated and depends on many factors. The discussion here focuses mainly on the melt state of polymers.

6.3.1.1 Random coil

The conformation of a polymer chain is a statistical distribution of many possibilities and, therefore, a random coil [18]. The random coil is characterized by its molecular weight (chain length) and chain stiffness. The melt viscosity is proportional to molecular weight. The greater the molecular weight,

230 Digital Manufacturing

the higher the melt viscosity. If the random coil is sided with branches, the branches will increase the melt viscosity as well.

The mobility of the random coil depends on temperature. Below T_g, the molecular mobility is limited to side chains or short-ranged movement for backbones of all the amorphous phase. Above T_g, the mobility of backbones increases to segmental flexibility except for the packed crystalline phase. Above T_m, the crystalline phase is unpacked and molten, and the polymer is fully amorphous as a melt.

6.3.1.2 Globular and stretched states

For long and flexible chains a globular (coiled) and stretched states can exist. During melt flow, be it shear flow or elongational flow, the globular state is deformed to a stretched state by stresses imposed on polymer chains. Likewise, the stretched state can relax and recoil after stresses are released [19]. Generally, the higher the molecular weight, the longer it takes to relax and recoil. The globular and stretched states of polymer chains can explain the expansion and contraction behaviors of a polymer.

6.3.1.3 Entanglement

In polymer melt, polymer chains are held together by not only intermolecular forces but also the entanglement of the chains. The higher the molecular weight, the higher the effect of entanglement. Entanglements are mechanical interlocks between polymer chains (Fig. 6.3). They are different from chemical cross-links in thermosets. Due to the mechanical restrictions, the polymer chains move within limited space defined as virtual tubes. The thermal movement of the entangled polymer chains within these restrictions is called reptation. Entanglement can be quantified by the number of entanglements per chain or the average molecular weight between two entanglements, from which the entanglement density can be calculated.

Entanglement has a significant effect on melt viscosity when the molecular weight of the polymer exceeds a critical value, such as when the chains are sufficiently long. The viscosity becomes proportional to the 3.4th power of molecular weight as shown in Eq. (6.10) [20]. Below this critical value, the melt viscosity is proportional to molecular weight. This explains the power-law model for describing the non-Newtonian behavior of polymer

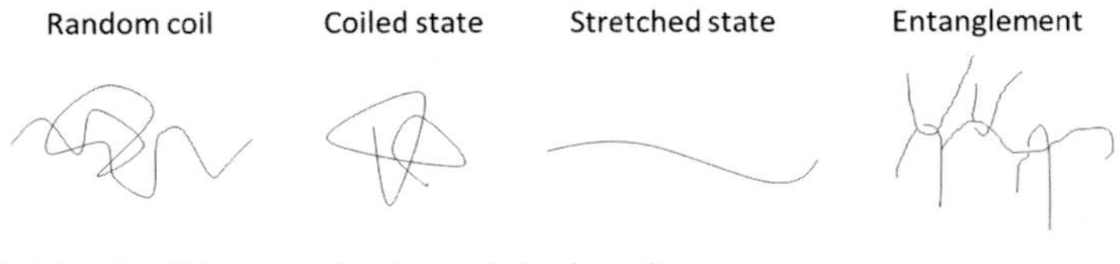

FIGURE 6.3 Possible states of polymer chains in melt.

melt. Besides, the effects of temperature and pressure on viscosity can largely be explained by entanglement as well.

$$\eta_0 \propto M \ (M < M_c)$$
$$\eta_0 \propto M^{3.4} \ (M \geq M_c)$$

(6.10)

6.3.2 Rheological properties

The resistance to deformation to flow under forces or stresses is called the rheological properties of a material. Generally, the modulus is used to measure the resistance to deformation and viscosity is used to measure the resistance to flow. The state of a material can be solid or liquid at room temperature, but the properties or behaviors of a material can be anywhere between the property of a solid and the property of a liquid when temperature change. Viscoelasticity is thereby defined from the combined properties of solid and liquid.

6.3.2.1 Viscosity

There are many types of viscometers for measuring viscosity, such as capillary, rotational, and vibrational. The key idea behind these techniques is to control either strain rate or stress and measure the other at a certain sampling frequency so that a graph of viscosity versus strain rate is obtained. The curve shown in the graph describes the viscosity of the material. Although the curve presents more accurate and meaningful data, a single number indicator such as the Melt Flow Index (MFI) is more popular in the industry, which measures how many grams of polymer melt flow out of a viscometer every 10 minutes under specified conditions. MFI is inversely related to viscosity or molecular weight. A higher MFI indicates a lower viscosity or a lower molecular weight.

6.3.2.2 Viscoelasticity

The polymer melt is a viscoelastic fluid, which shows both fluid viscosity and solid elasticity. Oscillatory testing with a small amplitude and angular frequency ω can be used to characterize viscoelasticity, such as a Dynamic Mechanical Analysis [21]. Viscoelasticity can be described by complex modulus or complex viscosity. Each is defined with a real component and an imaginary component. The word "complex" refers to a complex number in mathematics. Complex modulus and complex viscosity are related by the angular frequency. The tangent of phase angle θ measures the ratio between a material's elastic and viscous response, indicating whether the material, at the specified conditions, is more solid-like or fluid-like.

$$E^* = E' + iE'' = \sqrt{E'^2 + E''^2}$$

(6.11)

$$\eta^* = \eta' + i\eta'' = \sqrt{\eta'^2 + \eta''^2} \tag{6.12}$$

$$\omega = \frac{E^*}{\eta^*} = \frac{E'}{\eta'} = \frac{E''}{\eta''} \tag{6.13}$$

$$Tan(\theta) = \frac{E''}{E'} = \frac{\eta''}{\eta'} \tag{6.14}$$

where E' is storage modulus, E'' is imaginary loss modulus. η' is dynamic viscosity, η'' is imaginary storage viscosity. Both E^* and η^* can be a function of temperature, and curves of modulus or viscosity versus temperature are the rheological characteristic of the material.

6.3.3 Thermal properties

Thermal properties describe the behaviors of the material when heat passes through it during entering (heating) or leaving (cooling). Depending on the aspect of interest in a material, there are different measurements of thermal properties, such as phase transition temperatures, thermal diffusivity, and shrinkage.

6.3.3.1 Phase transition temperatures

The basic thermal properties of a thermoplastic polymer are glass transition temperature T_g, crystallization temperature T_c, and melt point T_m. They can be obtained from DSC curves and used as references for setting polymer processing parameters.

6.3.3.2 Thermal diffusivity

Thermal diffusivity measures the rate of heat transfer through a material. In an isotropic material and at constant pressure, thermal diffusivity is defined as:

$$\alpha = \frac{k}{\rho C_P} \tag{6.15}$$

where ρ is density, C_P is specific heat capacity, k is thermal conductivity. Each variable is temperature dependent. A material with a lower thermal diffusivity is more heat resistant.

6.3.3.3 Coefficient of thermal expansion

The thermal expansion reflects the energies of vibrating molecules. In general, polymers have a positive coefficient of thermal expansion (CTE). The coefficient of linear thermal expansion (CTE) is defined as:

$$\alpha_L = \frac{1}{L}\frac{dL}{dT} \tag{6.16}$$

where L is linear dimension and T is temperature. Polymer chains are more isotropic in the coiled state and more anisotropic in the stretched state. Hence, the thermal expansion of polymer may not be the same in all directions. For anisotropic thermal expansion the coefficients of linear expansion in perpendicular and parallel directions ($\alpha_\perp$ and $\alpha_\parallel$) are usually used.

6.3.3.4 Thermally reversible and irreversible shrinkage

These are the two modes of shrinkage during the cooling of a polymer, and they are thermally reversible. The first shrinkage mode is related to the energies of polymer molecules and it is the same as how other materials shrink during cooling. Upon cooling, polymer molecules become less vibrational and the polymer material returns from the thermally expanded state, thus shrinking in size. A different mode of shrinkage in polymer is crystallization. Crystallization can significantly contribute to shrinkage by packing polymer chain segments very tightly and densely. The higher crystallinity, the higher density, and hence the higher shrinkage.

In a polymer, shrinkage can also occur during heating or at an elevated temperature. It is termed thermal contraction. In contrast to the abovementioned reversible process, this thermal contraction at the high temperature is due to molecular orientation and it is thermally irreversible [22]. Polymer chains can be frozen in a stretched state during polymer processing. Upon heating the stretched state returns to a coiled state like the contraction of rubber, resulting in shrinkage of the polymer. It is possible that thermal expansion and thermal contraction coexist during heating, but which one is the dominant process depends on the extent of polymer chains in the stretched state. In highly oriented polymers such as fibers, the expansion effect is likely negligible compared to the contraction.

There are two implications from thermal contraction: (1) both heating and cooling can contribute to shrinkage and (2) the least shrinkable polymer is amorphous with no stretched chains. In other words, the shrinkage is purely a reverse of thermal expansion.

6.3.3.5 Thermal stress and warpage

Thermal stress is mechanical stress caused either by preventing thermal expansion/contraction or by differential expansion/contraction. When the accumulated thermal stress exceeds the yield stress of the polymer ($\sigma_T > \sigma_{yield}$), plastic deformation will occur.

In AM, when a hot layer is deposited, the bottom is usually in touch with a polymer or a substrate and the top is exposed to air. Due to different thermal diffusivities of air and polymer, the top will cool faster than the bottom, resulting in more shrinkage at the top. When the differential shrinkage induces thermal stress greater than the yield stress, warpage, or curling will occur (Fig. 6.4). Because the edges or corners usually cool faster than the

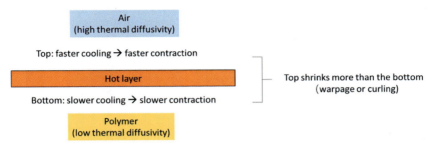

FIGURE 6.4 Warping due to thermal stress generated by differential shrinkage.

center, the warpage effect is more obvious at these locations. The warping issue in AM can be addressed either by mechanically stabilizing the layer via adhesion or support, or by optimizing the process parameters to achieve more uniform cooling and reduce the thermal stress below the yield stress.

6.4 Printability in 3D printing

The printability in 3D printing has two components: design printability and material-process compatibility. If a design is not printable, it must be revised to comply with design guidelines for an AM process or a particular AM printer. If a material is not compatible, the material may be modified, or a new printing system may be developed specifically for printing that material. Since the design for additive manufacturing can be a standalone subject by itself, the discussion here focuses on general requirements for polymer materials to be compatible and printable.

6.4.1 Layering

The seven process categories in the AM family can be further grouped into two approaches: processes that directly build a layer of the object and others that build a layer of the object from a preset layer of the material (Fig. 6.5). The former takes a single step to fabricate an object layer. The latter undergoes two steps, where a uniform layer of the material is essential before building an object layer.

6.4.1.1 Effect of layering on multimaterial printing

One inherent drawback of the layering approach is that it is less compatible with multimaterial printing. Multimaterial printing refers to printing of multiple materials within a single layer and not in different layers or builds. Therefore processes involving layering are more challenging for multimaterial printing since the deposition of material and formation of an object layer are separate. Yet, there are some successful attempts by increasing the design

Build directly	Layering and build
• Material jetting • Material extrusion • Directed energy deposition	Free flowing materials • Vat polymerization • Powder bed fusion • Binder jetting Sheets • Sheet lamination

FIGURE 6.5 Building a layer of an object.

complexity of the printing process [23,24]. Sometimes, printing multiple materials in alternative layers is loosely considered multimaterial printing, given the challenges in layering-based processes. In contrast, the processes without layering have a natural potential for multimaterial printing or digital material fabrication. It is because the deposition of material is controlled simultaneously, as the object layer is formed.

6.4.1.2 Layering of free flow materials

Free flow materials such as liquid and powder can be characterized by viscosity and flowability, respectively. For a liquid, depending on where the solidification occurs, layering of a liquid can be natural (floating layer) or require mechanical intervention such as recoating, suction, or vibration (top layer and bottom layer). If solidification is at the top, surface tension must be considered it affects the layering quality (Fig. 6.6).

For a powder bed the solidification occurs mostly at the top layer. Powder flowability is affected by many parameters such as particle size, distribution, morphology, roughness, charge, moisture, particle density, bulk density, and recycling. Powder flowability can be experimentally indicated by some simple numbers such as Hausner ratio, compressibility index, angle of repose (AOR), and powder cohesion (c). Particle size is the most important parameter and there is a good correlation between flowability and powder particle size d [25,26].

$$\tan(AOR) \propto \frac{1}{\rho d} \quad (6.17)$$

where ρ is particle density and d is particle size. Besides, it is also worth noting the differences between the flowability test conditions and the processing conditions of a particular printer.

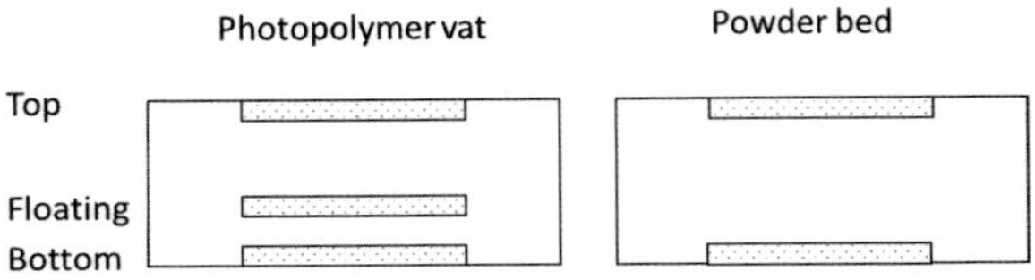

236 Digital Manufacturing

FIGURE 6.6 Free flow material can be solidified at different locations.

Solidification at the bottom layer in a powder bed is also possible, and inverted selective laser sintering is an example [23]. Theoretically, it may also be possible to solidify a floating layer inside a powder bed. However, it would be extremely challenging to continue the printing process, because unlike liquid, powder does not flow back to refill the cavity.

6.4.1.3 Layering of sheet materials

In commercial sheet lamination processes, the sheets are either continuous in a roll or a stack of laminates. The layer thickness is fixed by material, not adjustable as a machine parameter. Therefore it is not possible to print a part at a variable layer thickness. As a roll of sheet advances to the build platform, normal force from compression is needed to ensure bonding and smoothness. For a stack of sheets, besides compression, grabbing precisely one sheet each time is also critical.

6.4.2 Energy and material bonding

3D printing is about eliminating the surfaces of feedstock materials. Therefore there are two main approaches to material bonding: cohesion and adhesion (Fig. 6.7). Cohesion is the attraction between the same materials and adhesion is the attraction between dissimilar materials. Most AM processes take a cohesive approach, while sheet lamination and binder jetting are mostly adhesive based. Here, the interaction between a polymer material and the external energy is discussed. Printability is based on the phase change of the feedstock material induced by the transformation of energy.

6.4.2.1 Particle energy

Photo energy is more common than electron energy in AM because it does not require a vacuum environment. Photon energy is quantum energy carried by a single photon. It is used to trigger polymerization or to heat polymers. The power density is the product of photon energy and photon flux, which is the number of photons per area.

$$P = \emptyset E = \emptyset \frac{hc}{\lambda} \tag{6.18}$$

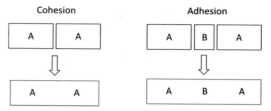

FIGURE 6.7 Material bonding in additive manufacturing.

where P is power density, $\varnothing$ is photon flux, E is photon energy, h is Plank's constant, c is speed of light, and λ is wavelength.

Polymerization is an exothermic process. Only a small amount of threshold energy is needed to trigger the spontaneous polymerization reaction. In AM photopolymers the energy input ranges from ultraviolet to infrared light in the light spectrum. Each light source determines the number of photons required to excite an initiator. In two-photon or multiphoton polymerization the absorption of two or more low energy photons (smaller frequency) is required, and the absorption must be simultaneous. Sequential absorption leads to decay of the first absorbed photon and does not excite the initiator. Therefore the efficiency decreases as the number of photons increases (Fig. 6.8). As heat is being released, initiators must also be thermally stable to avoid ablation. In addition, the transparency and scattering in the resin can affect absorption as well.

In AM thermoplastics the solidification process is endothermic. Hence, a higher power is required. In addition, due to the various properties of light, the absorbed energy is a fraction of the total energy (Fig. 6.9). Each powder particle is not isolated but a neighbor to each other. The energy leaving from one particle may arrive at another. In the case of sintering, the thermal properties of the polymer are critical, especially T_g, T_c, and T_m. Below T_g, the particle is a rigid solid. The polymer chains at the surface cannot diffuse to the surface of their neighbors to join the particles. Above T_m, the particle turns into a melt drop, and it will coalesce with adjacent melt drops and be deformed by melt surface tension. To maintain the stability of particle yet the mobility of the surface material, the absorbed energy of each particle must heat the polymer particle above its T_g but below its T_m. However, to avoid crystallization shrinkage during cooling, the temperature of the particle must be maintained above its T_c. The interval between the onsets of T_c and T_m is referred to as sintering window.

6.4.2.2 Mechanical and thermal energy

Most materials used in extrusion interact with mechanical energy or a combination of mechanical and other energies such as thermal energy (Fig. 6.10). In fused deposition modeling (FDM) the filament is mechanically

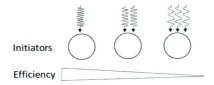

FIGURE 6.8 Absorption of photons by molecules.

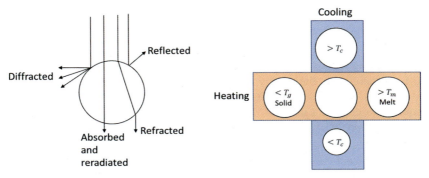

FIGURE 6.9 Absorption of photons by particles.

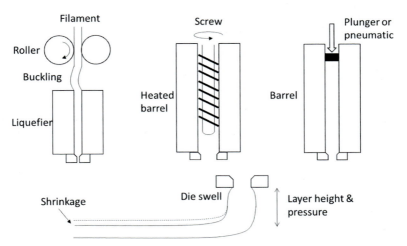

FIGURE 6.10 Extrusion by mechanical and thermal energy.

compressed at the drive wheel while being heated in the liquefier. Ideally, the compressive force should be completely transferred to the liquefier to push the melt out of the nozzle. However, buckling can occur before the entrance of the liquefier if the compressive stress exceeds the yield stress.

There is one approach to use the ratio of compressive modulus of the filament and the apparent viscosity of the polymer melt measured at a particular temperature to represent the ability of the filament to function without buckling, which is referred to as the printability of FDM [27].

$$\frac{E}{\eta_a} > \frac{8Ql(L/R)^2}{\pi^3 r^4 k} \tag{6.19}$$

where k is a scaling factor between the compressive pressure and the pressure driving the melt flow. The extrudate usually experiences a die swell upon exit of the nozzle due to polymer chains relaxing from the flow-induced stretched state. The filament gradually shrinks by thermal contraction and crystallization. The layer height affects the compressive pressure between the nozzle and the extrudate. There should be some pressure to compress the printed layer to avoid interlayer voids.

Another type of mechanical and thermal extrusion is by a screw extruder [28]. Typically, polymer pellets are fed into the heated barrel and pushed through a nozzle. Screw extrusion does not require the preparation of flexible filament, thus allowing a wider range of polymer materials and polymer blends. In addition, the solid-to-melt transition is inside the barrel with no buckling concern. However, the venting is a problem when melting the pellets because voids and volatiles can only be released after extrusion.

Many new polymers for extrusion AM are carried out by mechanical energy. They take advantage of the polymers' properties, such as thixotropic behavior or chemical reactions to stiffen the extruded filament so that the printed part can self-support [15,29,30]. A less explored advantage of this process is the ability to retract material back to the barrel, like erasing what is just printed. The combination of extrusion and retraction in a single barrel or separate barrels may be a powerful tool in the future.

6.4.2.3 Adhesive energy

Besides the cohesion of polymers, polymers can also interact with dissimilar adhesive materials to bond together to form a 3D object, such as binder jetting. Binder jetting involves the interaction of jetting drops and powder

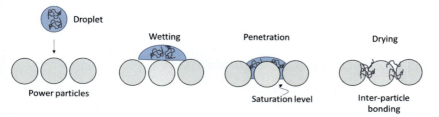

FIGURE 6.11 Binder jetting.

240 Digital Manufacturing

particles at the powder bed surface (Fig. 6.11). The selection of a binding agent is critical. The printed parts can be weak, mechanically, if the bonding strength is low or if the agents are subject to degradation under the postprint environment. The adhesive agents can be placed inside the liquid (in-liquid printing) or mixed in a powder bed (in-bed printing) [31]. Unlike ceramic or metal parts, where binders can be burnt away to improve purity and density, it is challenging to remove binders and purify and densify the polymer.

As a liquid is involved in jetting, it is necessary to make sure the liquid is breakable. Ohnesorge number is the ratio between two dimensionless numbers, Weber number and Reynolds number. Weber number is the ratio between fluid inertia and surface tension, while Reynolds number is the ratio between fluid inertia and viscosity. In general, Oh should be between 0.1 and 1. Above 1, the viscosity is too high to be jetted. The Oh number is also applicable to other AM processes that involve jetting, such as material jetting and multijet fusion [32].

$$Oh = \frac{\sqrt{W_e}}{R_e} = \frac{\sqrt{\frac{\rho d V^2}{\sigma}}}{\frac{\rho d V}{\eta}} = \frac{\sqrt{\eta}}{\sigma \rho d} \tag{6.20}$$

where η is dynamic viscosity of the liquid, ρ is density, d is diameter of droplet, and σ is surface tension.

The spreading of a droplet on a powder bed is governed by *Young−Dupre* equation [33], in which the energy of adhesion is defined as,

$$E = \gamma(1 + \cos \theta) \tag{6.21}$$

where γ is the surface tension of the liquid, and θ is contact angle.

Afterward, the voids among the particles serve as capillaries to allow the liquid to infiltrate. The saturation level (Eq. 6.22) is defined as the ratio between the infiltrated liquid and powder void space, and it depends on the powder bed packing [31]. In general, the denser the packing, the higher the saturation. Too low a saturation will leave insufficient connections among particles after drying and resulting in weak bonding. However, an exception is that the droplets are reactive with the powder particles.

$$\text{Saturation level} = \frac{V_{\text{binder}}}{V_{\text{void}}} \tag{6.22}$$

6.5 Characteristics of 3D printed parts

AM involves a wheel of four aspects: design, material, process, and application [34]. Implementation from design to application produces a physical part from a virtual twin. The difference between the virtual and physical twins is the characteristics of 3D printed parts. Besides dimensions and geometries, there are more differences regarding the spatial distribution of part content and part property, such as porosity, anisotropy, and heterogeneity.

6.5.1 Porosity

Additively manufactured polymer parts contain a certain level of default porosity due to imperfections in intralayer and interlayer bonding (Fig. 6.12). This default porosity is process-induced and should not be confused with the designed porosity, which is frequently used to study the behaviors of porous materials, such as energy absorbers [35,36]. The benefits and drawbacks of this default porosity depend on their applications. In most engineering applications the voids are undesirable and seen as defects decreasing mechanical properties and other functional performance. Optimization of process parameters, in-process void reduction, and postprocessing are usually conducted to eliminate the void fraction. However, in tissue engineering scaffolds, where a higher porosity or multiscale porosity is desired [37,38], this default porosity may help facilitate the diffusion of oxygen and nutrients. Furthermore, this default porosity may also be useful for liquid and gas filtration application, provided there is a control of pore size and pore-size distribution.

6.5.2 Anisotropy

A material is anisotropic if the property shows significantly different values when measured in different directions. As property and structure are corelated, there is usually a corresponding anisotropic microstructure in an additively manufactured part. Process−structure−property relationships are ongoing topics in AM research. The most common anisotropy comes from the property contrast in the $x-y$ plane and z-direction (Fig. 6.13). The degree of bonding in the z-direction is a function of layer thickness, while the degree of bonding in the $x-y$ plane is not. In general, the more the layers, the more the anisotropy. In the extreme case, if there is only one layer, then it is supposed to be isotropic. Moreover, within the $x-y$ plane, if the printing

FIGURE 6.12 Porosity by default.

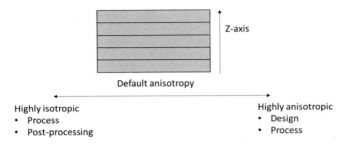

FIGURE 6.13 Anisotropy by default.

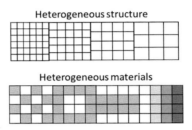

FIGURE 6.14 Heterogeneity by design.

is scan type, the property in the *x*-direction could be different from the *y*-direction as well.

One implication of the anisotropy is the effect of build orientation, which is fixed and cannot be changed during a build. A decision on which direction of a part accepts or allows anisotropy needs to be made. However, the decision can be difficult when the part geometry is complex. Again, the benefits, drawbacks, and relevance of this anisotropy depend on the applications, and there are always methods to increase or decrease anisotropy.

6.5.3 Heterogeneity

Unlike anisotropy, heterogeneity refers to the variation of properties from point to point, and the variation can be continuous or discrete. By default, there should be some degree of heterogeneity in the microstructure and property of 3D printed polymer parts, such as crystalline morphology [39,40]. This pattern can be typically periodic due to the layer-by-layer processing. The default heterogeneity may or may not be useful, depending on the application. The more meaningful and perhaps useful heterogeneity is the design of a heterogeneous structure/microstructure, as well as the design of a heterogeneous distribution of materials (Fig. 6.14). They can be used for infill or to tailor a material property by not changing its composition and other interesting applications, such as 4D (four-dimensional) printing [41,42]. Two enablers for designed heterogeneity are voxel modeling and multimaterial printing. Although this is an expected capability of AM, the full potential in industrial application is yet to be revealed in the future. Besides, the current concept of multimaterial printing is almost equal to material jetting due to a lack of alternative methods. More effective and efficient multimaterial approaches are needed in the future as well.

6.6 Summary

Polymer AM is a multidisciplinary field of polymer science, polymer engineering, polymer processing, and AM. Thermosets and thermoplastics are a classification of polymers with a good match with current AM processes. While the typical characteristic of thermosets is curing, the typical characteristic of thermoplastics is melting. Understanding of curing behaviors and melt behaviors is of paramount importance to understand the polymer behaviors in AM. The material printability in 3D printing depends on layering and the interaction between material and energy. Due to the diverse nature of AM processes, the printability is very much process-dependent or even printer-dependent.

The default characteristics of 3D printed polymer parts are porosity, anisotropy, and heterogeneity, which must be considered before application. More importantly, polymer materials for AM must satisfy the printability requirement, and the printed part property must meet the application requirement. Failing either one may land a particular polymer AM less relevant to the intended use.

6.7 Further recommendation

This chapter covers fundamental concepts in polymer materials in AM. For specific lists of AM processes and compatible polymer materials, the readers are recommended to explore further the Senvol Database (http://senvol.com/database/), which is the first and most comprehensive database for industrial AM materials, including polymers.

References

[1] V.R. Gowariker, N. Viswanathan, J. Sreedhar, Polymer Science, New Age International, 1986.

[2] S. Fakirov, Fundamentals of Polymer Science for Engineers, Wiley, 2017.

[3] H. Dodiuk, S.H. Goodman, 1 — Introduction, in: H. Dodiuk, S.H. Goodman (Eds.), Handbook of Thermoset Plastics (Third Edition), William Andrew Publishing, Boston, MA, 2014, pp. 1—12.

[4] D. Ratna, Handbook of Thermoset Resins, ISmithers Shawbury, 2009.

[5] O. Olabisi, K. Adewale, Handbook of Thermoplastics, Vol. 41, CRC Press, 2016.

244 Digital Manufacturing

[6] J. Fuh, et al., Curing characteristics of acrylic photopolymer used in stereolithography process, Rapid Prototyping Journal 5 (1999).

[7] D.A. Walker, J.L. Hedrick, C.A. Mirkin, Rapid, large-volume, thermally controlled 3D printing using a mobile liquid interface, Science (New York, NY) 366 (6463) (2019) 360–364.

[8] J. Stansbury, S.H. Dickens, Determination of double bond conversion in dental resins by near infrared spectroscopy, Dental Materials 17 (1) (2001) 71–79.

[9] A.M. Herrera-González, et al., Analysis of double bond conversion of photopolymerizable monomers by FTIR-ATR spectroscopy, Journal of Chemical Education 96 (8) (2019) 1786–1789.

[10] T.F. Scott, W.D. Cook, J.S. Forsythe, Photo-DSC cure kinetics of vinyl ester resins. I. Influence of temperature, Polymer 43 (22) (2002) 5839–5845.

[11] R.T. Alarcon, et al., Use of DSC in degree of conversion of dimethacrylate polymers: easier and faster than MIR technique, Journal of Thermal Analysis and Calorimetry 132 (2) (2018) 1423–1427.

[12] F. Rueggeberg, K. Tamareselvy, Resin cure determination by polymerization shrinkage, Dental Materials 11 (4) (1995) 265–268.

[13] A. Blume, J. Kiesewetter, Determination of the crosslink density of tire tread compounds by different analytical methods, KGK Kautschuk Gummi Kunststoffe 72 (9) (2019) 33–42.

[14] J.-Y. Lee, J. An, C.K. Chua, Fundamentals and applications of 3D printing for novel materials, Applied Materials Today 7 (2017) 120–133.

[15] L.J. Tan, W. Zhu, K. Zhou, Recent progress on polymer materials for additive manufacturing, Advanced Functional Materials 30 (43) (2020) 2003062.

[16] B. Zhang, et al., Reprocessable thermosets for sustainable three-dimensional printing, Nature Communications 9 (1) (2018) 1–7.

[17] T. Yuan, et al., 3D printing of a self-healing, high strength, and reprocessable thermoset, Polymer Chemistry 11 (40) (2020) 6441–6452.

[18] L.J. Smith, et al., The concept of a random coil: residual structure in peptides and denatured proteins, Folding and Design 1 (5) (1996) R95–R106.

[19] R.G. Larson, P.S. Desai, Modeling the rheology of polymer melts and solutions, Annual Review of Fluid Mechanics 47 (2015) 47–65.

[20] C. Kuo, W. Lan, Gel spinning of synthetic polymer fibres, Advances in Filament Yarn Spinning of Textiles and Polymers, Elsevier, 2014, pp. 100–112.

[21] H.A. Barnes, J.F. Hutton, K. Walters, An Introduction to Rheology, Vol. 3, Elsevier, 1989.

[22] M. Trznadel, M. Kryszewski, Thermal shrinkage of oriented polymers, Journal of Macromolecular Science, Part C: Polymer Reviews 32 (3–4) (1992) 259–300.

[23] J. Whitehead, H. Lipson, Inverted multi-material laser sintering, Additive Manufacturing 36 (2020) 101440.

[24] D. Han, et al., Rapid multi-material 3D printing with projection micro-stereolithography using dynamic fluidic control, Additive Manufacturing 27 (2019) 606–615.

[25] G. Xu, et al., Investigation on characterization of powder flowability using different testing methods, Experimental Thermal and Fluid Science 92 (2018) 390–401.

[26] Z. Guo, et al., Theoretical and experimental investigation on angle of repose of biomass–coal blends, Fuel 116 (2014) 131–139.

[27] N. Venkataraman, et al., Feedstock material property–process relationships in fused deposition of ceramics (FDC), Rapid Prototyping Journal (2000).

[28] S. Koga, et al., Thermodynamic modeling and control of screw extruder for 3D printing, in: 2018 Annual American Control Conference (ACC), 2018, IEEE.

[29] H. Li, et al., 3D bioprinting of highly thixotropic alginate/methylcellulose hydrogel with strong interface bonding, ACS Applied Materials & Interfaces 9 (23) (2017) 20086–20097.

[30] N.P. Levenhagen, M.D. Dadmun, Reactive processing in extrusion-based 3D printing to improve isotropy and mechanical properties, Macromolecules 52 (17) (2019) 6495–6501.

[31] M. Ziaee, N.B. Crane, Binder jetting: a review of process, materials, and methods, Additive Manufacturing 28 (2019) 781–801.

[32] F. Dini, et al., A review of binder jet process parameters; powder, binder, printing and sintering condition, Metal Powder Report 75 (2) (2020) 95–100.

[33] M.E. Schrader, Young-Dupre revisited, Langmuir: The ACS Journal of Surfaces and Colloids 11 (9) (1995) 3585–3589.

[34] C.K. Chua, K.F. Leong, 3D Printing and Additive Manufacturing: Principles and Applications Fifth Edition of Rapid Prototyping Fifth Edition, World Scientific Publishing Company, 2016.

[35] W. Yang, et al., Acoustic absorptions of multifunctional polymeric cellular structures based on triply periodic minimal surfaces fabricated by stereolithography, Virtual and Physical Prototyping 15 (2) (2020) 242–249.

[36] A. Alomarah, et al., Compressive properties of 3D printed auxetic structures: experimental and numerical studies, Virtual and Physical Prototyping 15 (1) (2020) 1–21.

[37] Z. Meng, et al., Design and additive manufacturing of flexible polycaprolactone scaffolds with highly-tunable mechanical properties for soft tissue engineering, Materials & Design 189 (2020) 108508.

[38] J. An, et al., Design and 3D printing of scaffolds and tissues, Engineering 1 (2) (2015) 261–268.

[39] A. Nogales, et al., Structure development in polymers during fused filament fabrication (FFF): an in situ small-and wide-angle X-ray scattering study using synchrotron radiation, Macromolecules 52 (24) (2019) 9715–9723.

[40] S. Petersmann, et al., Process-induced morphological features in material extrusion-based additive manufacturing of polypropylene, Additive Manufacturing 35 (2020) 101384.

[41] Z.X. Khoo, et al., 3D printing of smart materials: a review on recent progresses in 4D printing, Virtual and Physical Prototyping 10 (3) (2015) 103–122.

[42] A.Y. Lee, et al., Contactless reversible 4D-printing for 3D-to-3D shape morphing, Virtual and Physical Prototyping 15 (4) (2020) 481–495.

Chapter 7

Metal additive manufacturing

Chao Cai[1,2] and Kun Zhou[2]

[1]*State Key Laboratory of Materials Processing and Die and Mould Technology, Huazhong University of Science and Technology, Wuhan, China,* [2]*HP-NTU Digital Manufacturing Corporate Lab, School of Mechanical and Aerospace Engineering, Nanyang Technological University, Singapore*

Abbreviations

AM	additive manufacturing
AFM	atomic force microscope
ASTM	American Society for Testing and Materials
BJ	binder jetting
CNC	computer numerical control
DED	direct energy deposition
DLD	direct laser deposition
DMLS	direct metal laser sintering
EBAM	electron-beam additive manufacturing
EBM	electron-beam melting
EDS	Energy Disperse Spectroscopy
GA	gas atomization
GE	General Electric Company
HIP	hot isostatic pressing
ICP-OES	inductively coupled plasma optical emission spectroscopy
ISO	International Standards Organization
LCSM	laser confocal scanning microscope
LENS	laser engineered net shaping
LPBF	laser powder bed fusion
Micro-CT	micro-computed tomography
OM	optical microscope
PA	plasma atomization
PBF	powder bed fusion
PREP	plasma rotating electrode process
PSD	particle size distribution
SL	sheet lamination
SLM	selective laser melting
SEM	scanning electron microscope

Digital Manufacturing. DOI: https://doi.org/10.1016/B978-0-323-95062-6.00005-X
© 2022 Elsevier Inc. All rights reserved.

248 Digital Manufacturing

SLS	selective laser sintering
SP	stylus profilometer
UAM	ultrasonic additive manufacturing
WAAM	wire arc additive manufacturing
WLI	white-light interferometer
XPS	X-ray photoelectron spectroscopy
XRF	X-ray fluorescence

7.1 Introduction

Metal additive manufacturing (AM) is one of the most impactful AM categories. Metallic components are created through the accumulation of metallic materials in the form of powders, wires, or sheets via a layer-by-layer manner using various energy sources [1,2]. Over the last decades, the significant progress in the constituent techniques of metal AM, involving higher energy lasers, more rational software control systems, and more efficient powder preparation methods, boosts it to become a mature and prevalent digital processing technology [3]. This revolutionary technology enables the integration of design freedom, material diversity, and green manufacture, making it widely accepted as a new paradigm for the production of high-end components for critical applications, such as in aerospace, automotive, and health-care fields [3–5]. Grand View Research reported that the global metal AM market size was valued at USD 772.1 million in 2019 and is expected to grow at a compound annual growth rate of 27.8% from 2020 to 2027, due to this technology having great manufacturing flexibility, low waste, and high cost-effectiveness [6]. Increasing usage of metal AM in the overall manufacturing landscape further demonstrates their broad application perspective in engineering.

This chapter aims to provide a systematic review on metal AM technology regarding its technique classification and principles, feedstock preparation and characterization techniques, standard testing methods for the mechanical properties of printed metallic components, instinctive defects within printed components and corresponding detection methods, common postprocessing, and representative applications. It begins with a comprehensive overview of the classification of metal AM techniques in Section 7.2, and the principles, advantages, disadvantages, build capability, and renowned manufacturers of each technique are summarized. In Section 7.3 the preparation techniques for metal AM powder feedstock and powder characterization approaches are described in detail. Section 7.4 presents the applicable standard test methods for the mechanical properties of metallic AM components in terms of tensile, compressive, hardness, and fatigue properties. The printing and metallurgical defects within printed components and the corresponding inspection methods are unveiled in Section 7.5. Section 7.6 briefly reviews common postprocessing methods and their purposes for metallic AM components. Section 7.7 highlights typical influential application showcases of metal

AM in aerospace, automotive, and healthcare. Finally, a brief outlook on the scientific and technological challenges of metal AM and their respective potential resolutions is outlined in Section 7.8.

7.2 Classification of metal additive manufacturing technology

A considerable number of metal AM systems utilizing varied energy sources and feedstock forms are available in the market. In this chapter the metal AM technology is classified into four categories based on the American Society for Testing and Materials (ASTM) International Committee F4, which are powder bed fusion (PBF), direct energy deposition (DED), binder jetting (BJ), and sheet lamination (SL). The technological principle, material specifications, advantages, and limitations of each AM technique are presented thoroughly to help the reader establish a basic conceptual understanding of metal AM technology in this section.

7.2.1 Powder bed fusion

PBF creates products by selectively melting or fusing a powder-form raw material on a powder bed layer by layer under the assistance of a high energy source. Generally, there are five principal components in all PBF systems (Fig. 7.1A):

1. High energy source: A laser beam or electron beam is applied to consolidate the specific powder particles in a layer-wise approach.

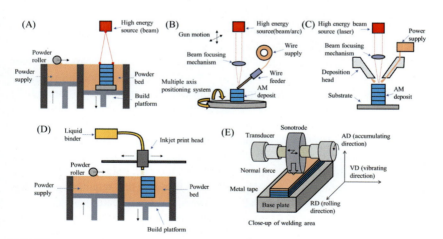

FIGURE 7.1 Schematic representations of metal AM systems: (A) PBF, (B) wire-fed DED, (C) powder-fed DED, (D) BJ, and (E) SL systems [4]. *AM*, additive manufacturing; *BJ*, binder jetting; *DED*, direct energy deposition; *PBF*, powder bed fusion; *SL*, sheet lamination. *Adapted from Materials & Design. 139 (2018) 565–586.*

250 Digital Manufacturing

2. Scanning motion system: This system focuses the heat source to the designated areas on the surface of the powder bed.
3. Powder supply and roller system: This mechanism supplies a layer of fresh powder on the top of the previously scanned layer(s).
4. Build platform: This mechanism is equipped with a metallic substrate plate that allows metallic components to be printed on it. The substrate plate also descends a certain height after each layer is scanned for the spreading of a new layer.
5. Enclosure chamber: This mechanism sustains an inert-atmosphere condition or a high vacuum environment to protect printed metallic objects from oxidation.

Compared with other metal AM techniques, the PBF technique has the distinct advantages of high processing resolution, good mechanical strength, and cost-effective customization for the printed metallic components. There are three mainstream types of PBF processes: selective laser sintering (SLS), laser powder bed fusion (LPBF), and electron-beam melting (EBM). SLS is mainly applied to manufacture polymeric components, and the latter two (LPBF and EBM) are the two predominant metal PBF processes.

7.2.1.1 Laser powder bed fusion

LPBF, also known as direct metal laser sintering and selective laser melting, was developed and commercialized in 1995 by EOS GmbH (Germany) based on the SLS system by using a high-power laser to melt metallic powder particles completely [7]. The initial LPBF system used a CO_2 laser that can only melt limited metal materials. These lasers are still prevalent in SLS machines because the laser powers are sufficient to sinter or fuse most polymeric materials. With the development of laser technology, all commercialized LPBF systems are equipped with either ytterbium or Nd-YAG fiber lasers with power ranging from 200 to 1000 W, which can melt most metallic materials. Laser beams are emitted from a laser source and go through an F-theta lens or dynamical mirror, generating an energy-focused laser spot in a diameter range of $50-100 \, \mu m$ on the surface of a powder bed.

LPBF process begins by converting the desired geometry from computer-aided design (CAD) to Standard Triangle Language (STL) which slices the objects into multiple layers. A laser beam is focused onto the powder bed, melting and scanning the powder bed, and forming the required geometry. Once a layer is completed, the platform lowers and the recoater spreads a fresh layer of powder onto the platform. The laser scanning is repeated and a component is formed. The enclosure chamber is filled with argon or nitrogen protective gas to avoid the oxidation of powder feedstock and metallic products. The specific powder particles, under the action of laser beams, experience rapid melting/solidification and repeated heat cycles during printing, which induces the formation of extensive thermal stresses within the products. To mitigate these thermal stresses, most commercialized LPBF machines are equipped with a preheating system that heats either the substrate plate or the entire building chamber. Preheating temperatures in LPBF are usually below 500°C. The applied layer thickness in LPBF does not exceed 100 μm, so the powder particles utilized in LPBF are preferably smaller than 100 μm. The recommended average particle size for LPBF is in the range of 15−63 μm [8].

Most printed objects require either sand or bead blasting posttreatment after printing to remove the adhesive powder on their surfaces. The surface finish of LPBF objects is generally rougher than that of machined counterparts. In special applications, LPBF components need to have a polishing treatment to improve their surface finish.

Compared with other metal AM techniques, LPBF can create objects with high resolution and good surface finish due to the small laser spot diameter, fine powder particle size, and thin layer thickness. The powder particles without the laser scanning can be reused directly after simply sieving the agglomerated powder clusters. The main disadvantage of LPBF is that the unavoidable high-temperature gradients in printed products stimulate thermal stress throughout the printed objects, which causes cracking and distortion or even printing interruption.

7.2.1.2 Electron-beam melting

EBM was commercialized in 1997 by Arcam AB Corporation in Sweden. The process principles of EBM and LPBF are similar with the main difference in their energy source. The energy source in the EBM system is an electron beam. EBM should be carried out under a high vacuum environment because electrons will suffer from deflections once they collide or interact with gas molecules, which results in the energy dissipation and decentralization of the electron beam. The application of a high vacuum environment can prevent the electrostatic charging of powder, alleviating printing process instability and the likelihood of smoking. Smoking is a phenomenon where powder particles are ejected from the surface of the powder bed owing to the mutual repulsion between charged metallic particles.

During EBM printing the powder bed is preheated by scanning the powder bed several times using unfocused electron beams. This preheating approach can endow the powder bed with more than 1000°C that is much higher than those in LPBF. On the one hand, such high preheating temperatures can significantly decrease the temperature gradient, mitigating thermal stresses, and cracks within the printed objects. On the other hand, some low-melting-point alloying elements (such as Al) in powder feedstock might undergo evaporation and loss under the high-temperature vacuum environment. After build completion the printed components are encapsulated by the slightly sintered powder bed resulting from the high preheating temperature. These encapsulated components have to be subjected to sandblasting using

252 Digital Manufacturing

an identical powder to separate the adhesive powder. The separated powder from the printed components can be reused after sieving.

Electron beams with high powers (over 3 kW) can penetrate a much thicker powder layer as compared with a laser beam source in LPBF. Usually, relatively coarse powder sizes ranging from 45 to 105 μm are preferred for EBM while the fraction of small particles in the feedstock should be as little as possible because the small particles would deteriorate the flowability and easy to cause smoking. The thicker layer thickness can shorten the printing time compared to LPBF but in exchange for a lower degree of printing accuracy.

7.2.2 Direct energy deposition

The typical apparatus for DED consists of a deposition arm or multiple deposition arms (movable along multiple axes) for materials supply, an energy source(s) (electron beam, laser beam, or arc beam) for melting the raw powder or wire materials, and a movable build plate for materials accumulation. DED can simultaneously deliver and melt feedstock on the build plate or a component, saving the time spent in the powder spreading in contrast to PBF. As a result, the deposition ratios of DED are about one order of magnitude faster than those of PBF [9]. The rapid deposition rate of DED is an appealing feature for the manufacture of large-scale components. However, the printing accuracy and surface quality of DED objects are much lower than those of PBF-printed counterparts because of the thick deposition thickness (0.1 to a few millimeters) and coarse energy beam spot size (up to several millimeters in diameter). Postmachining treatment is usually required for DED components before industrial applications. Many DED systems incorporate computer numerical control (CNC) machining and allow in situ machining during the printing process or machining in the same machine after the printing completion.

This technique closely resembles welding, and a flat initial surface is not necessarily required, implying that DED can add new materials onto an existing component. This unique characteristic makes the repairing of high-value components applicable for DED by selectively restoring the damaged portion of components, such as train wheels and turbine blades. DED also demonstrates the capability of processing functionally graded material components due to its flexibility to tune the material concentrations by adjusting the ingredient compositions via the multiple deposition heads.

Based on the feedstock form, the DED technique can be subclassified into powder-fed and wire-fed systems (Fig. 7.1B and C). Wire-fed DED systems can provide a more accurate and efficient deposition ratio, as well as a higher feedstock utilization rate compared to powder-fed DED systems. This is because the powder feedstock is jetted from the nozzle(s), and a portion of powder particles will deviate from the energy beam spot during the jetting process, resulting in the waste of feedstock. Besides, the price of wire materials is half of the equivalent powder materials. DED can also be subcategorized by the thermal source into direct laser deposition (DLD), electron-beam additive manufacturing (EBAM), and wire arc additive manufacturing (WAAM). Each kind of DED technique has its peculiarities, advantages, and limitations. More details will be discussed in the following sections.

7.2.2.1 Direct laser deposition

The early concept of DLD, which stated that an object is created in a layerwise additive deposition approach by powder or wire metallurgy using a laser beam source, was proposed by Brown et al. in the 1980s. In 1995 researchers at Sandia National Laboratories innovated this process that was commercialized later and registered to a trademark laser engineered net shaping (LENS) by Optomec [7]. LENS utilizes powder-fed materials as the feedstock and is furnished with multiple nozzles for more effective powder delivery. LENS is one of the first and most popular commercial DED brands, making it unarguably the most representative DED technique in academic and industrial organizations. In addition to LENS, there are various other DLD processes developed in the last decades, for instance, laser cladding, laser direct casting, direct metal deposition, freeform laser consolidation, laser metal deposition, and laser metal forming.

7.2.2.2 Electron-beam additive manufacturing

EBAM, also known as electron beam freeform fabrication, was invented by the Sciaky company of Chicago in 2004. Sciaky collaborated with Boeing to further develop and upgrade the EBAM system with the ability to manufacture commercial aviation components in 2005. During EBAM an electron beam gun melts metallic wire feedstock in a vacuum chamber to manufacture near-net-shape parts. After the printing process, additional surface machining is required to acquire a precise surface quality. The maximum material deposition rate is in the range of 3.18−11.34 kg/h, which is one of the fastest and most efficient metal deposition processes on the market to date [10]. Moreover, many EBAM systems can offer dual wire-fed heads to improve deposition efficiency further. Meanwhile, the dual wire-fed heads can simplify the feedstock transformation process of different deposition materials and supply two different metal alloys into a single melt pool to fabricate customized multimaterial parts.

7.2.2.3 Wire arc additive manufacturing

WAAM utilizes an electric arc as the heat source to melt metal wires onto a metal substrate. The origin of this process can be traced back to the patent written by Baker in 1925. There are three types of sources for arc generation in WAAM: gas metal arc, gas tungsten arc, and plasma

254 Digital Manufacturing

arc. The major advantages of WAAM are its cheap and open-source equipment due to the mature electric arc technology in the welding field. In comparison with DLD, WAAM can provide a more efficient thermal source and have high coupling efficiency even with materials such as Al, Cu, and Mg alloys, which are reflective metal alloys with poor laser coupling efficiency. WAAM does not require a vacuum environment, a prerequisite for EBAM, to operate. This endows the capability of WAAM to build large-scale objects as the build chamber has no volumetric limits. In practice, the heat sources and wire materials are always installed on multiaxis robotic arms to acquire higher printing flexibility. In addition, the depletion and evaporation of alloying elements, which easily occur during EBAM due to the vacuum condition, can be alleviated during the WAAM process. Nonetheless, the inert gas used for the arc formation is expensive, and the printed components are prone to distortion due to the high residual stress generated by the arc.

7.2.3 Binder jetting

Ely Sachs and Mike Cima invented binder BJ at the Massachusetts Institute of Technology in 1989 (then well known as three-dimensional printing), and later this technology was licensed to several different companies, including Z Corporation (acquired by 3D Systems in 2011) in 1994 and ExOne in 1996, for commercialization. Contemporarily, the main commercial BJ machines with the capability of printing metal materials are from HP, ExOne, Digital Metal, and Desktop Metal. The recently developed HP Metal Jet printers claim to be 50 times more productive than conventional BJ printers with a printing high voxel resolution of 1200×1200 dpi grid.

BJ process begins by spreading a layer of powder onto the platform while a roller evenly spreads and compacts the powder. It selectively deposits a liquid binding agent onto a bed of powder through the print head, which then infiltrates the specific powder surfaces to bond the powder particles together (Fig. 7.1D). A heater is used to dry/cure the layer after which the platform is lowered. A fresh layer of powder is then applied, and the process is repeated. The printed parts, also known as green parts, can be taken out from the powder bed directly after the liquid binding agent dries. The green parts are fragile and easily damaged since the powder particles are physically bonded without metallurgical reactions. To achieve higher mechanical performance, the printed parts are usually subjected to appropriate posttreatments, such as high-temperature sintering or resin infiltration.

Theoretically, any powdered material that can be spread and wetted by ejective binders is processible using this technique. The cost to set up a BJ system is far lower than PBF as there are no requirements of costly thermal sources and environmental control systems for shielding gas or vacuum condition. Unfortunately, the major flaws of BJ products are the uneven shrinkage during the sintering process and the inferior mechanical strength. The uneven shrinkage that occurs in the sintering process leads to significant deviations in the geometric accuracy of final objects. Internal stresses induced by shrinkage can also trigger delamination and cracking within the printed objects. The sintered parts still maintain a high level of porosity and cause their mechanical properties to be much lower than the counterparts printed by other metal AM.

7.2.4 Sheet lamination

Proposed by Helisys Inc. in 1986 and patented in 1987, SL is one of the first developed and commercialized AM techniques. During the SL process, thin sheets or rolls of materials are cut into precisely shaped laminations based on cross-sectional layers of the CAD model using a laser or a sharp blade. These laminations are then stacked or glued together via ultrasonic, adhesive, clamping, or thermal bonding layer by layer to create objects (Fig. 7.1E). The excess materials in each layer are left to use as a support structure and subsequently removed from the edge of the printed objects when the consolidation process is completed.

Ultrasonic additive manufacturing (UAM) is the primary SL technique for metal printing. UAM evolves from the ultrasonic seam welding technique and utilizes ultrasonic energy to join metallic sheets or ribbons with a layer thickness of $\sim 100\,\mu$m to manufacture complex metallic components. In UAM the operating temperatures required to join the metallic sheets or ribbons are as low as 30%−50% of the melting point of metals. The low processing temperature can mitigate thermal distortions in the printed components. Nevertheless, the UAM process is seldom applied in the industry owing to the disadvantages, such as the weak mechanical properties in the build orientation and limited lamination material categories available. The technical details of this process will not be discussed in this chapter.

In summary, each AM technique has its fundamental principle, advantages, disadvantages, and printing volume capability (Table 7.1). These technological features drive each technique to excel in certain specific applications, where their unique manufacturing advantages can be exploited. For example, PBF shows overwhelming superiority in the manufacturing of objects with geometrical complexity, high resolution, and good surface finish. DED can create large-scale near-net-shape parts with nearly full densities and strong metallurgical bonding rapidly. BJ is capable of fabricating refractory materials into complex-shaped components at a low cost and a faster rate as compared to PBF processes. SL enables different metal materials to be processed to form multimaterial structures, which are difficult to process using conventional technology.

TABLE 7.1 Summary of fundamental principle, typical technology, advantages, disadvantages, build volumes, and representative manufacturers of metallic additive manufacturing techniques.

ASTM category	Fundamental principle	Typical technology for metal manufacturing	Advantage	Disadvantage	Build volume (mm^3)	Representative manufacturer/ country
PBF	A thermal energy fuses a small designated region of the powder bed to build components	• EBM • LPBF	• High accuracy • High mechanical strength • Large range of material options	• Relatively slow process • High-cost equipment • High-cost materials	$X \leq 1000$ $Y \leq 1000$ $Z \leq 2500$	• ARCAM, Sweden • EOS, Germany • Concept Laser • Renishaw, UK • Matsuura, Germany • 3D Systems, USA
DED	A focused thermal energy melts metallic materials during deposition	• DLD • EBAM • WAAM	• Fabricate large components • Reduce manufacturing time and cost • Fabricate multiple materials simultaneously	• Low accuracy • Low surface finish • Low feature details • High-cost equipment	$X \leq 3000$ $Y \leq 3500$	• Optomec, USA • InssTek, USA • Sciaky, USA • Irepa Laser, France • Trumpf, Germany • BeAM, France

BJ	A liquid bonding agent is jetted onto thin layers of powder to glue products layer by layer.	• Indirect inkjet printing (binder 3DP)	• Free of support/ substrate • A wide range of materials • Large build volume • High print speed • Relatively low cost	• Fragile parts with inferior mechanical properties • Requirement of postprocessing • Shrinkage in the post sintering process	$X \leq 4000$ $Y \leq 2000$ $Z \leq 1000$	• ExOne, USA • HP Metal Jet, USA • Desktop Metal, USA • Digital Metal, Sweden
SL	Sheets or foils of materials selectively stack and consolidate into products	• UAM	• High speed • Low cost • Fabricate multiple materials simultaneously	• Weak mechanical properties in the build direction • Requirement of postprocessing • Limited available material	$X = 150-250$ $Y = 200$ $Z = 100-150$	• 3D Systems, USA • MCor, Ireland

ASTM, American Society for Testing and Materials; *BJ*, binder jetting; *DED*, direct energy deposition; *EBAM*, electron beam additive manufacturing; *EBM*, electron beam melting; *LPBF*, laser powder bed fusion; *PBF*, powder bed fusion; *SL*, sheet lamination; *UAM*, ultrasonic additive manufacturing; *UK*, United Kingdom; *USA*, United States; *WAAM*, wire arc additive manufacturing.

258 Digital Manufacturing

7.3 Preparation and characterization techniques for metal additive manufacturing feedstock

The two main forms of feedstock for metal AM are wire and powder feedstocks. For metal AM technology, particulate materials, which dominated the market of metal AM feedstock with a volume share of 92.6% in 2019, are the predominant form of feedstock [11]. Wire feedstocks are generally welding wires produced from casting, roll bonding, drawing, or extrusion techniques. Powder feedstocks are mainly produced using gas atomization (GA), plasma atomization (PA), and plasma rotating electrode process (PREP). AM feedstock characteristics play an important role in influencing the processing windows or the functionalities and properties of fabricated components. For wire feedstocks, contaminants and defects on the wire surface such as cracks and scratches may result in porosity in the fabricated components. Wires used for AM are typically range from 0.45 to 1.2 mm. For metal powders, key characteristics are particle size distribution (PSD), particle shape, flowability, and chemical composition. The characteristics of the metal powder, such as PSD, sphericity, flowability, and chemical composition, play a vital role in either the processing window establishment or the functionalities and properties of printed products. The characteristics of metal particles are highly affected by metal powder preparation processes. In this section, three common powder preparation processes for AM metal particulate feedstock are introduced and compared in detail. Subsequently, the typical powder measurement methods for AM powders in terms of PSD, flowability, and chemical composition are summarized.

7.3.1 Powder preparation techniques

There are diverse metal powder preparation processes, among which three methods, GA, PA, and PREP, are commercially applied in the production of AM metal powders. The metal powders prepared by these three methods have consistent characteristics and meet stringent requirements of metal AM techniques, ensuring stable product service performance under prescribed printing process parameters.

7.3.1.1 Gas atomization

In GA, metals or metal alloys are smelted in a protected crucible furnace for a period of time until the composition of the metal liquids becomes homogeneous. The metal liquids then flow down to a refractory metal nozzle from the crucible furnace and are atomized into tiny molten droplets by a high-pressure inert gas flow. Finally, the molten droplets solidify into spherical or near-spherical particles during the falling process in an inert gas chamber. The schematic representation of the GA process is illustrated in Fig. 7.2A.

Metal additive manufacturing Chapter | 7 **259**

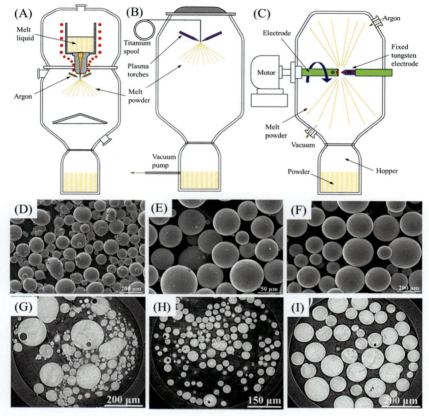

FIGURE 7.2 Schematic representation of these three powder preparation methods and representative morphologies of Ti-6Al-4V particles prepared by each method: (A), (D), and (G) GA schematic and corresponding prepared particles; (B), (E), and (H) PA schematic and corresponding prepared particles; and (C), (F), and (I) PREP schematic and corresponding prepared particles [12]. *GA*, Gas atomization; *PA*, plasma atomization; *PREP*, plasma rotating electrode process.

Even though GA is a mature process in the industry, there are still two unavoidable issues in GA-prepared powder particles. The first issue is that a substantial amount of satellite particles are generated when solidified metal particles rebound from the bottom of the chamber and collide with the falling molten particles. The satellite particles, which have adverse effects on the flowability and packing density, can be observed in Fig. 7.2D and G. The other issue is the formation of gas pores or bubbles within the metal particles when the partial high-pressure gas applied in GA is trapped and retained in the molten particles. These gas-trapped pores will remain and decrease the densities of printed components, which are detrimental to the mechanical properties of printed components, especially fatigue properties.

260 Digital Manufacturing

7.3.1.2 Plasma atomization

PA uses argon plasma torches to melt and atomize prequalified metal wires into fine droplets (Fig. 7.2B). These torches generally adopt tungsten electrodes, ensuring minimal erosion and contamination of the powder. The metal powders can possess higher purity because the molten wires need not contact refractory metals in a crucible. Moreover, PA can achieve a higher yield of small powders, a better quality of sphericity, a lower fraction of satellite particles, and a more homogenous PSD (Fig. 7.2E and H) compared with GA. The major drawback of PA is the high cost of processing the feeding materials into wire form.

7.3.1.3 Plasma rotation electrode process

Metal or metal alloys are made into consumable electrodes as feeding materials in PREP. The end side of the target materials is heated by a plasma torch and melted into molten liquids (Fig. 7.2C). Following that, the molten liquids are ejected and split into small molten droplets by strong centrifugal forces through the rotation of the target material at high speeds of 3000–15,000 rpm. The small molten droplets are subsequently solidified into powder particles during the flight process in an inert gas environment. The advantages of PREP are the capability of preparing high purity powders with no or minor gas-trapped pores and fewer satellite particles (Fig. 7.2F and I). The main limitation of PREP is its broad PSD ranging from 45 to 800 μm, making PREP-prepared powders unsuitable for the PBF technique. A systematic review and comparison of these three methods are summarized in Table 7.2.

7.3.2 Powder characterization techniques

Determination of AM metal powder properties is a necessary process for industrial production or new materials exploration. It facilitates the selection of powders with consistent physiochemical properties and the manufacture of products with consistent service performance. PSD, flowability, and chemical composition are the paramount properties for AM metal powder particles, and details on standardized measurements for these properties will be outlined in this section.

7.3.2.1 Powder size distribution

The PSD of powder feedstock has a significant influence on the layer thickness and packing density. It determines the printing precision, surface quality, and density of printed components. Microscope observation, sieving analysis, and dynamic light scattering are the common measurement approaches for the PSD detection of AM metal powders.

TABLE 7.2 A systematic review and comparison of gas atomization (GA), plasma atomization (PA), and plasma rotating electrode process (PREP) methods.

Method	Feed material	Powder size range (μm)	Advantage	Disadvantage
GA	Rod	0–200	• Relative low cost • Higher fine particles yield	• Satellite particles • Gas-trapped pores • Impurity contamination
PA	Wire	25–125	• High purity • Fewer satellite particles • Higher fine particles yield	• Costly wire feedstock • Gas-trapped pores • Limited feedstock options
PREP	Rod	50–800	• High purity • High sphericity • Fewer satellite particles • Fewer gas-trapped pores	• Low fine powder yield • High cost

Microscope observation uses microscope equipment, such as optical microscope (OM) and scanning electron microscope (SEM), to observe and measure the morphology and size of powder particles. Sieving analysis evaluates PSDs by allowing powder particles to pass through a series of sieves of progressively smaller mesh sizes. Subsequently, the weight of blocked powder particles in each sieve is recorded and calculated as a fraction of the whole mass. ASTM B214 specifies the standard operational procedure for this method.

Dynamic light scattering is the most common technique used to measure PSDs of AM metal powders. The principle of this method is based on the Brownian motion of dispersed particles in an aqueous solution. Before measurements, an aqueous solution is applied to disperse the target powder particles. The Brownian motion of particles in the solution causes the imposed light to be scattered at different intensities. The PSDs of target powders can be determined by measuring the random changes in the intensity of light scattered from the solution according to the Stokes–Einstein relationship. ASTM B822-20 provides a standard guideline for this testing method.

262 Digital Manufacturing

7.3.2.2 Flowability

Powder flowability is a critical characteristic for metal AM, particularly for the PBF technique, because it can strongly impact the uniformity of powder spreading. Hall and Carney flowmeter funnel approaches are frequently used to measure powder flowability owing to their operational simplicity. The principle of the two methods is the same: a certain weight of powder particles flow through a funnel and the time taken to flow through the funnel is used to calculate the flow rate. The funnel has a standard size and geometry with a 60-degree cone angle, but the outlet holes in Hall and Carney flowmeter funnel approaches are Φ 2.54 mm and Φ 5.08 mm, respectively. The Carney flowmeter funnel is applied to measure powder particles that have difficulty passing through the Hall flowmeter funnel. The relevant standard testing details can be found in ASTM B213-20 (for Hall) and ASTM B964-16 (for Carney).

7.3.2.3 Chemical composition

In practice, the chemical composition of powder feedstock has a slight deviation from the nominal composition. The composition deviation could deteriorate the microstructures and performance of printed components. Powders can be polluted or oxidized when exposed in a nonvacuum condition for a long time or repeatedly exposed in an AM building chamber. The common techniques used for inspecting the chemical composition of AM powders are energy dispersive spectroscopy (EDS), X-ray photoelectron spectroscopy (XPS), X-ray fluorescence (XRF), and inductively coupled plasma optical emission spectroscopy (ICP-OES).

EDS takes advantage of the theory that each element has a unique atomic structure allowing a unique set of peaks on its electromagnetic emission spectrum. It is a universal method to identify and quantify the elemental composition in a small area of the surface of a sample. EDS detectors are commonly installed on SEMs.

XPS, also known as electron spectroscopy for chemical analysis, is a surface-sensitive quantitative spectroscopic technique that can detect not only the element composition but also the chemical state of each element. During XPS measurements, monoenergetic X-rays are excited and irradiated onto target sample surfaces to trigger the emission of photoelectrons from targeted samples. Sequentially, the emitted photoelectrons are collected by an electron energy analyzer. Finally, the element identification, quantity, and chemical state of the detected samples can be determined by analyzing the binding energy and intensity of the collected photoelectron peaks.

XRF analyzers detect the chemistry of a sample by measuring excited characteristic "secondary" (or fluorescent) X-rays from target materials that are being bombarded with high-energy X-rays or gamma rays. Element and chemical analyses can be conducted from the collected fluorescent X-ray. From the viewpoint of their measurement principles, the difference between XRF and XPS is the form of emitted energy from target materials.

ICP-OES, also regarded as inductively coupled plasma atomic emission spectroscopy, exploits the fact that excited electrons emit energy at a given wavelength when they return to the ground state after excitation by high-temperature plasma. Each element emits energy at specific wavelengths peculiar to its atomic character, making the detected elements identifiable.

7.4 Mechanical properties standard testing for metallic additive manufacturing components

Metallic AM components are generally applied in high-end industries that demand components with robust mechanical properties. It is crucial to evaluate the mechanical properties of metallic AM components in a rational approach to determine whether these components conform to the requirements of specific applications. The scope of this section reviews applicable standard test methods from ASTM and International Standards Organization (ISO) to evaluate the predominant mechanical properties (tense, compression, hardness, and fatigue) of metallic AM components.

7.4.1 Tension

Tensile testing is a basic mechanical measurement to detect how a material resists loaded tensile forces until it fractures. Properties measured directly via a tensile test are ultimate tensile strength σ_{UTS}, tensile yield strength σ_{TY}, and elongation at break ε_T. Uniaxial tensile testing is the most commonly used for obtaining the mechanical characteristics of isotropic materials. ASTM E8 is a standard test method on the tensile behavior of metallic materials under uniaxial tensile stresses at room temperatures ($10°C-38°C$). ISO 6892-1 is the equivalent tensile standard, but it covers more types of sample geometries such as sheet and wire. ASTM E21-20 and ISO 6892-2 both specify the operational details for the tensile tests of metallic materials at elevated temperatures (above $38°C$). ISO 15579 provides the guideline for tensile testing of metallic materials at temperatures between $-196°C$ and $10°C$. For cryogenic-temperature (below $-196°C$) tensile tests, the information is specified in ASTM E1450-16 and ISO 19819. ISO 26203-1 and ISO 26203-2 give specific requirements for testing metallic materials at high strain rates ranging from 10^{-2} to 10^3/s at room temperature. A preliminary understanding of the tensile properties of common metallic AM components can be achieved from Table 7.3.

7.4.2 Compression

During compressive testing a material experiences opposing uniaxial forces that push inward upon the specimen from opposite sides. The loading forces in compressive testing are the opposite of those in typical tensile tests, but

TABLE 7.3 Tensile property comparison of common metallic components fabricated by additive manufacturing and conventional methods.

Method	Materials	Heat treatment	Orientation	σ_{TY} (MPa)	σ_{UTS} (MPa)	ε_T (%)	Refs.
LPBF	316 L stainless steel	As-built	H	438	528	10	[13]
		As-built	V	435	504	16	
	Inconel 718	As-built	H	643	991	13	[14]
		As-built	V	572	904	19	
		1100°C/1 h	H	1159	1377	8	
		1100°C/1 h	V	1074	1320	19	
	AlSi10Mg	As-built	H	250	350	2.5	[15]
		As-built	V	240	280	1.2	
		T6	H	285	340	4.5	
		T6	V	290	330	2.2	
	Ti-6Al-4V	As-built	H	1137	1206	7.9	[16]
		As-built	V	962	1166	1.7	
		1050°C/1 h	H	913	1019	8.9	
		1050°C/1 h	V	836	951	7.9	
		950°C/1 h	H	944	1036	8.5	
		950°C/1 h	V	925	1040	7.5	
		700°C/1 h	H	900	1000	1.9	
		700°C/1 h	V	965	1046	9.5	

EBM	316 L stainless steel	As-built	H	334	571	29.3	[17]
		As-built	V	396	437	30.6	
	Inconel 718	As-built	H	894	1061	11.5	[18]
		As-built	V	925	1138	15.7	
	Ti-6Al-4V	As-built	H	899	978	9.5	[19]
		As-built	V	869	928	9.9	
DED	Ti-6Al-4V (powder feedstock)	As-built	H	916	1032	19	[20]
		As-built	V	961	1072	17	
	Inconel 718 (powder feedstock)	As-built	H	590	845	11	[21]
		1050°C/1 h + 980°C/1 h	H	1133	1240	9	
	316 L stainless steel	As-built	H	330−345	540−560	35−43	[22]
Cast	316 L stainless steel	As-built	−	365	596	69	[13]
	Inconel 718	As-built	−	758	802	5	[18]
	AlSi10Mg	As-built	−	160−185	300−350	3−5	[23]
	Ti-6Al-4V	As-built	−	750	875	4.5	[24]
Wrought	316 L stainless steel	As-built	−	170−290	170−290	∼40	[25]
	Inconel 718	As-built	−	1034	1241	10	[18]
	Ti-6Al-4V	As-built	−	795	860	10	[26]

DED, Direct energy deposition; *EBM*, electron beam melting; *LPBF*, laser powder bed fusion; *H*, horizontal orientation that is parallel to the substrate plane; *V*, vertical orientation that is perpendicular to the substrate plane.

266 Digital Manufacturing

both experimental measurements are carried out with the same machine. The goal of compression tests is to detect the behavior or response of materials and obtain the corresponding compressive properties (such as ultimate compressive strength σ_{UCS}, yield strength σ_{CY}, elongation at break ε_C) when it experiences compressive loads.

The generally applicable standards for compressive testing are ASTM E9−19 and ASTM E209−18, which are specific to the compression of cylinder-shaped metallic materials at room temperatures and elevated temperatures (beyond 538°C), respectively. There is no relevant test standard in ISO. A comparison of the compressive properties of common metallic components fabricated by AM and conventional methods is summarized in Table 7.4.

7.4.3 Hardness

The hardness property determines a material's capability to resist plastic deformation, penetration, indentation, and scratching. It is a crucial indicator of the material's resistance to wear and abrasion in engineering. Generally, there is a proportional relationship between hardness and strength, which means that tougher metallic materials can possess higher strength.

Several methods have been developed for hardness testing, including Brinell, Rockwell, Vickers, Knoop, Tukon, Mohs scale, and Scleroscope. The first five methods are indentation-based testing that measures the deformation resistance of a material by loading a constant compression force via a sharp indenter. Mohs scale is used to obtain scratch hardness by measuring the fracture or permanent plastic deformation of a material that is subjected to friction from a sharp object. Scleroscope is a rebound method where the height of the "bounce" of a diamond-tipped hammer dropped from a fixed height onto the targeted material is detected.

For the indentation-based hardness measurements, ASTM E10-18 and ISO 6506-1 provide the details of the Brinell hardness measurement at room temperature, including testing indenter requirements and operational procedures. Rockwell hardness measures the hardness at specific locations on a part, which may not be representative of the whole part. Its standard testing guidelines are specified in ASTM E18-20 and ISO 6508-1. Vickers and Knoop hardness test methods can be acquired in ASTM E384-17 (for Vickers and Knoop), ISO 6507-1 (for Vickers), and ISO 4545-1 (for Knoop). Since the various types of hardness tests reflect the hardness of a material in different aspects, ASTM E140 was developed to relate and convert one hardness scale to another.

The equipment and operations for determining the scratch hardness of a material using a diamond stylus are introduced in ASTM G171-03. Regarding the rebound hardness, ASTM E448-82 covers the verification of Scleroscope hardness instruments and the calibration of standardized hardness test metallic blocks.

TABLE 7.4 Compressive property comparison of common metallic components fabricated by additive manufacturing and conventional methods.

Method	Materials	Heat treatment	Orientation	σ_{CY} (MPa)	σ_{UCS} (MPa)	ε_C (%)	Refs.
LPBF	316 L stainless steel	As-built	H	~300	~600	–	[27]
	Inconel 718	As-built	H	1248	–	–	[28]
		As-built	V	1120	–	–	
	AlSi10Mg	As-built	H	317	–	–	[29]
		T6	H	169	–	–	
	Ti-6Al-4V	As-build	V	1292	1602	13.0%	[30]
EBM	Ti-6Al-4V	As-built	H	879	1412	3.0	[31]
		As-built	V	953	1587	3.0	
DED	316 L stainless steel	As-built	H	330–345	540–560	35–43	[32]
	Ti-6Al-4V	As-built	H	916	1032	19.0	[33]
	Inconel 718	As-built	H	590	845	11.0	[34]
		1050°C/1 h + 980°C/1 h	H	1133	1240	9.0	

DED, Direct energy deposition; *EBM*, electron beam melting; *LPBF*, laser powder bed fusion.

268 Digital Manufacturing

Among these measurement methods, Vickers is the universal testing approach for metallic AM components. The Vickers hardness of common metallic AM components is listed in Table 7.5.

7.4.4 Fatigue performance

Fatigue is a structural failure of a material that is subjected to a cyclic load. This type of structural damage occurs when the cyclic loading forces are far below the static strength of materials. The cyclic forces can induce the initiation of a fatigue crack, and the crack will grow gradually after each cycle until the structure fractures. Fatigue is usually used as a guide to select metallic materials for service under conditions of repeated direct stress.

ASTM E466-15 is a standard practice regarding conducting axial, constant-amplitude, and force-controlled fatigue tests at ambient temperature for metallic specimens. This axial force fatigue test is used to determine the effect of variations (such as geometry, surface condition, stress) on the fatigue resistance of metallic materials subjected to direct stress for relatively large numbers of cycles. ISO 1099 is similar to ASTM E466-15, but it mainly highlights the relation between applied stress and the number of cycles to failure for a given material condition (such as hardness and microstructure) at various stress ratios. ASTM E606 is identical to E466-15, except for adopting strain control instead of force control.

ASTM E2368-10 and ISO 12111 are the standard measurements for the thermomechanical fatigue of metallic specimens under uniaxial loading and strain control. A thermomechanical fatigue cycle is performed under a condition where uniform temperature and strain fields over the specimen gage section are simultaneously varied and independently controlled.

ASTM E2714-13 contains the determination of creep-fatigue deformation or crack formation in nominally homogeneous materials subjected to uniaxial forces under isothermal conditions. ISO 12106 is the equivalent standard.

Fatigue crack growth rate expresses materials resistant to stable crack extension under cyclic loads. Both ASTM E647 and ISO 12108 provide guidelines to characterize the fatigue crack growth rate of preprepared samples that are notched to have a precrack. During testing the samples are under linear-elastic stress conditions with applied forces perpendicular to the crack plane and a constant stress ratio. ASTM E2760 is the standard test method for creep-fatigue crack growth testing. It concerns fatigue cycling with sufficiently long loading/unloading rates or hold times, or both, to cause creep deformation at the crack tip.

7.5 Defects in metallic additive manufacturing components

Due to the layer-by-layer accumulation and rapid melting/solidification of material processing features, there are many unique and unavoidable metallurgical

Metal additive manufacturing Chapter | 7 **269**

TABLE 7.5 Hardness comparison of common metallic components fabricated by additive manufacturing and conventional methods.

Method	Materials	Heat treatment	Orientation	Hardness	Refs.
LPBF	316 L stainless steel	As-built	H	219−239	[35]
		As-built	V	228	
	316 L stainless steel	As-built	−	209	[36]
		650°C/2 h	−	202	
		950°C/2 h	−	182	
		1100°C/2 h	−	173	
	Inconel 718	As-built	H	304	[37]
		1100°C/1 h	H	258	
		1100°C/7 h	H	217	
		1250°C/1 h	H	210	
		1250°C/7 h	H	207	
	AlSi10Mg	As-built	H	133	[38]
		450°C/2 h	H	96	
		500°C/2 h	H	88	
		550°C/2 h	H	64	
	Ti-6Al-4V	As-build	H	396	[39]
			V	405	
		705°C/3 h/AC	H	350	
			V	374	
		850°C/2 h/AC	H	393	
			V	374	
		1015°C/0.5 h/AC + 850/2 h/FC	H	400	
			V	385	
EBM	316 L	As-built	V	164−178	[40]
	Inconel 718	As-built	H	435−442	[41]
	Ti-6Al-4V	As-built	H	366	[42]
		925°C/1 h/WC + 450/4 h/AC	H	401	
		925°C/1 h/AC + 450/4 h/AC	H	358	

(*Continued*)

270 Digital Manufacturing

TABLE 7.5 (Continued)

Method	Materials	Heat treatment	Orientation	Hardness	Refs.
DED	316 L stainless steel	As-built	H	174−205	[43]
	Ti-6Al-4V (powder feedstock)	As-built	H	333−351	[44]
	Ti-6Al-4V (wire feedstock)	As-built	H	327	[45]
	Inconel 718 (powder feedstock)	As-built	H	228−263	[46]
			V	251−281	
Cast	316 L stainless steel	As-built	−	165	[47]
	Inconel 718	As-built	−	266	[48]
	AlSi10Mg	As-built	−	95−105	[23]
	Ti-6Al-4V	As-built	−	320	[49]
Wrought	316 L stainless steel	As-built	−	215−225	[50]
	Inconel 718	As-built	−	318	[51]
	Ti-6Al-4V	As-built	−	342	[50]

DED, Direct energy deposition; *EBM,* electron beam melting; *LPBF,* laser powder bed fusion.

defects in AM metallic components, which are absent in the traditionally processed metallic components. The common defects include excessive residual stresses, pores, cracking, distortions, and poor surface qualities in metal AM components. These defects significantly deteriorate the mechanical and physical performance of metallic AM components and hamper their full potential for advanced applications. The formation mechanisms and detection methods for these defects are introduced systematically in this section.

7.5.1 Defect categories

The existence of defects within AM metallic components is a matter of considerable concern for engineering applications, especially for components suffered from cyclic loads. Understanding the formation mechanisms of defects during the printing process is significant to suppress or eliminate defects generation by utilizing proper printing parameters or strategies.

7.5.1.1 Excessive residual stresses

Residual stresses are stresses that are remained in the components after the completion of mechanical processes. Residual stresses are an inherent characteristic in metallic AM products, especially in PBF and DED metallic products. This is because metallic AM products experience repeated rapid melting/solidification and heating/cooling processes. Hence, it leads to a spatially varied thermal distribution, as well as nonuniform expansion and contraction during printing.

Metallic AM components usually retain excessive residual stresses with high magnitudes. When residual stresses are higher than yield stress of materials, warping or plastic deformation may take place. If the local residual stresses are beyond the ultimate tensile stress of printed materials, it may induce cracking within the printed parts. The magnitude of residual stresses is closely related to that of the yield strength of printed materials. The materials with higher yield strength can generate a higher magnitude of residual stresses. For example, the residual stresses in DED Inconel 718 samples with yield strength of ~ 1100 MPa are ~ 1.5 times higher than that in DED AISI 316 stainless steel specimens with yield strength of ~ 450 MP [52,53].

The type and magnitude of residual stresses in the printed components are varied and inhomogeneous during the material accumulation process (Fig. 7.3) [54]. Within a printed component, different positions have different types and magnitudes of residual stresses. For instance, residual stresses show tensile and compressive types at the edges and center of the printed objects, respectively. The magnitude of residual stresses always presents dynamic changes during printing. The residual stresses generally improve with the increase in the printing height. After cracking the stresses are partially relieved and redistributed within the printed objects.

7.5.1.2 Pores

Pore is an unavoidable defect in metallic AM objects, and the quantities of pores in metallic AM products are much more than that in the counterparts fabricated by conventional methods. Pore, as one of the most detrimental defects, has negative effects on the mechanical performance of final parts (especially for fatigue properties due to the pores acting as crack initiators). The pores can be primarily classified into lack-of-fusion pores and metallurgical pores according to their shape morphologies and formation mechanisms (Fig. 7.4) [55].

Lack-of-fusion pores, as the name signifies, are the result of incomplete metallurgical bonding between layers or adjacent melt pools, leading to irregular morphologies (usually observed to have a flat disk-like morphology) with relatively large sizes (hundreds of microns). These pores are processing defects, which means that their formation is dependent on the printing parameters. A low energy input (such as lower input power, high

272 Digital Manufacturing

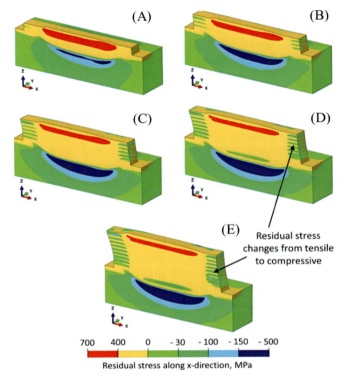

FIGURE 7.3 Distribution of residual stresses along the longitudinal direction after the deposition of (A) 2nd, (B) 4th, (C) 6th, (D) 8th, and (E) 10th layer of IN 718 powder on IN 718 substrate using a laser source [54].

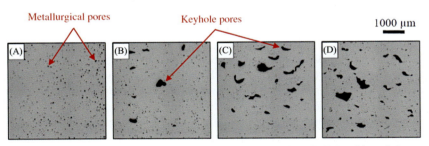

FIGURE 7.4 Lack-of-fusion and metallurgical pores in LPBF AlSi10Mg with varied scan speeds: (A) 250, (B) 500, (C) 750, and (D) 1000 mm/s [55]. *LPBF*, Laser powder bed fusion.

scan speed, and large hatching space) is more likely to cause insufficient melt and then create the lack-of-fusion pores.

Metallurgical pores possess spherical or near-spherical morphologies with a small size (lower than tens of microns). In contrast to lack-of-fusion pores the formation mechanisms of metallurgical pores are more sophisticated. These are related not only to the printing parameters but also to the raw materials. Presently, several formation mechanisms have been uncovered, such as pore transfer from feedstock powder, instability of depression zones during the printing process, vaporization of volatile elements, and gas precipitation.

7.5.1.3 Cracking

Metallic AM components suffer from cracking, which is a common defect in PBF and DED processes and can be simplified into cold cracking, hot cracking, and liquid cracking [56].

Cold cracking is a process that denotes the formation of cracks induced by residual stresses. Once the residual stresses exceed the ultimate tensile stress of the printed material, cracking will occur. Extensive experimental research discovers that the existence of pores can intensify cold cracking as residual stresses tend to accumulation around pores. Additionally, cold cracks prefer to generate between layers, which leads to delamination (Fig. 7.5A) [57]. The lengths of cold cracks in metallic AM components usually reach a millimeter scale.

Hot cracking, also referred to as solidification cracking, is generated during the solidification process. The rapid melting/solidification process during printing causes the trapping of liquid between already solidified dendrites. These weak mushy liquid/solidified regions rupture and tear under the action of residual stresses. Solidification cracking is material dependent and appears in nickel-base superalloys, 6000 and 7000 series Al alloys usually [58]. These cracks show jagged morphology with lengths of ~ 200 μm.

Liquid cracking refers to cracks that are caused by the liquidation of already solidified regions. Low-melting-point phases or elements are enriched at grain boundaries due to the constitutional undercooling during the rapid solidification process. When laser energies penetrated from the top layer to the solidified layers, these zones are easy to be remelted. The remelted liquated zones are torn by contracting the surrounding material and then pull apart the weakened grain boundaries. Fig. 7.5D shows that a low-melting-point carbide remelted for the LPBF process of a nickel-based alloy gives rise to liquidation cracking formation and propagation along the grain boundary [56].

7.5.1.4 Distortions

Metallic AM components are commonly subjected to distortions that are induced by excessive residual stresses. Shape distortions are detrimental to dimensional accuracy and have negative effects on the performance of printed components, particularly for large, thin-walled structures. Distortions can occur not only during the printing process but also during the removal

274 Digital Manufacturing

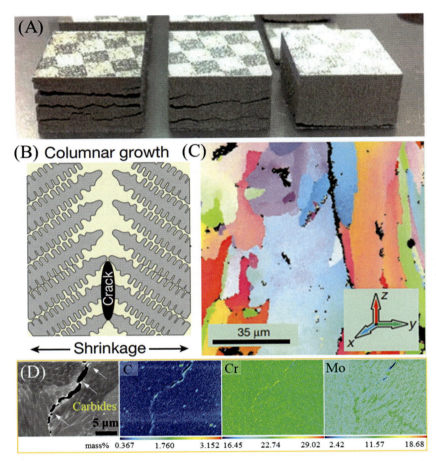

FIGURE 7.5 Cold and hot cracking in metallic AM products: (A) delamination in EBM stainless steel [57], (B) schematic representation of the columnar growth of dendrites during the solidification process, resulting in cracks due to solidification shrinkage, and (C) hot cracking at the interface between columnar grains in LPBF Al7075 [58], and liquid cracking caused by the liquidation of already solidified regions in the LPBF nickel-based alloy [56]. *AM*, Additive manufacturing; *EBM*, electron beam melting; *LPBF*, laser powder bed fusion.

process of components from the substrate. Severe distortions can bring about printing failure or the fabrication of inferior components.

There are different types of distortions, including warpage, shrinkage, bending, angular, and rotational distortions in AM objects. Warpage is easy to arise at the bottom of printed components because the maximum residual stresses concentrate at the interface between the bottom of printed components and substrates (Fig. 7.6A) [59]. These residual stresses are compressive and tensile at the center and edge of the bottom, respectively, which promote

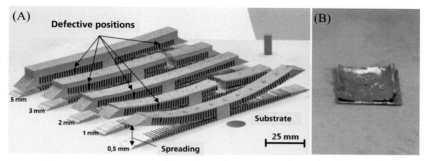

FIGURE 7.6 Two representative distortions in metallic AM components: (A) cantilever distortion after support removal for LPBF AlSi10Mg [59] and (B) flaking during LPBF of pure Al [60]. *AM*, Additive manufacturing; *LPBF*, laser powder bed fusion.

the warpage from the edges to the center. Shrinkage is an intractable issue in metallic AM components since shrinkage is not homogeneous in metallic AM components because the amount of contraction is different in each direction. Flaking is another type of distortion and results from the melt thin layer being peeled off from the precious layers or substrate due to their weak metallurgical bonding strength (Fig. 7.6B) [60].

7.5.1.5 Large surface roughness

The surface quality of components is crucial to certain aspects of the components' mechanical performance, such as fatigue limitation, wear resistance, mating, and sealing. Usually, metallic AM components cannot meet surface roughness requirements in engineering applications. This creates a need to undergo secondary postprocessing machining or polishing for most printed metallic components before engineering applications (Fig. 7.7A) [61]. The summary of the surface roughness of metallic AM components is listed in Table 7.6. Three major aspects determining the resultant surface roughness of printed components are STL file, material feedstock, and processing parameters.

STL is the industry standard file type for AM. 3D model of components needs to be exported as STL files from their native CAD software. During the export process, their 3D models are converted into the machine language of G-code through a process called "slicing," which makes STL files printable. STL uses a series of triangles to represent solid 3D models of the components. A bunch of triangles makes up the contours of models, making the surface of components rougher.

The morphology (size and shape) and type (powder and wire) of the material feedstock play vital roles in determining the surface roughness of printed components. For powder-fed AM, powder feedstock will adhere to the surface of components, so their size and shape directly influence the

276 Digital Manufacturing

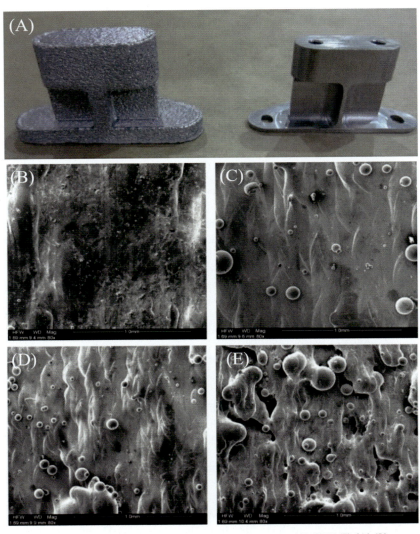

FIGURE 7.7 Surface topography of metallic AM components: (A) EBM Ti-6Al-4V components before and after machining [61] and LPBF TiC/Ti composite parts with (B) smooth surface (C and D) rough surface with a small number of fine balls, and (E) rough surface with large-sized balls using scan speeds from 0.1 to 0.4 m/s [62]. *AM*, Additive manufacturing; *EBM*, electron beam melting; *LPBF*, laser powder bed fusion.

surface roughness of the workpiece. The quality of the powder is another factor because low-quality powders with nonspherical shapes readily agglomerate, exacerbating the surface quality of printed components. For the wire-fed AM the diameter of wire feedstock significantly impacts the surface

Metal additive manufacturing Chapter | 7 **277**

TABLE 7.6 Summary of the surface roughness for metallic additive manufacturing components.

Process	XY surface roughness (μm)	Z surface roughness (μm)
PBF	20–200	20–200
Powder-fed DED	100–1000	100–1000
Wire-fed DED	2000–5000	1000–10000
BJ	20–65	50–100

BJ, Binder jetting; *DED*, direct energy deposition; *PBF*, powder bed fusion.

finishing of printed components. Generally, the surface quality of components manufactured from powder-fed feedstock is much better than that printed from wire-fed feedstock.

Nonoptimal printing parameters can cause the occurrence of the balling phenomenon, which dramatically deteriorates the surface quality of printed components (Fig. 7.7B−E) [62]. The balling effect stems from the spheroidization of the melt pool caused by the unstable molten pool using inappropriate printing parameters. There are two types of balling: ellipsoidal balls with a mean diameter of ~ 500 μm and spherical balls with an average diameter of ~ 10 μm. The coarse balls originate from discontinuous melt tracks, which arise from inferior wettability between molten pools and substrates (or previous layer). The formation of fine balls is attributed to the unstable molten pools, leading to the splashing of the molten metal. Balling is susceptible to scan speed, and balling will become more severe with an increase in scan speed, which triggers the increasingly unstable molten pools. These balls will hamper the uniform spreading of a new layer of powder on the previous layer and aggravate the surface roughness of components.

7.5.2 Defects detection techniques

There are many available techniques to detect defects in metallic AM components. The principle of techniques suitable for metallic AM components is listed in the subsection, and these techniques are classified based on the measurement features and principles. The advantages and drawbacks of each technique are elucidated to help readers find effective methods for specific defects measurement.

7.5.2.1 Residual stress measurement

Residual stress detection can be classified into destructive and nondestructive experimental measurement methods. The destructive techniques remove material from the detected specimens to release partial stresses, to break the equilibrium state of residual stresses within specimens. The remaining specimens then experience deformation due to the redistribution of residual stresses to form a new equilibrium state. The resultant deformation in the remaining specimens is recorded and used to back-calculate the initial residual stresses. The destructive methods can be subcategorized into complete destructive and semidestructive techniques. The semidestructive techniques remove only a small amount of material from components while retaining their structural integrity.

The nondestructive techniques measure the physical or crystallographic properties of the specimens and then obtain the near-surface and volumetric residual stresses of components via theoretical calculation. Neutron diffraction, synchrotron diffraction, X-ray diffraction, magnetic, and ultrasonic are common techniques for nondestructive methods. The first three utilize diffraction technology to inspect spacing changes in the lattice of measured components and deduce their residual stresses. The ultrasound technique detects the residual stresses by measuring variations in the speed of ultrasound waves passing through the specimens. The magnetic technique depends on the interaction between magnetization and elastic strain in ferromagnetic materials. The categories and measurement size capacities of destructive and nondestructive techniques are summarized in Fig. 7.8.

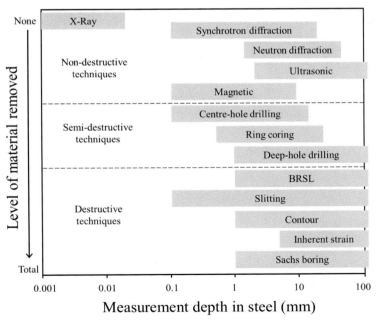

FIGURE 7.8 Summary of categories and measurement size capacities of residual stress detection techniques.

7.5.2.2 Porosity detection

Porosity in metallic AM components can be detected quantitively via nondestructive methods such as Archimedes' principle, gas pycnometry, and micro-computed tomography (micro-CT), and destructive methods (microscopic cross-sectional analysis).

Archimedes' principle and gas pycnometry are density-based testing methods, which cannot measure porosity characteristics, such as pore distribution, sizes, and morphologies. In Archimedes' principle the porosity of printed objects can be deduced by comparing measured real density utilizing the physical law of buoyancy to their theoretical densities. The principle of gas pycnometry, in which the density of objects can be obtained by measuring their mass and volume (using gas displacement), is similar to Archimedes' principle. The major advantages of both density-based methods are time-saving, easy to implement, and nondestructive.

The direct porosity measuring method, micro-CT, can reconstruct a 3D voxel model of the target objects with high resolution of internal structures via X-ray. The merit of micro-CT is its capability of providing visualized pore distribution, quantitative pore sizes, and overall porosity values of entire objects. The weaknesses of micro-CT are its high equipment cost, intricate data processing, and limited measurement size range.

The principle of the microscopic cross-sectional method is dependent on the capture of the cross-sectional morphologies of ground and polished target samples via an OM or SEM. Subsequently, the porosity can be determined through computing ratios between areas of black pixels and white pixels based on the microscopic cross-sectional morphologies using an image processing software. This method can merely obtain rough porosity values, as the quantitative analyses are easily influenced by the grinding and polishing preparation and thresholding algorithms set in the image processing software.

7.5.2.3 Printing accuracy determination

There are two main kinds of 3D measurement techniques: contact and noncontact.

- Contact 3D measurement techniques probe objects by collecting position points of their surfaces through physical touch. Contact 3D measurement can collect higher resolution and more accurate geometrical data when compared with the noncontact techniques. Unfortunately, the wide application of this process in industries is impeded by long measurement times, its limited capability of measurement sizes, and geometrical complexity.
- Noncontact 3D measurement captures the geometrical shape of objects through a media (such as laser, structural light) without physical contact. The detected geometrical information is then converted into digital data. Laser, structured light, and photogrammetry solutions are the representative and common noncontact 3D measurement techniques in engineering applications.

280 Digital Manufacturing

Laser 3D scanners generally comprise three main elements, a laser transmitter, a camera receiver, and a target object. The laser transmitter casts a laser beam upon the target object, and then the camera receiver can record where the beam and object intersect. The advantages of the laser-based technique are its low price, fast measurement speed (less than 10 minutes on average for an object), and desirable precision level (of the order of 0.01 mm). As for the disadvantages, it should be noted that it is difficult to measure transparent objects or reflective surface, and limited measurement size ranges (only a few meters).

Structured-light (or white-light) 3D scanners project light patterns onto target objects. The position information of light patterns is subsequently read by a camera. Several images will be captured and acquired from the various positions in real time until the reconstruction is completed. In contrast with the laser-based technique that gathers data via a laser line or a point at a time, the structured-light technique exploits white- or blue-light technology to capture the entire area using multiple patterns and images. Thus it makes the structured-light measurement faster and more efficient. The structural light is also safe for people, even to the naked eye.

Photogrammetry captures the images of inspected objects from multiple angles using digital cameras. The software can merge the captured images and generate a point cloud based on the relative positions of the pixels in captured images. The advantages of this method are easy manipulation and the capability of offering colorful information. However, the scans are limited to low resolution and accuracy.

7.5.2.4 Surface topography measurement

There are four common surface topography measurement methods to detect the surface roughness of the printed metallic components: stylus profilometer (SP), atomic force microscope (AFM), white-light interferometer (WLI), and laser confocal scanning microscope (LCSM). The former two are contact techniques, and the latter two belong to noncontact techniques.

- SP method uses a tip probe to detect the surface roughness of a sample by physically moving it along the surface of the sample to acquire the vertical displacements of the probe. The vertical displacements of the probe can then be used to reconstruct the surface topography of the measured sample.
- AFM employs a cantilever equipped with a sharp tip (probe) at its end to scan the specimen surface. During measurements, when the tip approaches the surface of the sample, forces between tip and sample will lead to a displacement of the cantilever. A monitor gives deflection feedback to the piezo scanner that is used to maintain a constant contact force. The cantilever displacement provides feedback to the piezo scanner that is measured to obtain the vertical information of the sample surface, representing the surface topography of the target sample.

- WLI adopts the interference effects that occur when the light reflected from the sample is superimposed with the light reflected by a high-precision reference mirror to measure the surface roughness of a sample.
- LCSM uses a laser as the light source to emit a laser beam and scan the sample surface. The laser receiver then detects the reflection information of the position in the focus of the confocal optical system. The 3D profile of the sample surface is created by accumulating the in-focus position information in the vertical direction.

The technological characteristics, advantages, and disadvantages of these four common surface topography measurement methods are summarized in Table 7.7.

7.6 Postprocessing

Postprocessing is often required after the completion of printing processes for metallic AM components to meet specific specifications or achieve enhanced properties, such as surface quality, geometrical accuracy, and mechanical properties. The general postprocessing approaches for metallic AM components can be classified into (1) removal of surficial adhesive excess powders, support structures, and substrate plates; (2) heat treatment; and (3) surface finishing.

7.6.1 Removal of adhesive powders, support structures, and substrate plates

After printing, excessive powder particles that are adhered to the surface of printed parts and piled up in the internal channels or structural holes of parts must be cleaned off before subsequent processing, particularly for the PBF parts. Sand or bead blasting is commonly applied to remove adherent powder particles and is also an effective surface finishing technique for metallic AM parts. Nevertheless, sand or bead blasting is not efficient for many complex-shaped parts, especially for parts with internal channels or structural holes. An alternative for cleaning such parts should combine an ultrasonic with an appropriate cleaning solvent.

Support structures are one of the most important elements for creating freeform and intricate geometries in AM technology. Parts with complex design features like overhangs, holes, and bridges are likely to collapse during the printing process if such features are not supported. Support structures can ensure these features remain attached to the main body of the printed parts and prevent structural collapse during printing. Support structures in metal parts are typically removed via sand or bead blasting and machining.

Substrate plates provide a platform to adhere to and immobilize the printed metallic objects. After printing the metallic objects are usually removed from the substrate plate using wire electrical discharge machining.

TABLE 7.7 Technological characteristics, advantages, and disadvantages of the four common surface topography measurement methods.

Method	Resolution (nm)	Height measurement capability	Advantage	Disadvantage
SP	1	< 1 mm	• Clear wave profile • Capability of long-distance measurement	• Time-consuming • Inability to measure viscous samples • Stylus wear
AFM	0.01	< 10 μm	• High resolution • Capability of analyzing physical properties (electrical property, magnetic property, friction, viscoelasticity, etc.)	• Inability to measure large samples • Difficulty in operations
WLI	0.1	A few mm	• Rapid measurement • Capability of measuring large-size objects	• Limitation in certain objects with a good reflection • Low resolution for XY stage measurements
LCSM	0.1	A few mm	• Colorful observation • Easy operation	• Incapability of high-precision measurements below 1 nm • Incapability of measuring materials that absorb laser beam wavelength

AFM, Atomic force microscope; *LCSM*, laser confocal scanning microscope; *SP*, stylus profilometer; *WLI*, white-light interferometer.

7.6.2 Heat treatment

Heat treatments are essential and critical postprocessing for metallic AM structural components in industrial applications. The thermal processing aims to relieve residual stresses, modify microstructures, and close pores for metallic AM components to obtain stable mechanical performance and longer service life.

In general, a relatively low-temperature (below the phase transformation temperature) treatment, followed by furnace cooling, is employed for the residual stress relief of metallic AM components. There is no standardized thermal treatment that can be applied to every material. The selection of thermal parameters is strongly dependent on the types of materials and sizes of components. Moreover, residual stress cannot be relieved completely even with a long postprocessing time because it is not able to eliminate all crystal defects within printed components.

The rapid cooling rates in the metal AM processes result in the formation of undesirable nonequilibrated phases or microstructures in components. These nonequilibrated phases or microstructures severely deteriorate the mechanical properties of metal components. Thermal postprocessing is a common approach to resolve these metallurgical defects through the transformation of nonequilibrium phases or microstructures to equilibrate phases or microstructures. Temperatures for microstructure modification are always higher than the phase transformation temperature of target materials. However, available standard thermal postprocessing procedures are limited to conventionally manufactured components but are not suitable for the metal AM components. At the moment, there are no unified criteria for microstructure—modification heat treatments for metal AM components. The heat treatment process needs to be designed according to the initial microstructure and resultant microstructure to suit the mechanical properties required in specific applications.

Hot isostatic pressing (HIP) is one of the material processing and postprocessing methods. HIP can compress materials by combining high temperatures (from several hundred to 2000°C) and isostatic gas pressures (from several tens to 200 MPa) in a high-pressure containment vessel. This technique is widely applied to eliminate pores in metal AM parts and thus increases the ductility and fatigue resistance of parts [63].

7.6.3 Surface finishing

Media blasting, laser shock peening, vibratory/rotary/tumbling finishing, and CNC machining are the prevalent methods for improving the surface finish of metal AM components.

Media blasting is a process that uses pressurized air to shoot pieces of abrasive materials (sand, grit, and ceramic beads) out of a nozzle. The process is a simple way to reduce surface roughness. However, it is labor-intensive and hard to achieve a uniform and precise finish on entire parts.

Laser shock peening has emerged as an alternative surface treatment capable of overcoming the limitations of conventional media blasting. During this process a solid-state laser beam is pulsed on a metallic surface. Shock waves are then generated, traveling throughout the component and causing plastic deformation.

Vibratory/rotary/tumbling finishing methods oscillate a chamber of variously shaped abrasives to create continuous friction between abrasives and submerged parts. These approaches can achieve an even smooth finish without needing special jigs or fixtures, even for large-scale components.

CNC machining is also a common approach to improve tolerances and surface finish of metallic AM parts.

7.7 Applications

AM can provide revolutionary processing strategies that enable the cohesive integration of design freedom, material diversity, and green manufacturing. AM is a disruptive technology, but it is not expected to replace conventional technology used to manufacture large quantities of low-cost identical products. It is a desirable alternative approach to complement traditional methods in the fabrication of high-value, customized, and small-scaled products that can only be produced at the expense of high-cost, time-consuming, or complicated processing procedures via traditional technology. AM is one of the most important cornerstones for Manufacturing 4.0 and widely applied in many advanced domains, such as aerospace, automotive, healthcare, marine, and energy, as shown in Fig. 7.9 [64].

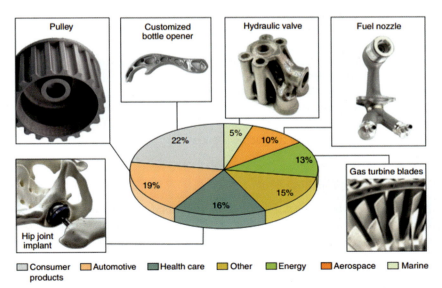

FIGURE 7.9 Metal AM applications in various industries and the revenue distribution among these industries [64]. *AM*, Additive manufacturing.

This section presents typical innovative applications of metallic AM components in the aerospace, automotive, and health-care domains in detail.

7.7.1 Aerospace

AM can realize flexibility in design, size, and material, as well as high material utilization rate, fast production, and delivery. These features make it ideal for aerospace engineering and stimulate the aerospace industry to become one of the earliest adopters of AM technology starting from 1989 [65]. AM has evolved into a strategic technology in the aerospace field and brought about high profit throughout the aerospace supply chain to date. With the rapid development of AM technologies, applications in aerospace have transformed from initial rapid prototyping to current component manufacturing and repair. During the manufacture of critical components in aircraft, AM manifests significant advantages like reductions in weight, cost, and delivery time and improvement in the fuel efficiency and service life of aircraft.

7.7.1.1 Direct component manufacturing

The high-precision PBF techniques are always adopted for the manufacture of small but geometrically complex components. The manufacturing flexibility drives engineers to design the components with an integral structure and optimized topology. For instance, engineers fully exploit the design freedom driven by LPBF to redesign a bracket connector in the Airbus A350 XWB with a porous bionic structure. The redesigned connector made of Ti alloy via a LaserCUSING system is 30% lighter than its milled or cast predecessor made of Al alloy (Fig. 7.10A) [66]. These variations reduce fuel consumption and improve the load capacity of the aircraft. More importantly, due to the tool-less manufacturing by employing LPBF, it reduces the production cost and shortens the research and development time by up to 75% compared to conventional processing routes.

General Electric Aviation (GE) in collaboration with EOS redesigned and manufactured fuel nozzles for their next-generation LEAP jet engines using LPBF decades ago [70]. The number of the modified fuel nozzle was reduced from 18 to a whole piece, which decreased the weight of the nozzle by 25% relative to that of the predecessor. LPBF enables the modified nozzle with more intricate cooling pathways and support ligaments, endowing the nozzle with a fivefold increase in durability and a $\sim 15\%$ improvement in fuel efficiency. Besides, GE acquired Avio Aero in 2013 and developed TiAl blades (Fig. 7.10B) fabricated by EBM for GE9X engines applied in Boeing's next-generation 777X jets [67]. The TiAl blades, with about half of the weight of conventional counterparts made of nickel alloy, can bear a spin

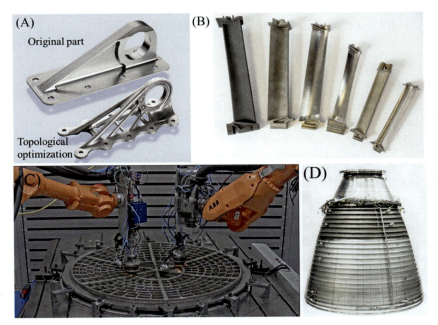

FIGURE 7.10 Applications of direct component manufacturing in aerospace: (A) original and topologically optimized bracket connectors for the Airbus A350 XWB [66], (B) TiAl blades produced by EBM for GE9X engines [67], (C) a demonstrator of mirror components for satellite applications with a diameter of 1.5 m [68], and (D) printed advanced the Ariane 6 nozzle (SWAN) with a diameter of 2.5 m [69]. *DED*, Direct energy deposition; *EBM*, electron beam melting.

inside the engine at 2500 times per minute under searing heat and titanic forces.

Unlike PBF, DED is adopted to fabricate large-scale structural components due to the poor printing resolution and fast deposition rate [71]. In 2003 Boeing produced a Ti alloy pylon rib for F-15 fighter jets using a self-developed DED system [72]. The Ti alloy pylon rib is the first 3D-printed metal part that is qualified and installed on a military aircraft. In 2017 Boeing collaborated with Norsk Titanium to print structural Ti alloy components using a wire-based DED (named RPD by Norsk Titanium) system [73]. The structural Ti alloy components were certified with FAA certification deliverables on Boeing 787 Dreamliner through a rigorous testing program. This prompted Boeing 787 Dreamliner to become the first commercial airplane equipped with certified AM Ti alloy parts in a structural application. In the same year, according to the requirements of Thales Alenia Space, Norsk Titanium completed and passed mechanical evaluation testing for their DED structural components. Adoption of DED makes Thales Alenia Space successfully reduce buy-to-fly ratios by half and product lead times by 6 months for their parts.

Fraunhofer IWS, ESA, and Airbus made a demonstrator of mirror components for satellite applications with a diameter of 1.5 m (Fig. 7.10C) using their developed hybrid AM system, which merged the high flexibility of DED with the precision of cryogenic machining and showed the feasibility of the whole process chain for the manufacturing of large-volume titanium components [68]. GKN Aerospace printed the first advanced Ariane 6 nozzle (SWAN) with a diameter of 2.5 m (Fig. 7.10D) for the Vulcain 2.1 engine manufactured by Airbus Safran Launchers in France [69]. The cost of the nozzle made by the DED process was reduced by 40%, and the number of parts was decreased from approximately 1000 to 100.

7.7.1.2 Damage component repair

High-performance components in the aerospace industry, such as compressors, blisks, turbines, and blades, are made from high-value materials and produced through complex processing procedures. These components have to serve in extremely harsh environments (like corrosion, impacts, and variable thermal cycles), which easily cause the damage and failure of the components. When compared with the manufacture of new components, the repair and remanufacturing processes can be considered a more effective alternative for the damaged components in terms of manufacturing cost and delivery time. It has been reported that the energy consumption and labor used for repairing an engine are 55% and 67% of those required for fabricating a new engine, respectively [74].

The fundamental principle of AM repair technology is to utilize a laser source DED system that adds repair powder material to build the damaged geometry and then generate a metallurgical bonding interface between the damaged component and the restored part.

In cooperation with seven universities and research institutes, three aeroengine and component manufacturers, five industrial partners, and three repair centers as end users, Fraunhofer Institute launched the project FANTASIA in the funding support of the European Commission in 2006 [75]. FANTASIA, which stands for flexible and near-net-shape generative manufacturing chains and repair techniques for complex-shaped aeroengine parts, aims to decrease repair cost and turnaround time for aeroengine parts by at least 40%. Many damaged components, which had to be discarded formerly owing to the lack of a feasible repair approach, can be successfully restored with the assistance of DED currently. Damaged bosses, brackets, and flanges (Fig. 7.11A) in the BR715 high-pressure turbine case are reconditioned using a DED

FIGURE 7.11 Applications of damage component repair in aerospace [76]: (A) repairing the damaged flange of a BR715 high-pressure turbine case and (B) repairing damaged damping wire walls of the Br715 HPC front drum.

process [76]. Damaged damping wire grooves of the Br715 HPC front drum are also remanufactured by restoring a local groove wall (Fig. 7.11B) [76]. It is reported that 15 different repair components have obtained the application certification from Rolls-Royce.

Optomec repaired a T700 blisk, which was subjected to corrosion on the airfoil's leading edges during service, using the DED process [77]. RPM Innovations showcased that a Ti-6Al-4V bearing housing in a gas turbine engine, the bearing seat of which was in an out-of-tolerance condition caused by the over abrasion, was restored using DED and completed an evaluation trial in an engine. The cost of bearing housing repair was reduced by half, and the delivery time was decreased from several weeks to a few days in contrast with manufacturing a new one.

7.7.2 Automotive industry

AM technology has opened a new avenue for the design and manufacture of lighter, stronger, and safer products with lower costs and shortened lead times for the automotive sector. Many renowned and innovative automotive companies (such as BMW, Bugatti, Audi) have been in cooperation with the leaders of the AM domain (such as HP, Stratasys, 3D Systems, EOS, and Envision TEC) to explore more efficient AM supply chains for the mass production of end-use parts [3,78]. Furthermore, many CAD software companies (Autodesk, Dassault Systèmes, and Siemens Software) participated in reducing the weight of components during the product design phase. The metal AM applications in the automotive industry have developed and evolved from initial rapid prototyping to current high-performance structural components, such as chassis, transmission, and engine components.

7.7.2.1 Commercial vehicles

MIMO Technik exploits the advantage of fabrication flexibility to design and manufacture two typical components in the Porsche 911 turbo via LPBF. The first case is a turbo exhaust manifold with a wall of 1 mm in thickness that is printed from Inconel 625 powder (Fig. 7.12A) [79]. The new design endows the manifold with a reduced assembly and optimized gas flow characteristics in contrast to its conventionally processed counterparts. The usage of heat-resistant Inconel 625 meant that the manifold could be designed thinner relative to the previous equivalent made of stainless or cast steel, which reduces the weight as well. The second one is an LPBF printed turbo intercooler end

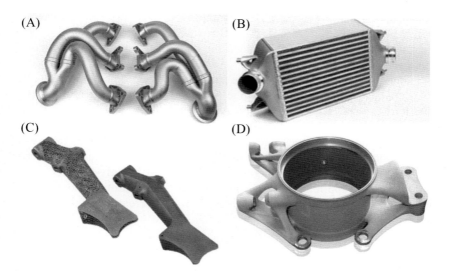

FIGURE 7.12 Applications of metal AM in the automotive industry: (A) the LPBF turbo exhaust manifold [79], (B) the LPBF turbo intercooler end tank [79], (C) the conventional and developed brake pedal [80], and (D) the LPBF AlSi10Mg knuckle [81]. *LPBF*, Laser powder bed fusion.

290 Digital Manufacturing

tank with a 1.8-mm wall thickness from AlSi10Mg powder (Fig. 7.12B) [79]. The modified tanks are capable of a faster cooling rate and lighter weight than the predecessor. Volkswagen, one of the world's largest and most innovative vehicle makers, has adopted HP Metal Jet technology to produce high-performance functional parts with specific structural requirements, such as gearshift knobs. In addition, Volkswagen's long-term plans with HP's technology involve accelerating manufacturing processes for mass-customizable parts like key rings and exterior name plates.

7.7.2.2 Racing vehicles

Motorsports is the cradle of many ground-breaking technological developments due to the high requirements of innovation and security. This spurred AM to play a paramount role in this field. In the Formula 1 (F1) race, Williams Martini Racing cooperated with EOS to develop a brake pedal via LPBF for their F1 race car as a conventional Al alloy brake pedal is brittle (Fig. 7.12C) [80]. The team succeeded in creating several fast iterations to achieve topology optimization with a lightweight design for the brake pedal. The newly designed brake pedal with lighter weight and lower costs is produced from Ti-6Al-4V powder by LPBF. In 2014 engineers from Rennteam Uni Stuttgart used LPBF to revamp an AlSi10Mg knuckle that connects the vehicle's axle, bearings, and brake system for their Formula Student Germany car (Fig. 7.12D) [81]. Based on loading forces during usage, lightweight design and topological optimization were developed for the knuckle with a hollow structure. The new knuckle was 20% stronger and 35% lighter than the cast equivalent, which aided the team to win the Formula Student Germany competition.

English Racing team in Oregon encountered a problem regarding its Mitsubishi Evo race car, the engine of which always failed immediately due to the overly high oil pressure. A new under-pulley gear with a larger diameter that would turn slower to lower the oil pressure was proposed. However, long lead times and high casting mold costs led to failure in production. With the application of LPBF the team overcame the processing challenges in which the gear was fabricated in only 5 hours. After the successful application of the printed gear, 35 additional components were successfully produced using LPBF for different race applications across the country.

7.7.3 Healthcare

Patient-specific products can be more beneficial for the patient's recovery, compared with using standardized products designed for the general public. These products have also attracted considerable attention from the biomedical industry. The conventional techniques can not realize the mass production of custom-made products with geometrical complexity demanded in biomedical engineering applications. Based on patient-specific demands, AM technology can create geometrically complex structures with a wide range of materials without restrictions. Rapid and accurate manufacturing of on-demand products can simplify the reconstruction of the biostructures and expedite the therapy implementation.

7.7.3.1 Dentistry

SmarTech Publishing forecasts the annual growth of AM applications in the dental market will reach $\sim 35\%$, and the total amount will increase to USD 9.5 billion by 2027 [82]. The applications of metal AM in dentistry involve the manufacture of jawbones and dentures (including crowns and bridges).

In some cases, patients need to implant a new chin and jaw to replace their damaged chin and jaw. Belgian company LayerWise (required by 3D System) produced an entire replacement Ti jaw (Fig. 7.13A) for an elderly woman [83]. The complex-shaped patient-specific implant was manufactured integrally via LPBF and omitted the traditional assembly process. After surgery the printed Ti jaw allowed the patient to speak and swallow as usual.

Metal AM is also prevalent in producing partial and entire dentures. Dentures are usually made by conventional investment casting [87]. The casted products generally suffer from casting metallurgical defects, shape distortion, high costs, and long processing time. The AM fabricated dentures must be suitable for patients without compromise in shape and quality. With the development of 3D scanning and image processing techniques, 3D reconstruction of dentures enables AM technology to become a feasible and applicable method for the manufacture of dentures with high accuracy and repeatability. As many as 10,000,000 metallic AM dental parts are produced each year [84]. CoCr-based alloys are generally used as the materials for the dentures since their mechanical properties, corrosion resistance, and biocompatibility meet the requirements of dental implants. Many AM equipment manufactures devote their work to exploring specialized metallic materials and printers for dental manufacturing. For example, EOS developed a Cobalt Chrome SP2 material for dental applications. EOS stated that they could produce 450 units within 24 hours and 80,000 units within a year with one machine at a low cost. Fig. 7.13B showcases the building platform of an EOS M 100 machine with dental crowns and bridges [84].

7.7.3.2 Orthopedic implants

The human skeleton is composed of two types of bones, compact (cortical) bone and cancellous (trabecular or spongy) bone. Both are made of porous structures beneficial for the growth of osteocytes, blood vessels, and so on. Hence, metallic orthopedic implants with appropriate pore sizes and porosity

292 Digital Manufacturing

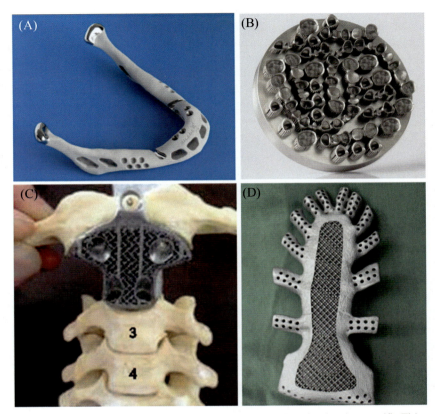

FIGURE 7.13 Applications of metal AM in healthcare: (A) the printed patient-specific Ti jaw implant [83] and (B) the printed CoCr-based dental crowns and bridges [84], (C) the printed customized porous Ti alloy vertebral implant inserted between C1 and C3 vertebras [85], and (D) the printed custom-made titanium prosthesis [86].

are considered to be ideal bone substitutes. First, porosity can make implants penetrable for cell migration, tissue, and vessel in-grown, facilitating the coalescence between implants and surrounding bones. Second, the porosity design can ensure implants with compatible mechanical properties that can significantly reduce stress-shielding between implants and natural bones when compared to traditionally nonporous implants. Nowadays, AM has evolved into a desirable alternative for rapidly producing patient-specific implants with adjustable pore sizes and porosity in medical applications.

Peking University Third Hospital designed and printed a self-stabilizing and customized porous Ti alloy vertebral implant (Fig. 7.13C) inserted between C1 and C3 vertebras via EBM for a 12-year-old boy who suffered from Ewing sarcoma [85]. The patient was able to ambulate after the 7-day postoperative recovery. Additionally, it was evidenced that the implant osseointegration was excellent, and the implant was stably fixed without subsidence or displacement through CT inspection 1 year later after the surgery.

A large anterior chest wall defect was cured successfully by transplanting a custom-made porous Ti64 chest wall implant (Fig. 7.13D) made via LPBF [86]. The implanted artificial sternum showed good osseointegration, and there were no complications in the following 3 months. Customized metallic implants used in hip, knee, and foot joints were also reported. These cases reflect that patients can benefit from the quick and one-time construction of the personalized chest wall implant using AM in terms of time-saving and efficient therapy.

7.8 Conclusion and perspectives

In this chapter the common metal AM techniques such as PBF and DED were presented with a focus on the relevant fundamental concepts and typical applications. Indeed, metal AM has been widely used in various high-end manufacturing industries. However, as an emerging technology, many scientific and technological challenges are not yet understood and addressed thoroughly. The theoretical such as heat and mass transfer between high energy sources and metal materials, defect evolution, and their effect on the performance of printed components, ultrafast solidification behavior of molten pools, nonequilibrate grain structure, and heterogeneous microstructure need to be investigated in depth. In industrial production the qualification of products is established by trial-and-error testing methods that are time-consuming, laborious, and costly due to the lack of AM material property database and standards. The printing parameters or strategies optimized from small coupons are difficult to be applied to the manufacture of large and complex components because the processing parameters and scanning strategies are strongly dependent on the sizes and geometries of components. Thermal treatment to modify the microstructure of as-printed metallic components is imperative. However, the relevant studies are still at an initial stage, resulting in the lack of criteria in the industrial field.

There are several helpful solutions or research directions to overcome these scientific and technological challenges potentially. Mechanistic models, which can correctly predict the solidification behavior, distribution, and magnitudes of residual stresses, evolutions of defects, microstructure, and properties for metallic AM components, should be developed and utilized. A combination of emerging digital tools regarding machine learning and big data to establish optimized processing windows and material property databases for metallic AM components should also be taken into consideration in priority. Furthermore, the advancement of metallurgical science and technology and the maturity of computer hardware and software can play a synergistic role in solving these scientific and technological issues.

294 Digital Manufacturing

References

[1] J. Gonzalez-Gutierrez, S. Cano, S. Schuschnigg, C. Kukla, J. Sapkota, C. Holzer, Additive manufacturing of metallic and ceramic components by the material extrusion of highly-filled polymers: a review and future perspectives, Materials 11 (5) (2018) 840.

[2] D. Bourell, J.P. Kruth, M. Leu, G. Levy, D. Rosen, A.M. Beese, et al., Materials for additive manufacturing, CIRP Annals 66 (2) (2017) 659−681.

[3] T. DebRoy, H.L. Wei, J.S. Zuback, T. Mukherjee, J.W. Elmer, J.O. Milewski, et al., Additive manufacturing of metallic components−process, structure and properties, Progress in Materials Science 92 (2018) 112−224.

[4] Y. Kok, X.P. Tan, P. Wang, M.L. Nai, N.H. Loh, E. Liu, et al., Anisotropy and heterogeneity of microstructure and mechanical properties in metal additive manufacturing: a critical review, Materials & Design 139 (2018) 565−586.

[5] D. Gu, X. Shi, R. Poprawe, D.L. Bourell, R. Setchi, J. Zhu, Material-structure-performance integrated laser-metal additive manufacturing, Science 372 (6545) (2021) 1−15.

[6] M. Rosenberg, K. Ilić, K. Juganson, A. Ivask, M. Ahonen, I.V. Vrček, et al., Potential ecotoxicological effects of antimicrobial surface coatings: a literature survey backed up by analysis of market reports, PeerJ 7 (2019) e6315.

[7] T. Wohlers, T. Gornet, History of additive manufacturing, Wohlers Report 24 (2014) 118.

[8] C. Meier, R. Weissbach, J. Weinberg, W.A. Wall, A.J. Hart, Critical influences of particle size and adhesion on the powder layer uniformity in metal additive manufacturing, Journal of Materials Processing Technology 266 (2019) 484−501.

[9] A. Gisario, M. Kazarian, F. Martina, M. Mehrpouya, Metal additive manufacturing in the commercial aviation industry: a review, Journal of Manufacturing Systems 53 (2019) 124−149.

[10] C. Gao, S. Wolff, Eco-friendly additive manufacturing of metals: Energy efficiency and life cycle analysis, Journal of Manufacturing Systems 60 (2021) 459−472.

[11] I. Gibson, D. Rosen, B. Stucker, Materials for additive manufacturing, in: Additive Manufacturing Technologies, Springer, Cham, 2021, pp. 379−428.

[12] G. Chen, S.Y. Zhao, P. Tan, J. Wang, C.S. Xiang, H.P. Tang, A comparative study of Ti-6Al-4V powders for additive manufacturing by gas atomization, plasma rotating electrode process and plasma atomization, Powder Technology 333 (2018) 38−46.

[13] A. Röttger, K. Geenen, M. Windmann, F. Binner, W. Theisen, Comparison of microstructure and mechanical properties of 316 L austenitic steel processed by selective laser melting with hot-isostatic pressed and cast material, Materials Science and Engineering: A 678 (2016) 365−376.

[14] E. Chlebus, K. Gruber, B. Kuźnicka, J. Kurzac, T. Kurzynowski, Effect of heat treatment on the microstructure and mechanical properties of Inconel 718 processed by selective laser melting, Materials Science and Engineering: A 639 (2015) 647−655.

[15] D. Buchbinder, Generative fertigung von aluminiumbauteilen für die serienproduktion, in: AluGenerativ Abschlussbericht BMBF 01RIO639A-D, 2010.

[16] T. Vilaro, C. Colin, J.D. Bartout, As-fabricated and heat-treated microstructures of the Ti-6Al-4V alloy processed by selective laser melting, Metallurgical and Materials Transactions A 42 (10) (2011) 3190−3199.

[17] C. Wang, X. Tan, E. Liu, S.B. Tor, Process parameter optimization and mechanical properties for additively manufactured stainless steel 316L parts by selective electron beam melting, Materials & Design 147 (2018) 157−166.

[18] M.M. Kirka, F. Medina, R. Dehoff, A. Okello, Mechanical behavior of post-processed Inconel 718 manufactured through the electron beam melting process, Materials Science and Engineering: A 680 (2017) 338–346.

[19] H. Rafi, N. Karthik, H. Gong, T.L. Starr, B.E. Stucker, Microstructures and mechanical properties of Ti6Al4V parts fabricated by selective laser melting and electron beam melting, Journal of Materials Engineering and Performance 22 (12) (2013) 3872–3883.

[20] J.S. Keist, T.A. Palmer, Role of geometry on properties of additively manufactured Ti-6Al-4V structures fabricated using laser based directed energy deposition, Materials & Design 106 (2016) 482–494.

[21] X. Zhao, J. Chen, X. Lin, W. Huang, Study on microstructure and mechanical properties of laser rapid forming Inconel 718, Materials Science and Engineering: A 478 (1–2) (2008) 119–124.

[22] L. Costa, R. Vilar, Laser powder deposition, Rapid Prototyping Journal 15 (4) (2009) 264–279.

[23] R.N. Lumley, Technical data sheets for heat-treated aluminum high-pressure die castings, Die Casting Engineer (2008) 32.

[24] M. Koike, P. Greer, K. Owen, G. Lilly, L.E. Murr, S.M. Gaytan, et al., Evaluation of titanium alloys fabricated using rapid prototyping technologies—electron beam melting and laser beam melting, Materials 4 (10) (2011) 1776–1792.

[25] I. Yadroitsev, M. Pavlov, P. Bertrand, I. Smurov, Mechanical properties of samples fabricated by selective laser melting, in: 14èmes Assises Européennes du Prototypage & Fabrication Rapide, 2009.

[26] J. Alcisto, A. Enriquez, H. Garcia, S. Hinkson, T. Steelman, E. Silverman, et al., Tensile properties and microstructures of laser-formed Ti-6Al-4V, Journal of Materials Engineering and Performance 20 (2) (2011) 203–212.

[27] B. AlMangour, D. Grzesiak, J.M. Yang, Rapid fabrication of bulk-form $TiB_2/316L$ stainless steel nanocomposites with novel reinforcement architecture and improved performance by selective laser melting, Journal of Alloys and Compounds 680 (2016) 480–493.

[28] C. Pei, D. Shi, H. Yuan, H. Li, Assessment of mechanical properties and fatigue performance of a selective laser melted nickel-base superalloy Inconel 718, Materials Science and Engineering: A 759 (2019) 278–287.

[29] N.T. Aboulkhair, I. Maskery, C. Tuck, I. Ashcroft, N.M. Everitt, The microstructure and mechanical properties of selectively laser melted AlSi10Mg: the effect of a conventional T6-like heat treatment, Materials Science and Engineering: A 667 (2016) 139–146.

[30] X.Y. Zhang, G. Fang, S. Leeflang, A.J. Böttger, A.A. Zadpoor, J. Zhou, Effect of subtransus heat treatment on the microstructure and mechanical properties of additively manufactured Ti-6Al-4V alloy, Journal of Alloys and Compounds 735 (2018) 1562–1575.

[31] R. Alaghmandfard, C. Dharmendra, A. Hadadzadeh, B.S. Amirkhiz, A. Odeshi, M. Mohammadi, Dynamic compressive response of electron beam melted Ti−6Al−4V under elevated strain rates: microstructure and constitutive models, Additive Manufacturing 35 (2020) 101347.

[32] M.A. Bevan, A.A.H. Ameri, D. East, D.C. Austin, A.D. Brown, P.J. Hazell, et al., Mechanical properties and behavior of additive manufactured stainless steel 316L, Characterization of Minerals, Metals, and Materials 2017, Springer, 2017, pp. 577–583.

[33] T. Gualtieri, A. Bandyopadhyay, Laser engineering net shaping of microporous Ti6Al4V filters, Frontiers in Mechanical Engineering 2 (2016) 9.

[34] K. Yuan, W. Guo, P. Li, J. Wang, Y. Su, X. Lin, et al., Influence of process parameters and heat treatments on the microstructures and dynamic mechanical behaviors of Inconel 718 superalloy manufactured by laser metal deposition, Materials Science and Engineering: A 721 (2018) 215–225.

296 Digital Manufacturing

[35] Y. Zhong, L. Liu, S. Wikman, D. Cui, Z. Shen, Intragranular cellular segregation network structure strengthening 316L stainless steel prepared by selective laser melting, Journal of Nuclear Materials 470 (2016) 170−178.

[36] M. Kamariah, W. Harun, N. Khalil, F. Ahmad, M. Ismail, S. Sharif, Effect of heat treatment on mechanical properties and microstructure of selective laser melting 316L stainless steel, in: IOP Conference Series: Materials Science and Engineering, 257(1), 2017.

[37] W.M. Tucho, P. Cuvillier, A. Sjolyst-Kverneland, V. Hansen, Microstructure and hardness studies of Inconel 718 manufactured by selective laser melting before and after solution heat treatment, Materials Science and Engineering: A 689 (2017) 220−232.

[38] W. Li, S. Li, J. Liu, A. Zhang, Y. Zhou, Q. Wei, et al., Effect of heat treatment on AlSi10Mg alloy fabricated by selective laser melting: microstructure evolution, mechanical properties and fracture mechanism, Materials Science and Engineering: A 663 (2016) 116−125.

[39] L.A. Chicos, S.M. Zaharia, C. Lancea, M.A. Pop, I. Cañadas, J. Rodríguez, et al., Concentrated solar energy used for heat treatment of Ti6Al4V alloy manufactured by selective laser melting, Solar Energy 173 (2018) 76−88.

[40] I. Segura, J. Mireles, D. Bermudez, C. Terrazas, L. Murr, K. Li, et al., Characterization and mechanical properties of cladded stainless steel 316L with nuclear applications fabricated using electron beam melting, Journal of Nuclear Materials 507 (2018) 164−176.

[41] S. Goel, M. Ahlfors, F. Bahbou, S. Joshi, Effect of different post-treatments on the microstructure of EBM-built alloy 718, Journal of Materials Engineering and Performance 28 (2) (2019) 673−680.

[42] H. Galarraga, R.J. Warren, D.A. Lados, R.R. Dehoff, M.M. Kirka, P. Nandwana, Effects of heat treatments on microstructure and properties of Ti-6Al-4V ELI alloy fabricated by electron beam melting (EBM), Materials Science and Engineering: A 685 (2017) 417−428.

[43] J. Yu, M. Rombouts, G. Maes, Cracking behavior and mechanical properties of austenitic stainless steel parts produced by laser metal deposition, Materials & Design 45 (2013) 228−235.

[44] B.E. Carroll, T.A. Palmer, A.M. Beese, Anisotropic tensile behavior of Ti−6Al−4V components fabricated with directed energy deposition additive manufacturing, Acta Materialia 87 (2015) 309−320.

[45] E. Brandl, A. Schoberth, C. Leyens, Morphology, microstructure, and hardness of titanium (Ti-6Al-4V) blocks deposited by wire-feed additive layer manufacturing (ALM), Materials Science and Engineering: A 532 (2012) 295−307.

[46] E.L. Stevens, J. Toman, A.C. To, M. Chmielus, Variation of hardness, microstructure, and laves phase distribution in direct laser deposited alloy 718 cuboids, Materials & Design 119 (2017) 188−198.

[47] F. Bartolomeu, M. Buciumeanu, E. Pinto, N. Alves, O. Carvalho, F. Silva, et al., 316L stainless steel mechanical and tribological behavior—a comparison between selective laser melting, hot pressing and conventional casting, Additive Manufacturing 16 (2017) 81−89.

[48] <http://americancastingco.com/materials/>.

[49] K. Beyl, K. Mutombo, C. Kloppers, Tensile properties and microstructural characterization of additive manufactured, investment cast and wrought Ti6Al4V alloy, in: IOP Conference Series: Materials Science and Engineering, IOP Publishing, 2019, p. 012023.

[50] M. Ziętala, T. Durejko, M. Polański, I. Kunce, T. Płociński, W. Zieliński, et al., The microstructure, mechanical properties and corrosion resistance of 316 L stainless steel fabricated using laser engineered net shaping, Materials Science and Engineering: A 677 (2016) 1−10.

[51] S. Olovsjö, A. Wretland, G. Sjöberg, The effect of grain size and hardness of wrought alloy 718 on the wear of cemented carbide tools, Wear 268 (2010) 1045−1052.

[52] R. Barros, F.J. Silva, R.M. Gouveia, A. Saboori, G. Marchese, S. Biamino, et al., Laser powder bed fusion of Inconel 718: residual stress analysis before and after heat treatment, Metals 9 (12) (2019) 1290.

[53] I. Yadroitsev, I. Yadroitsava, Evaluation of residual stress in stainless steel 316L and Ti6Al4V samples produced by selective laser melting, Virtual and Physical Prototyping 10 (2) (2015) 67−76.

[54] T. Mukherjee, W. Zhang, T. DebRoy, An improved prediction of residual stresses and distortion in additive manufacturing, Computational Materials Science 126 (2017) 360−372.

[55] N.T. Aboulkhair, N.M. Everitt, I. Ashcroft, C. Tuck, Reducing porosity in AlSi10Mg parts processed by selective laser melting, Additive Manufacturing 1 (2014) 77−86.

[56] S.S. Sun, Q. Teng, Y. Xie, T. Liu, R. Ma, J. Bai, et al., Two-step heat treatment for laser powder bed fusion of a nickel-based superalloy with simultaneously enhanced tensile strength and ductility, Additive Manufacturing (2021) 102168.

[57] W.J. Sames, F.A. List, S. Pannala, R.R. Dehoff, S.S. Babu, The metallurgy and processing science of metal additive manufacturing, International Materials Reviews 61 (5) (2016) 315−360.

[58] J.H. Martin, B.D. Yahata, J.M. Hundley, J.A. Mayer, T.A. Schaedler, T.M. Pollock, 3D printing of high-strength aluminium alloys, Nature 549 (7672) (2017) 365−369.

[59] C. Li, J.F. Liu, X.Y. Fang, Y.B. Guo, Efficient predictive model of part distortion and residual stress in selective laser melting, Additive Manufacturing 17 (2017) 157−168.

[60] H.S. Park, D.S. Nguyen, Study on flaking behavior in selective laser melting process, Procedia CIRP 63 (2017) 569−572.

[61] S. Rawal, J. Brantley, N. Karabudak, Additive manufacturing of Ti-6Al-4V alloy components for spacecraft applications, in: 2013 6th International Conference on Recent Advances in Space Technologies (RAST), IEEE, 2013, pp. 5−11.

[62] D. Gu, Y.C. Hagedorn, W. Meiners, K. Wissenbach, R. Poprawe, Nanocrystalline TiC reinforced Ti matrix bulk-form nanocomposites by selective laser melting (SLM): densification, growth mechanism and wear behavior, Composites Science and Technology 71 (13) (2011) 1612−1620.

[63] A. Du Plessis, E. Macdonald, Hot isostatic pressing in metal additive manufacturing: X-ray tomography reveals details of pore closure, Additive Manufacturing (2020) 101191.

[64] T. Debroy, T. Mukherjee, J.O. Milewski, J.W. Elmer, B. Ribic, J.J. Blecher, et al., Scientific, technological and economic issues in metal printing and their solutions, Nature Materials 18 (10) (2019) 1026−1032.

[65] J. Edgar, S. Tint, Additive manufacturing technologies: 3D printing, rapid prototyping, and direct digital manufacturing, Johnson Matthey Technology Review 59 (3) (2015) 193−198.

[66] D. Dimitrov, E. Uheida, G. Oosthuizen, R. Blaine, A. Laubscher, A. Sterzing, et al., Manufacturing of high added value titanium components. A South African perspective. In IOP Conference Series, Materials Science and Engineering 430 (1) (2018) 012009.

[67] A. Sinha, B. Swain, A. Behera, P. Mallick, S.K. Samal, V. HM, A review on the processing of aero-turbine blade using 3D print techniques, Journal of Manufacturing and Materials Processing 6 (1) (2022) 16.

[68] B. Blakey-Milner, P. Gradl, G. Snedden, M. Brooks, J. Pitot, E. Lopez, M. Leary, F. Berto, A. du Plessis. Metal additive manufacturing in aerospace: A review. Materials & Design 209 (2021) 110008.

298 Digital Manufacturing

[69] T.D. Ngo, A. Kashani, G. Imbalzano, K.T. Nguyen, D. Hui, Additive manufacturing (3D printing): a review of materials, Methods, Applications and Challenges, Composites Part B: Engineering 143 (2018) 172−196.

[70] T. Kellner, Fit to Print: New Plant Will Assemble World's First Passenger Jet Engine With 3D Printed Fuel Nozzles, Next-Gen Materials, 29, General Electric, Boston, MA, 2014, p. 2016.

[71] J.C. Najmon, S. Raeisi, Review of additive manufacturing technologies and applications in the aerospace industry, Additive Manufacturing for the Aerospace Industry (2019) 7−31.

[72] National Research Council, Accelerating Technology Transition: Bridging the Valley of Death for Materials and Processes in Defense Systems, National Academies Press, 2004.

[73] Hybrid additive and subtractive manufacturing processes and systems: a review, Journal of Machine Engineering (2018) 18.

[74] A. Saboori, A. Aversa, G. Marchese, S. Biamino, M. Lombardi, P. Fino, Application of directed energy deposition-based additive manufacturing in repair, Applied Sciences 9 (16) (2019) 3316.

[75] R. Liu, Z. Wang, T. Sparks, F. Liou, J. Newkirk, Aerospace Applications of Laser Additive Manufacturing, Laser Additive Manufacturing, Laser Additive Manufacturing, Woodhead Publishing, 2017, pp. 351−371.

[76] A. Gasser, G. Backes, I. Kelbassa, A. Weisheit, K. Wissenbach, Laser metal deposition (LMD) and selective laser melting (SLM) in turbo-engine applications, Laser Technik Journal 2 (2010) 58−63.

[77] G. Piscopo, L. Iuliano. Current research and industrial application of laser powder directed energy deposition. The International Journal of Advanced Manufacturing Technology, (2022) 1−25.

[78] The rise of 3-D printing: The advantages of additive manufacturing over traditional manufacturing, Business Horizons 60 (5) (2017) 677−688.

[79] J.C. Vasco, Additive manufacturing for the automotive industry[M], Additive Manufacturing, Elsevier, 2021, pp. 505−530.

[80] E. Karayel, Y. Bozkurt, Additive manufacturing method and different welding applications, Journal of Materials Research and Technology 9 (5) (2020) 11424−11438.

[81] Additive manufacturing helps racing team finish first. Metal Powder Report. 68 (6) (2013) 32−33.

[82] L. Benfratello, S. Conte. Additive manufacturing adoption in dental practices: state of the art and future perspectives. Master of Science Thesis, Politecnico di Turin, 2019.

[83] World's First Patient-Specific Jaw Implant, Metal Powder Report. 67 (2) (2012) 12−14.

[84] A. Dawood, B.M. Marti, V. Sauret-Jackson, A. Darwood. 3D printing in dentistry. British dental journal, 219 (11) (2015) 521−529.

[85] N. Xu, F. Wei, X. Liu, L. Jiang, H. Cai, Z. Li, et al., Reconstruction of the upper cervical spine using a personalized 3D-printed vertebral body in an adolescent with Ewing sarcoma, Spine (Philadelphia, Pa: 1986) 41 (1) (2016) 50−54.

[86] J.L. Aranda, M.F. Jiménez, M. Rodríguez, G. Varela, Tridimensional titanium-printed custom-made prosthesis for sternocostal reconstruction, European Journal of Cardio-Thoracic Surgery 48 (4) (2015) 92−94.

[87] A. Bhargav, V. Sanjairaj, V. Rosa, L.W. Feng, Y.H. Jerry Fuh, Applications of additive manufacturing in dentistry: a review, Journal of Biomedical Materials Research, Part B: Applied Biomaterials 106 (5) (2018) 2058−2064.

Chapter 8

The emerging frontiers in materials for functional three-dimensional printing

Jia Min Lee[1], Swee Leong Sing[2,3], Guo Dong Goh[2], Guo Liang Goh[2], Wei Long Ng[1] and Wai Yee Yeong[1]

[1]*HP-NTU Digital Manufacturing Corporate Lab, School of Mechanical and Aerospace Engineering, Nanyang Technological University, Singapore,* [2]*Singapore Centre for 3D Printing, School of Mechanical and Aerospace Engineering, Nanyang Technological University, Singapore,* [3]*Department of Mechanical Engineering, National University of Singapore, Singapore*

List of abbreviations

ABS	acrylonitrile butadiene styrene
AM	additive manufacturing
CBAM	composite-based additive manufacturing
CIP	cold isostatic pressing
CNT	carbon nanotube
DLP	digital light processing
FDA	food and drug administration
FDM	fused deposition modeling
FFF	fused filament fabrication
GelMA	gelatin methacryloyl
HA	hyaluronic acid
HAP	hydroxyapatite
HV	vickers hardness
LAM	laser additive manufacturing
LCM	lithography-based ceramics manufacturing
LENS	laser engineered net shaping
LOM	laminated object manufacturing
LSD	layerwise Slurry Deposition
MA	methacryloy
MWCNT	Multiwalled carbon nanotube
NPs	nanoparticles
PA	polyamide
PCB	printed circuit board

Digital Manufacturing. DOI: https://doi.org/10.1016/B978-0-323-95062-6.00008-5
© 2022 Elsevier Inc. All rights reserved.

PEEK	polyether ether ketone
PEDOT	PSS poly(3,4-ethylenedioxythiophene) polystyrene sulfonate
PU	polyurethane
R2R	roll-to-roll
RGD	arginylglycylaspartic acid
SLA	stereolithography
SLB	selective laser burn-out
SLCOM1	selective lamination composite object manufacturing 1
SLM	selective laser melting
SLS	selective laser sintering
SWCNT	single-walled carbon nanotube
VOC	volatile organic compound
YSZ	yttrium-stabilized zirconia
1D	one-dimensional
2D	two-dimensional
3D	three-dimensional

8.1 Introduction

Since the conception of three-dimensional (3D) printing, or additive manufacturing (AM), 30 years ago, the use of 3D printing technologies has evolved from being just a prototyping tool in product development to making actual parts used in aerospace, automotive, biomedical, and electronic industries. 3D printing provides manufacturing advantages such as reduced production cost, reduced lead time, customizability, and increased design freedom. To encourage the adoption of 3D printing for manufacturing, it is important to ensure that 3D printed parts can perform on par or better than parts made from conventional manufacturing processes. In this context, enhanced performance in 3D printing parts can be understood as injecting functionality into 3D printed parts. Functionality in 3D printed parts is introduced at different stages of the fabrication process and can be generally described in four main aspects (Fig. 8.1).

- Developing functional materials where printing inks/powders have properties such as electrical conductivity [1,2], shape-memory [3−5], or biological compatiblity [6,7] which add dimensionality into 3D printed parts. Additives are added to the inks to improve the overall performance and printability of otherwise hard to process materials [8,9]. Ink formulation is dependent on the printing process, where properties such as yield stress, viscosity, laser absorptivity are characterized and evaluated before printing [10,11].
- Introducing functionality during printing such as process-induced microstructural changes (i.e., intrinsic effects) [6,12,13] and external stimulation during printing (e.g., extrinsic effects from acoustic focusing [14,15], rotational [16], magnetic field [17,18], plasma treatment [19]) resulted in 3D printed parts with anisotropic properties.

The emerging frontiers in materials **Chapter | 8** **301**

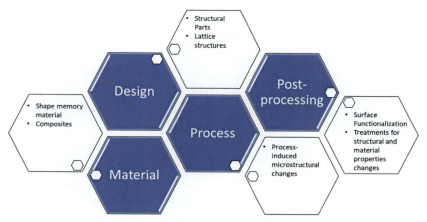

FIGURE 8.1 Introducing functionality in 3D printed parts through developing functional materials, designing lattices/parts that serve specific purposes, using 3D printing processes to induce microstructural changes to the printed parts, and functionalizing the printed parts using postprocessing treatments. *3D*, Three-dimensional.

- Designing functional elements in printed parts such as actuators, valves, pumps and channels in microfluidic [20,21], structures for heat exchanger [22], and bioinspired structures [18,23] that uses the advantage of 3D printing to produce parts with complex geometries. In this aspect, simulation helps users to understand the influence of different parameters on the overall parts' performance with minimal iterations.
- Post-processing of 3D printed parts is performed for purposes such as removal of support material and secondary curing, surface modification through polishing to remove surface roughness, surface functionalization, and treatments for inducing changes to structural and material properties. Two distinct approaches are used to introduce functionality at the post-processing stage. First, top-down processes such as etching and polishing create micro/nanostructural changes to the surface of the printed parts [24]. Second, bottom-up processes such as physical/chemical deposition, grafting, self-assembly, and coating are introduced to improve thermal and mechanical properties [25] and biological and chemical functions [26] of 3D printed parts.

The above-mentioned approaches have been used across different industries, where users are able to introduce functionalities at different stages of the 3D printing process. The following sub-sections will look at the advancement in materials for 3D printing as demonstrated in three main applications. These industries, namely, aerospace, biomedical, and electronics, have seen immerse adoption of 3D Printing technologies. In each sub-section, we will begin by looking at the overview of the industry. Following which, the different printing technologies used in the industry are summarized and compared. Within each industry the advancement and challenges for material development and 3D printing are discussed.

8.2 Composites materials for aerospace industry

Composite materials are widely used in the aerospace industry. They are slowly replacing conventional metal alloys owing to their high strength-to-weight and high stiffness-to-weight ratios. The weight saving using composite materials means the aircraft can carry more payloads and fuels to cover a longer range. Conventionally, composite parts are simple in shape due to the limitation of the fabrication technique. Fiber sheets crumble at corners and sharp edges making them not suitable for parts with complex geometry. The fabrication of mold is expensive and time-consuming. The fabrication process requires highly skilled operators and it is labor-intensive (Fig. 8.2).

The development of AM techniques for composite materials would enable the production of high performance, more advanced parts and components. Meanwhile, the development of the manufacturing process normally comes together with the development of composite materials suitable for AM techniques.

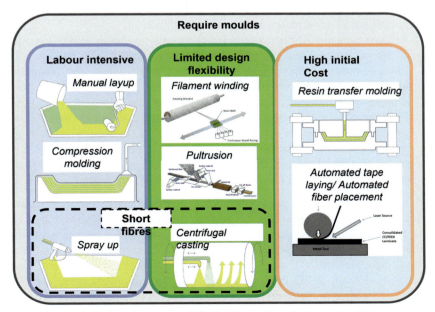

FIGURE 8.2 Limitations of conventional fabrication technique for composites [27].

8.2.1 Overview of the composite industry

The development of the composites in AM has taken two approaches. Fig. 8.3 shows the timeline of the development of the 3D printing composites. The first is the development of composite materials to suit the existing 3D printing techniques. It is normally done by premixing the reinforcement materials with the polymer matrix to form the composite feedstocks for the 3D printers. The material used is usually nano- or milli-scaled short fibers. 3D printing material company, Colorfabb, developed several composite filaments using various fibrous materials such as carbon fiber, wood and bamboo, and metallic materials such as copper, brass, and bronze as reinforcements [28]. 3D printer companies like Stratasys, EOS, and Ultimaker developed their nozzles and print parameters for printing the short fiber-reinforced thermoplastics [29].

The second approach is the development of new techniques or modified versions of existing techniques to realize the fabrication of the composite parts using 3D printing techniques. Markforged has pioneered the development of continuous fiber composite printers in 2014. The first generation of the composite printer, Markone, is a dual nozzle extrusion printer capable of extruding nylon and prepreg continuous carbon, Kevlar, and glass fibers [30]. EnvisionTEC introduced its composite printer that comprises woven fiber composites preimpregnated with thermoplastics—the selective lamination composite object manufacturing 1(SLCOM1) [31]. Impossible objects developed a composite-based additive manufacturing (CBAM) technique that uses both sheet lamination and inkjet printing techniques to fabricate composite parts [32]. Randomly oriented long carbon and glass fibers have been used as reinforcement and aerospace-grade polyether ether ketone (PEEK) and nylon 12 have been used as the thermoplastic matrix. Continuous composites developed printhead that can extrude copper and nichrome wires, in addition to continuous carbon and glass fibers [33], which gives the 3D printing technique the ability to create multifunctional composites. Fortify

FIGURE 8.3 Timeline of the commercialization of various AM techniques for composites. *AM*, Additive manufacturing.

304 Digital Manufacturing

developed digital composite manufacturing, which uses a magnetic field to align short glass and carbon fibers in the resin to form different patterns in each layer using digital light processing (DLP) stereolithography (SLA) [34]. Using this technique, high-resolution composites can be produced.

In short, the technological progression for composite 3D printing has been on the development of 3D printable composite materials for existing 3D printing techniques. Researchers have focused on trying to align the short fiber reinforcements during the print processes via self-induced alignment and external forces to improve the mechanical properties further. Continuous fiber reinforcement is made possible by modifying the existing printhead or coming up with a new 3D printing technique for composite materials. Multifunctional composites and large-scale composite printing are the trends moving forward.

8.2.2 Composites for three-dimensional printing

3D printing has seen a large rise in the composite industry for both the production of prototypes and tooling parts of the manufacture of customizable parts itself. Most of the manufacturing is done using short, chopped carbon fibers, but there are examples of continuous fibers (which does have parallels with automated fiber placement). Composite materials are mixtures of two or more materials that have superior properties as compared to their constituent materials. Composite materials normally consist of two phases: reinforcement and matrix. The reinforcement phase is typically fibers or particulates with excellent stiffness to provide the strength to the composite materials. The matrix holds the fiber reinforcement, distributes loads to the fibers, and gives the composite component its shape.

In 3D printing, polymeric materials have limitations in terms of mechanical and electrical properties. To improve the mechanical and electrical properties, reinforcement materials, which have superior mechanical and/or electrical properties, are added to form composite materials.

Reinforcement materials can come in various forms: particulate, short fibers, long fibers, and continuous fibers. Generally, continuous fibers can take higher loads, which results in better mechanical properties compared to short fibers. However, short fibers and particulate reinforcements offer better printing flexibility as compared to long and continuous fibers. Apart from that, reinforcement loading also plays an important role in determining the mechanical and electrical performance of the composites. The mechanical properties of the composite materials are dependent on the types of reinforcement and matrix used as well as the reinforcement loading. Short fibers can provide a small degree of improvement in mechanical properties and the typical range of tensile strength is about 40−80 MPa [32]. Composites with continuous fibers, on the other hand, can produce composite parts with higher tensile strength in the range of 100−800 MPa. Carbon fibers offer a superior high-strength-to-weight ratio making them suitable for use in the aerospace industry. Kevlar and glass fibers are some of the cheaper alternatives that also provide good strength-to-weight ratios [35]. Natural fibers have also been used as reinforcement for composite materials due to their good sustainability and renewability [36]. Surface modification of the reinforcement is performed, and additives such as surfactants and plasticizers [37] are added to improve the fiber−matrix bonding. Table 8.1 shows various fiber/matrix compositions that have been used to develop 3D printed composites, together with their tensile properties.

Matrix materials utilized in most commercial composites can be divided into two general categories: polymeric, including some thermosetting and thermoplastic resins, and metallic, consisting of pure metals and alloys. The function of the matrix in a composite material is usually multifold. Matrices are designed to protect the reinforcing phase from structural damage, corrosive attack, and reactions that would degrade the reinforcement properties. The matrix phase also serves to transmit applied stresses to reinforce constituents and stabilize them against buckling in situations where compressive stresses are axially applied to the cylindrical fibers.

In most scenarios, the matrix may share load-carrying responsibility with the reinforcing phase. This aspect is important in particulate strengthened composites. Furthermore, the matrix may be selected for its physical properties, such as density, thermal, electrical conductivity (or electrical sensitivity), thermal expansivity, melting or softening temperature, and translucency or opacity.

In general, there are a few factors that need to be taken into consideration in the development of 3D printable composite materials, namely, the material properties and the processability of the raw materials for the AM technique, in particular, (1) moisture resistance, (2) toughness, (3) improved processability, and (4) elevated temperature performance.

Moisture degrades the resins by plasticizing the epoxy resin, resulting in swelling of resin and decreasing its glass transition temperature [53]. As a result, the composite mechanical strength is reduced at higher temperatures. Hence, matrix resin with high moisture resistance reduces degradation at a higher temperature and can be used for high-temperature applications.

Epoxy resins tend to be brittle due to the high cross-link density of the polymers. It results in the brittle nature of the composites as characterized by the low transverse strain-to-failure, low impact strength, low shear strength, etc. The brittle nature can be reduced by using a more ductile and toughness matrix resin. There are generally two ways to improve the toughness of composites: (1) addition of an elastomeric, secondary-phase toughening agent; (2) the use of tougher matrix resins such as thermoplastic resins and some high-performance thermosets such as polybenzoxazine and cyanate ester [54].

The elevated temperature performance is a critical factor that determines whether the composite system is suitable for high-temperature applications. The highest operating temperature for thermoplastics is 240°C (PEEK) [55],

306 Digital Manufacturing

TABLE 8.1 Overview of additive manufacturing technologies processing composite materials for aerospace industry.

Technology	Process	Material	Refs.
Material extrusion	FFF	Discontinuous:	
		• SWCNTs (5%)/Carbon fiber (5%)/MAGNUM 213 ABS	[38]
		• Graphene nanoplatelets (4 wt%)/ABS	[39]
		Continuous:	
		• Carbon fiber (0.2−0.4 mm) (40%)/ABS	[40]
		• Carbon fiber (34.5%)/nylon	[41]
	Liquid deposition modeling	Discontinuous:	
		• SiC whiskers-carbon fibers/epoxy (Epon 826), nano-clay platelets, and DMMP	[42]
		• A-MWCNT (7 wt%)/ PVP (17 wt%)	[43]
Vat photopolymerization	SLA	Discontinuous:	
		• SiO_2 nanoparticles (20 nm)/ acrylate functionalized oligomers Genomer 4302 (G4302)/Sartomer CN922 with the monomer tris(2-hydroxyethyl)isocyanurate triacrylate	[44]
		• Electrical conductive whisker (TiC) (50 µm)/photopolymer	[45]
		• Al59Cu25.5Fe12.5B3 (25 µm)/ Accura Si40 resin	[46]
		Continuous:	
		• 800 e-glass fibers (4.6%)/Somos 3100 resin	[47]
		• Thornel T-300 tow of 3000 fibers (5%)/XB5081 Ciba Geigy	[48]
Sheet lamination	LOM	Mat/Sheet	
		• Aerospace-grade prepreg containing e-glass fibers (52−55 vol%) / epoxy	[49]
	CBAM	Carbon fiber/Kevlar/fiberglass/high-density polyethylene	[50]

(Continued)

The emerging frontiers in materials Chapter | 8 **307**

TABLE 8.1 (Continued)

Technology	Process	Material	Refs.
Powder bed fusion	SLS	Discontinuous:	
		• CNT (0.5 wt%)/PA-12 and PU	[51]
		• YSZ (5 wt%)/PA-6	[52]

ABS, Acrylonitrile butadiene styrene; *CBAM*, composite-based additive manufacturing; *CNT*, carbon nanotube; *DMMP*, dimethyl methylphosphonate; *FFF*, fused filament fabrication; *LOM*, laminated object manufacturing; *MWCNT*, multiwalled carbon nanotube; *PA*, polyamide; *PU*, polyurethane; *SWCNT*, single-walled carbon nanotube; *YSZ*, yttrium-stabilized zirconia.

whereas the maximum operating temperature for thermoset is 300°C (phenolic-triazine resin) [54].

Processability is also an important factor that determines whether a combination of fiber/matrix is suitable for the 3D printing process. Properties that need attention in the extrusion-based fused filament fabrication (FFF) technique include rheology, melt flow index, and the melting point of the matrix resin [56]. In SLA, rheology [44] and material optical absorption [57] on a specific wavelength need to be taken into consideration. In selective laser sintering (SLS), powder flowability and spreadability require special attention, especially when fiber content is high [58].

In general, all 3D printed techniques have been used to fabricate short fiber-reinforced composites. They are normally done by premixing the short fiber with the raw material thermoplastic feedstock to form composite feedstocks. Continuous fiber can only be processed by a few AM processes, namely, FFF [59], SLA [48], laminated object manufacturing (LOM) [60]. Thermoplastics are commonly used to fabricate 3D printed composites, although 3D printed thermoset composites are slowly growing. The mechanical properties are generally lower compared to that of the composites with similar fiber loadings manufactured using conventional fabrication techniques [61]. The poorer strength is attributed to the high porosity due to the lack of consolidation or compaction process. Nonetheless, new processes are being developed to overcome the limitations in terms of low fiber loading and high porosity.

8.2.3 Challenges and potentials in composites materials for aerospace industry

3D printing research on aerospace applications has been primarily on metals [62]. Some examples have 3D printed titanium waveguide brackets on the National Aeronautics and Space Administration Juno spacecraft, aluminum brackets on the Telstar 18 Vantage communication satellite, and thermal

308 Digital Manufacturing

blanket retention brackets on Boeing 702MP satellites [62]. Although 3D printed composite parts are still not widely adopted in the aerospace industry, 3D printed composites have many advantages over 3D printed metal. First, composites have a high strength-to-weight ratio. Second, the fabrication cost of 3D printed composites is lower. As the price to produce the feedstock material for composites is cheaper, the lower fabrication cost is particularly crucial to the cost-sensitive aerospace applications. Third, the lower processing temperature of the composites fits into a wider range of 3D printing techniques, which allows the fabrication of multifunctional multimaterial composite parts. For instance, real-time structural health monitoring of the airfoil can be achieved by embedding continuous copper wire onto a 3D printed continuous fiber-reinforced airfoil truss structures and laying fiber optics on the upper skin of the airfoil [63].

The layer-by-layer fabrication approach of 3D printing techniques makes the fabrication of highly optimized complex structures with intricate geometries possible. The complex structures can be directly manufactured using a 3D printing technique from the design file. It greatly shortens the lead time, reduces the number of parts required for assembly, and makes prototyping very economical [64]. The ability to selectively deposit materials in some of the 3D printing techniques enables the creation of multimaterial composite parts. Thus parts can be further optimized by tuning the material properties based on the usage and requirement.

Composite 3D printing is being used as a tool to create molds for the helicopter blades. Using the large-scale AM technology developed by Thermwood, the United States, the 6-m molds created have a good surface finish, tight tolerances, and the ability to withstand autoclave processing [63]. This shows the new opportunities that composite 3D printing could explore for large and intricate aerospace components.

With the ability of 3D printing techniques to selectively deposit materials, the fabrication of the composite material could benefit from better control of the material properties within part dimensions [65]. Instead of focusing on improvement in mechanical properties, multifunctionality through the selective deposition of nanomaterials can broaden the potential of composites to exhibit unique properties, like the spatial variation in thermal and electrical conductivity, photonic emissions tunable for wavelength [66]. This precision in control would provide a powerful toolbox in realizing the multifunctionality of the 3D printed composites.

Prior arts have shown encouraging findings on integrating nanomaterials to 3D printing. In general, there are two ways of realizing 3D printing of multifunctional composites [66], (1) premixing of the nanomaterials into the hose matrix, and then followed by 3D printing of the nanocomposite mixture as a complete part, and (2) 3D printing of composite structures with midway pauses of the print job, and the introduction of nanomaterials through manual deposition or automatic deposition via certain printing technique such as aerosol jet printing and inkjet printing.

Table 8.2 shows some potential applications that the 3D printed multifunctional composites can be used.

8.2.3.1 Scaling up to big area additive manufacturing

Scaling up the 3D printing of composites to suit industry needs poses some challenges that need to be dealt with. First, one of the biggest hurdles to expand composite 3D printing to large format printing is the low fabrication speed [68]. New advancements in large format 3D printing, like big area AM, eliminate the drawback by using a large diameter feeding screw in place of the FFF's filament heating mechanism. The large nozzle diameter can extrude 50 kg of material in an hour [69]. However, the high extrusion rate comes at the expense of lower print resolution. As a result, the surface finish of the printed parts is poor and requires postprocessing.

8.2.3.2 Lack of three-dimensional printable composite materials

The material range for 3D printable composites should continue to be widened. Current 3D printing matrix materials are relying on several thermoplastics materials, such as acrylonitrile butadiene styrene and nylon [68]. Apart from that, commercially available composite materials are restricted to specific 3D printing techniques and have fixed fiber/matrix contents. More fiber/matrix combinations need to be explored for various 3D printing techniques. In particular, the process—structure—property relationship in 3D

TABLE 8.2 Nonstructural applications of composites [67].

Properties of composites	Applications
Electrical properties	• Smartphones, touch screens, light-emitting diodes, and thin-film solar cells
	• Stretchable antenna, sensor, and light-emitting diodes
	• Actuators and sensors
	• 3D connectors of an electrical circuit
Thermal properties	• Thermally conducting materials
	• Heat sink and cooling system
	• Rocket motors
Optical properties	• Luminescent light-converting thin films for AC
	• Full-color AC-driven displays
	• Electrochromic devices

3D, Three-dimension.

310 Digital Manufacturing

printing will expedite the invention of lightweight, stronger, and multifunctional materials. For instance, multiscale-reinforced composites can be developed by combining nanofibers with continuous fiber in composite to create multifunctional composites with enhanced mechanical performance.

8.2.3.3 Poor out-of-plane strength

One of the drawbacks of the existing extrusion-based 3D printing technique is the inability to have z-direction reinforcement due to the three-axis configuration of the 3D printing technique [33]. This means that the fiber reinforcement can only exist in the $X-Y$ plane but not in the Z-direction. Although a more advanced 3D printing method using robotic arms [70] overcome the plane-wise reinforcement limitation, the lack of consolidation of the extruded composite filament remains. Thus delamination resistance is lower due to the poor interlayer adhesion.

3D printing of composite structures can be a turning point for AM technology. The potential of fabricating functional devices directly from commercial 3D printers with controllable properties has created a huge rush for new developments and research in this field. While many limitations still exist now, 3D printing of composites develops fast in recent years. The attractive combination of endless possibilities in the range of composite materials and extra customization of AM offers a unique new area in the manufacturing field for researchers and developers to explore.

8.3 Biomaterials for bioprinting

In the following section, we will highlight 3D printing advances as well as the challenges and potential of functional biomaterials in biomedical applications based on two aspects. First is the use of 3D printing technologies for biological applications, specifically bioprinting. The technologies can be classified according to the material processing techniques such as material jetting [71,72], vat photopolymerization [73,74], material extrusion [75,76], and free-form spatial printing [77]. Second, we will look at the printing of ceramics for biomedical implants.

8.3.1 Overview of bioprinting

Bioprinting refers to the use of computer-aided transfer processes for patterning and assembling living and nonliving materials with a prescribed two-dimensional (2D) or 3D organization to produce bioengineered structures [78,79]. The biomaterials that are commonly used in the bioprinting process can be categorized into natural and synthetic biomaterials [80], whereas the inks that are used in the bioprinting process can be classified as bioinks or biomaterial inks [81].

The emerging frontiers in materials Chapter | 8   **311**

8.3.2   Bioinks for bioprinting

A bioink consists of cells (mandatory) and biomaterials, while the biomaterial-ink only consists of the biomaterials. In the latter approach the biological component of bioprinting, that is, cells, can be introduced post-printing through seeding on the printed constructs (i.e., cell seeding approach).

8.3.2.1   Natural bioinks

- Alginate is a water-soluble polymer that undergoes a reversible cross-linking mechanism in the presence of divalent cations such as Ca^{2+} to form an ionic-cross-linked bioink. Hence, the ionic-cross-linked alginate bioinks can dissolve easily due to the displacement of divalent ions by monovalent ions under physiological conditions [82]. Furthermore, it is necessary to perform arginylglycylaspartic acid (RGD) modification to improve cell adhesion [83]. The alginate hydrogels are commonly printed using the extrusion-based bioprinting approach, in laser-assisted bioprinting systems and inkjet-based bioprinting. The cell-laden alginate precursor solution can either be printed into a $CaCl_2$ cross-linker pool [84] or used as a receiving substrate that is subsequently cross-linked by the cross-linker ($CaCl_2$) solution [85].
- Collagen is widely used for tissue engineering applications due its specific peptide sequences that are recognized by adherent cells [86,87]. Collagen precursor has been printed using microvalve and extrusion bioprinting system [88,89]. Collagen solution can be neutralized before printing [90] or neutralized during printing by alternating printing precursor solution along with pH neutralizer such as NaOH [91,92].
- Gelatin is a partially denatured form of collagen [93]. The hydrolysis of collagen into gelatin is affected by temperature, pH, and extraction duration. Modifications of gelatin to its derivatives are necessary to improve gelatin's stability for in vivo applications [94]. Gelatin has been printed using extrusion-based bioprinting, laser-assisted bioprinting, inkjet printing. Printed gelatin constructs are stabilized using enzymes [95], which improves the overall gel strength [96].
- Hyaluronic acid (HA) is a linear, nonsulfated glycosaminoglycan commonly found in most connective tissues. HA chains form flexible coils at low concentrations but entangle each other and form a temporary network at high concentrations [97]. Although HA is not favorable for cell attachment due to its highly negative charge and hydrophilicity, it can still bind to proteins and cells through cell surface receptors such as CD44, RHAMM, and ICAM-1 [98]. Modified HA can be printed using extrusion-based bioprinting, two-photon polymerization, laser-assisted bioprinting, and microvalve-based bioprinting.

312 Digital Manufacturing

8.3.2.2 Synthetic bioinks

- Gelatin methacryloyl (GelMA) is a photopolymerizable hydrogel, which has been chemically modified from water-soluble gelatin. GelMA is a synthetic hydrogel that contains both cell-binding and cell-degrading motifs. Only 5% of the amino acid residues in gelatin are chemically modified by methacryloyl (MA) [99]. The lack of reaction between the RGD motifs and MA ensures good cell-adhesion properties [100]. Photoinitiator (PI) is critical for the bioink formulation of photopolymerizable hydrogels such as GelMA. Some of the PIs used to initiate the reaction include water-soluble 2-hydroxy-1-[4-(2-hydroxyethoxy) phenyl]-2-methyl-1-propanone (Irgacure 2959), lithium phenyl-2,4,6-trimethylbenzoylphosphinate (LAP), 2,2'-azobis[2-methyl-n-(2-hydroxyethyl) propionamide] (VA086), eosin Y, ruthenium (Ru) and sodium persulfate, and benzylidene cycloketone-based PIs G2CK and P2CK (for two-photon polymerization). GelMA has been printed using extrusion bioprinting, inkjet-based bioprinting, SLA-based bioprinting, and two-photon.
- Polycaprolactone (PCL) is a hydrophobic, semicrystalline aliphatic polyester with a slow degradation rate ranging from 2 to 4 years, dependent on the molecular weight of the polymers [101]. It is considered nonbiodegradable due to the absence of relevant enzymes in the body for enzymatic degradation. Surface modifications, such as coating with amine layers [102], NaOH treatment [103], and immobilization of RGD peptides [104], are required to improve the biocompatibility. PCL-based scaffolds have been fabricated using SLA- or inkjet-based printing approaches. Cells are seeded onto the fabricated scaffolds postprinting. The PCL filaments/fibers were usually printed within the cell-laden constructs using an extrusion-based bioprinting approach to improve the overall mechanical integrity of the printed cell-laden constructs.
- Polyethylene glycol (PEG) is a linear synthetic polyether that is nontoxic and has been approved by the U.S. Food and Drug Administration (FDA) for use as excipients for different pharmaceutical formulations. PEG-based scaffolds undergo slow hydrolytic degradation both in vitro and in vivo [105]. Its tunable properties enable the grafting of enzyme-sensitive peptides to accelerate their degradation rates [106]. The most common cross-linking approach for PEG-based hydrogels is via photopolymerization; some of the reported PIs for PEG-based printing include Irgacure 2959, LAP, VA086, eosin Y, and riboflavin (vitamin B2) with triethanolamine coinitiator. PEG-based constructs have been printed using extrusion-, inkjet-, and SLA-based bioprinting.
- Pluronic F127 (PF127) is a synthetic polymeric compound, which has a central hydrophobic poly (propylene oxide) (PPO) block along with two hydrophilic poly (ethylene oxide) (PEO) sides (PEO−PPO−PEO). It is a water-soluble bioink that undergoes a reversible temperature-dependent

cross-linking process with a sol−gel transition temperature ranging from 15°C to 35°C depending on the polymer's concentration [107]. PF127 is commonly used as a "fugitive" bioink due to its high printability and easy removal. To ensure PF127 structures do not dissolute upon immersion in solution, the terminal hydroxyl moieties of PF127 are modified with photo-cross-linkable groups to form chemically cross-linked structures [108]. PF127 is commonly used in extrusion-based bioprinting [109]. However, its high intrinsic viscosity and thermosensitive properties make it challenging for jet formation when using droplet-based bioprinting approaches. PF127 has limited application in laser-based bioprinting as it remains as a solid coating on the quartz support [110].

8.3.3 Challenges and potential in bioprinting of biomaterials

Selecting biomaterials for bioprinting should be specific based on the needs of intended tissue types and the printing technologies used for depositing cells/materials (Table 8.3). The following are some dilemmas faced when formulating the appropriate bioink for bioprinting.

8.3.3.1 Natural material has relatively weak strength

Natural material, such as collagen and gelatin, forms hydrogel with relatively weaker mechanical properties with limited structural stability under pristine conditions. Increasing the storage modulus of bioink in the extrusion process improves the printability [128]. Methods such as semicross-linking bioink are applied to improve its mechanical properties [129]. However, this can shorten the printable window. Recently, a team used riboflavin combined with collagen to reduce gelation time and improve the stability of constructs [128]. However, it is important with any future modifications to verify the cell viability to ensure that the bioink is still biocompatible and maintains the essential biocompatibility and structural functionalities intended for its application.

8.3.3.2 Added components, added complexity

When additional components such as cells are added into the bioink formulation, challenges such as controlling the spatial location of cells and the lack of control over cell density within each deposited unit are apparent [130]. The lack of control over cell distribution is prevalent in the preprinting stage, resulting in the inconsistent printing [111,131]. One main contributing factor is the cell adhesion along the printing cartridge because of the presence of van der Waals interactions between the cells and the interior surface of the printing cartridge [132].

314 Digital Manufacturing

TABLE 8.3 Overview of additive manufacturing technologies used for processing biomaterials in bioprinting.

Technology	Process	Material	Refs.
Material extrusion	Pneumatic pressure	• Alginate	[111]
		• Gelatin	[112]
		• Hyaluronic acid	[113]
		• GelMA	[114]
		• PEG	[115]
		• Pluronic F127	[109]
	Melt extrusion	• PCL	[116]
Material jetting	Inkjet	• Alginate	[117]
		• GelMA	[118]
		• PEG	[119]
	Laser assisted	• Alginate	[120]
		• Collagen	[91]
		• Gelatin	[121]
		• Hyaluronic acid	[122]
	Microvalve based	• Collagen	[123]
		• Hyaluronic acid	[124]
Vat polymerization	Stereolithography	• GelMA	[125]
		• PEG	[126]
	Two-photon polymerization	• Hyaluronic acid	[127]

GelMA, Gelatin methacrylate; *PCL*, polycaprolactone; *PEG*, polyethylene glycol.

8.3.3.3 Limitation in print resolution

Print resolution and accuracy of the bioprinted construct is largely determined by the smallest material unit formed based on the printing processes. The print resolution for current users is still faced with several technical issues involving limitation in diffusion limit (approximately 200 μm) for culturing of printing tissue [133]. The printing process imposes mechanical stresses on cell behavior [134]. As such, there is a conflict between obtaining higher resolution of cell printing while ensuring viability and potency of cells are maintained postprinting.

The current biomaterials for bioprinting face multifaceted challenges in terms of biophysical properties. They can exhibit good bulk mechanical properties while providing dynamic biochemical properties like the extracellular matrix. Therefore research in biosynthetic ink that is application and process specific will serve as key drivers to unleash the full potential of bioprinting functional engineered tissue.

8.4 Ceramics for biomedical implants

Ceramics used for biomedical implants can be broadly categorized based on their reactivity with living tissues after implantation such as (1) bioinert (e.g., alumina and zirconia), (2) bioactive (e.g., bioglasses and glass-ceramics), and (3) bioresorbable (e.g., calcium phosphates, calcium phosphate cements, calcium silicate). The following section will look at bioinert ceramics, which have high mechanical and corrosion resistance, that have been successfully used in orthopedics. Specifically, the section will identify AM technologies for the processing of these bioinert ceramics.

8.4.1 Overview of three-dimensional printed ceramic implants

AM of ceramic has much potential for applications in the biomedical field. Samples of the ceramic dental restorations are shown in Fig. 8.4. All-ceramic dental restorations fabricated by SLS/cold isostatic pressing (CIP) technology (A) digital tooth models (B) ceramic products [135].

During laser processing of silica-based powder mixture, silica was melt to form a liquid glass phase and linked other particles, forming a porous structure [136]. Such processes, like SLS, that make use of materials that consist of at least one glass phase and one crystal phase, can be used for biomedical implants. The microstructure and density of the ceramic parts depend on the contents and interaction between constituents of the powder mixture, as well as the rate of heating and cooling [137]. Glass systems such as $SiO_2/Al_2O_3/P_2O_5/CaO/CaF_2$ [138] and $SiO_2/CaO/Na_2O/P_2O_5$ [139] have been studied. The glass-ceramic system $SiO_2/Al_2O_3/P_2O_5/CaO/CaF_2$ has

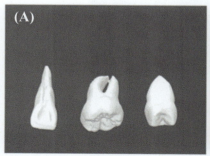
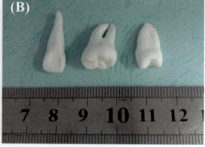

FIGURE 8.4 All-ceramic dental restorations fabricated by SLS/CIP technology (A) digital tooth models and (B) ceramic products [135]. *SLS*, Selective laser sintering.

good mechanical strength and fracture toughness due to the apatite and mullite phases produced [138]. Porous scaffolds with good fracture toughness were made from $SiO_2/CaO/Na_2O/P_2O_5$ using SLS [139]. Dental restoration framework made of $ZrO_2-Al_2O_3$ by selective laser melting (SLM) [140] and dental bridge made of ZrO_2 by SLA [141].

For use in bone implants and scaffolds, Tesavibul et al. used SLA to process 45S5 Bioglass with biaxial strength of 40 MPa [142]. A sample of a customized bone implant is shown in Fig. 8.5.

AM techniques have been used to manufacture ceramic parts with complex shapes as they reduce customization costs and lead time for individualized parts [143]. The traditional manufacturing processes, such as casting with abrasive machining, result in large tool expenditure and longer processing time [144]. Other advantages of AM include design freedom and precise control over the sizes and shape of complex ceramic structures. However, ceramics do not fuse in the same way as metals or polymers, and their extremely high melting point adds challenges to process them by AM. Nonetheless, several attempts have been made using all the seven categories of AM technologies, based on ISO/ASTM classification, for ceramic manufacturing, as shown in Table 8.4.

8.4.2 Ceramic materials by three-dimensional printing for biomedical implants

The results of ceramics processed using these techniques, such as silica (SiO_2), zirconia (ZrO_2), and alumina (Al_2O_3), are presented. Critical challenges encountered during AM of ceramics are discussed.

- Zirconia, such as yttrium-stabilized zirconia (YSZ), is an oxide ceramic material that has good corrosion resistance in addition to high flexural

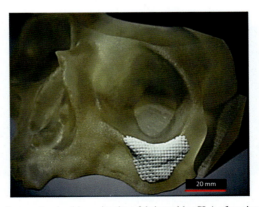

FIGURE 8.5 Bioglass customized bone implant fabricated by SLA after sintering [142].

The emerging frontiers in materials Chapter | 8 **317**

TABLE 8.4 Overview of additive manufacturing (AM) technologies used for processing ceramics.

Technology	Process	Material	Ref.
Material jetting	Cold spray AM	• SiC • Al_2O_3	[145]
Sheet lamination	SLB	• Al_2O_3	[146]
Material extrusion	FDM + robocasting	• Al_2O_3 • ZrO_2	[147]
	DIW	• Al_2O_3	[148]
	CODE	• ZrO_2	[149]
Vat polymerization	LCM	• Al_2O_3 • ZrO_2	[150,151]
	DLP	• Al_2O_3 • HA	[152,153]
	SLA	• Al_2O_3 • SiO_2	[141,154]
Powder bed fusion	SLM	• $ZrO_2-Al_2O_3$	[140]
	LSD	• SiO_2	[155]
	SLS	• SiC	[156]
Binder jetting		• Al_2O_3 • $BaTiO_3$	[157,158]
Directed energy deposition	LENS	• Al_2O_3	[144]
	LAM	• Al_2O_3-YAG	[159]

CODE, Ceramic on-demand extrusion; *DIW*, direct ink writing; *DLP*, digital light processing; *FDM*, fused deposition modeling; *LAM*, laser additive manufacturing; *LCM*, lithography-based ceramics manufacturing; *LENS*, laser engineered net shaping; *LSD*, layerwise slurry deposition; *SLA*, stereolithography; *SLB*, selective laser burn-out; *SLM*, selective laser melting; *SLS*, selective laser sintering.

strength and high fracture toughness. The fracture properties of YSZ can be enhanced via a transformation toughening mechanism [160].

In dental application, zirconia offers superior mechanical properties, which can be used to fabricate implants made of only ceramic [161]. Zirconia and its powder mixtures have been explored in SLM and SLA. Lian et al. obtained zirconia parts with 98.58% relative density after sintering. The parts also have a Vickers hardness of 1398 HV and surface roughness of 2.06 µm [141]. Ferrage et al. obtained components made from YSZ that have a relative density of 96.5% using SLS equipped with an Nd:YAG laser. However, a small amount of graphite is added to the

318 Digital Manufacturing

powder to improve laser absorptivity [162]. Scheithauer et al. were able to achieve parts with 99% relative density using a combination of fused deposition modeling and robocasting [147]. Harrer et al. used LCM to fabricate YSZ that has been reported to have theoretical density of 99.6% and a strength of 650 MPa [151]. Using material extrusion, Yu et al. were able to obtain YSZ samples with 94.6% relative density, with a flexural strength of 488.96 MPa and compressive strength of 1.56 GP [163].

- Alumina (Al_2O_3) can maintain its strength in high temperature and has high hardness, superior wear and corrosion resistance, low density and bioinertness, making it one of the most used engineering ceramics in biomedical applications.

  Fan et al. used an SLM system with a high-energy diode laser to form alumina single tracks, which have microhardness values within the range of 15.4−19.9 GPa [164]. Fayed et al. achieved a minimum porosity of 4.34% and Vickers microhardness of 1682 HV using SLS to process alumina [165]. Zhang et al. managed to melt Al_2O_3 using SLM fully. However, orderly cracks are also observed in the parts due to large temperature gradients that are inherent due to the process [166].

  Schwentenwein and Homa used lithography-based manufacturing to process Al_2O_3 and obtained 99.3% relative density compared to the theoretical density of alumina with four-point bending strength of 427 MPa and an associated Weibull modulus of 11.2. The parts formed also have a relatively low surface roughness (Ra) of between 0.36 and 1.08 μm [150]. With the use of binder jetting technology, Gonzalez et al. were able to obtain Al_2O_3 parts with 96% relative density, and with additional 16 hours of sintering, the compressive strength of the parts is 131.86 MPa [157]. Scheithauer et al. were able to achieve parts with 99% relative density using a combination of fused deposition modeling and robocasting [147].

  Rueschhoff et al. explored using continuous filament of a highly loaded colloidal suspension of Al_2O_3 in direct ink writing (DIW). However, the flexural strengths of the parts, between 134 and 157 MPa were lower than expected due to the porosity induced by the suspensions [148]. Tang et al. used a slurry to replace the typical sheet materials in LOM in the process termed selective laser burn-out to fabricate Al_2O_3 parts [146]. Li et al. achieved a relative density of 99.1%−99.5% for zirconia parts using CODE and the parts have an HV and fracture toughness (K_{Ivb}) of 13.1 GPa and 4.6 MPa m$^{1/2}$, respectively. The flexural strength obtained from a four-point bending test of the parts was 563 MPa [149].

- Other ceramics. There are still many ceramics that have the potential for fabrication using AM. The development of materials has been a key focus for biomedical applications using AM.

Exner et al. worked on micro laser sintering of silica−alumina (SiO_2/ Al_2O_3) using fine powder of less than 1 μm [167]. The produced components

The emerging frontiers in materials Chapter | 8 **319**

had good geometric resolution and surface quality. However, the material is still not suitable for high-strength components as the tensile strength of 120 MPa is still low, even after thermal treatment.

Shishkovsky et al. [168] explored the use of direct laser sintering of YSZ and aluminum/alumina powder mixture. The ceramic parts they developed were porous. Bertrand et al. [169] built zirconia–yttria (ZrO_2/Y_2O_3) components that have a density of 56% by partially melting the ceramic powder via SLS/SLM. Yves-Christian et al. reported SLM of Al_2O_3/ZrO_2 that can produce components that are almost fully dense and have flexural strength above 500 MPa. This is achieved by completely melting the ceramic powder with preheating at 1600°C during the process [170]. However, the samples have a dimensional accuracy of 150 μm and surface roughness of $Rz = 60$ μm. Wu et al. fabricated zirconia-toughened alumina using SLA. The fabricated parts have a density of 4.26 g/cm^3, HV = 17.76 GPa, bending strength = 530.25 MPa, and fracture toughness of 5.72 MPa m$^{1/2}$ [171].

Gaytan et al. used binder jetting to fabricate $BaTiO_3$ parts with the maximum relative density of 63% with postsintering at 1400°C for 4 hours [158]. Gan and Wong used SLM for spodumene glass ceramic. While internal cracks are observed, they managed to obtain a flexural strength of 2.10 MPa [172]. Minasyan et al. used SLS to fabricated parts using Si_3N_4 and are able to obtain full density, HV = 12 GPa, and compressive strength of 432 MPa [173]. Chen et al. fabricated dental restoration from 3Y-TZP ceramics using SLS and cold isostatic pressing. Cold isostatic pressing is used to improve the density and hence, the mechanical properties of the parts. The final products achieved the highest flexural strength of 279.5 MPa and a relative density of 86.65% [135]. Danezan et al. used SLS to process porcelain that consists of kaolinite, quartz, and potassium feldspar. However, they are only able to obtain a relative density of 40% [174]. Samples of the SLS-formed porcelain are shown in Fig. 8.6.

- Wang et al. used vat polymerization to fabricate SiOC and SiC from precursor preceramic polymers. The SiOC components achieved 97% theoretical density [175]. Revelo and Colorado used DIW to fabricate parts from kaolinite clay. However, the compressive strength has a wide range of 20−50 MPa [176].
- In biomedical research, Liu used direct laser sintering on hydroxyapatite (HAP) and silica slurry mixture to produce bone scaffolds [177]. Also working on HAP, Liu et al. used DLP to fabricate scaffolds and achieved a relative density of 94.9%, with bending strength of 41.3 MPa [153].

8.4.3 Challenges and potential in ceramics for three-dimensional printing

AM has shown its potential in manufacturing metal and polymer parts with complex shapes directly with minimal processing steps. However, many

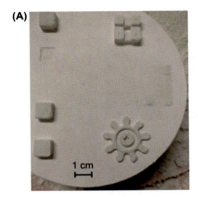

FIGURE 8.6 Photographs of the SLS porcelain objects (A) on the porcelain plateau and (B) separated from the porcelain plateau [174]. *SLS*, Selective laser sintering.

challenges remain in fabricating fully dense ceramic parts using these technologies. Successful integration of ceramics into AM processes will spur significant developments in the field of technical ceramics.

8.4.3.1 Flowability of ceramics

For powder bed fusion processes, good flowability of powder is critical for forming a thin and even layer during the powder spreading onto the substrate or the preceding layer [178]. Powder parameters such as particle-size distribution, morphology, and flow characteristics are critical and must be controlled [179]. The small powder particles tend to agglomerate and adhere to the surface of the coater and printing platform due to electrostatic charges. This created a challenge in achieving a homogeneous powder layer deposition [155]. It is important to note that the powder bed density has an influence on the final part quality produced by powder bed fusion processes. Gaytan et al. improved the powder flowability for binder jetting by heating them before every build. The heating also removes any excess binder and agglomerations [158].

Even for material extrusion processes, abrasion of the equipment components is very high due to the highly loaded suspensions or slurry used to create highly dense ceramic parts [147].

8.4.3.2 Thermal and residual stresses

Rapid solidification is typical in laser processing that results in full melting cause significant residual stresses and strains. Residual stresses were even reported for partial melting without infiltration [180]. The temperature gradients during the process and material physical properties affect the residual stresses. Ceramics generally allow only low plastic deformation at room temperature [181], with typical elongation at break of about 0.2%−0.6% for oxide ceramics.

Rapid melting and cooling of ceramics, which are typically characterized by high-temperature gradients in directed energy deposition and powder bed fusion processes [144], will also impact the grain size of the ceramic part produced negatively. As the laser system in such machines is unable to provide the capability to reach a high temperature that allows solid-state sintering and slow diffusion, high-temperature gradient occurs during ceramic processing. This undesired condition often results in cracks in the ceramic parts [182]. Wu et al. found that using water-cooled substrate in LAM can lower the heat accumulation in Al_2O_3−YAG during laser processing [159]. This method has the potential to reduce the thermal and residual stresses.

8.4.3.3 Dimensional accuracy

Mitteramskogler et al. found that light scattering within the ceramic-filled slurry for vat polymerization causes a certain amount of widening of dimensions in the final geometry. This overgrowth is both sensitive to the overall exposure area and exposure time. They suggest splitting the images used for this technique into core and contour (perimeter) to reduce such inaccuracies [152]. Sintering of the ceramic parts is usually accompanied by nonuniform dimensional changes such as reduction of axial dimensions, also known as shrinkage anisotropy [158].

Shrinkage in the parts fabricated can happen after sintering, as shown in Fig. 8.7. Liu et al. observed and concluded that to avoid defects caused by

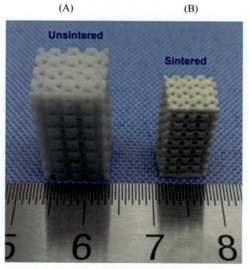

FIGURE 8.7 As fabricated HA bone scaffold (A) and after sintering (B) [153].

322 Digital Manufacturing

shrinkage, the slurry used for DLP must have as much solid content as possible. However, this reduces the viscosity of the slurry [153].

Further research on the technical issues and challenges in the processing of ceramics by AM is required. When these challenges are resolved, the future of direct AM ceramics will be disruptive to various industries, especially in biomedical sectors.

8.5 Conductive materials for electronic printing

Conventionally, electronics are predominantly fabricated by silicon wafer technology and printed circuit board (PCB) technology. However, the processes involved in these technologies are often complex and tedious. For instance, silicon chip fabrication normally involves numerous stages such as lithography, etching, and deposition [183]. Similarly, the PCB fabrication process consists of several steps such as masking, aligning, etching, drilling, and plating [184]. Some of these processes can produce harmful chemicals and require high energy consumption due to the high processing temperature. Besides, the "subtractive" approach (e.g., etching and drilling) used in these techniques inevitably brings forth excessive wastage of materials. Despite the disadvantages that these conventional techniques have brought, they are still accepted widely for the fabrication of electronics because of their compatibility with the mass production of high-quality electronic components. However, this has slowly changed since the inception of printed electronics.

8.5.1 Overview of three-dimensional printed electronics

Printed electronics is an emerging field of research that sees large markets because it circumvents the traditional expensive and inflexible silicon-based electronics and allows for the fabrication of a variety of devices on flexible substrates using high-throughput printing approaches. It is now possible to fabricate simple circuits embedded with several kinds of passive components, for example, resistors, capacitors, and inductors, which are the most fundamental building blocks of any electrical circuit.

Within additive printing techniques, it can be categorized into two main groups, contact and noncontact printing techniques. Examples of contact printing techniques include screen printing, gravure printing, and flexographic printing that allow fabrications of not only complex physical prototypes but also functional electronic components and circuitries. The comparisons of the printing techniques can be found in Table 8.5.

- Screen printing technique uses a stencil to prevent the ink from reaching the unnecessary areas of the substrate, resulting in ink patterns on the substrate. This roll-to-roll (R2R) printing technique involves feeding a continuous roll of flexible substrate. Adding viscosity modifiers will

TABLE 8.5 Comparisons of different printing techniques [185].

	Viscosity (Pa s)	Surface tension (mN/m)	Line width (μm)	Line thickness (μm)	Print speed (m/min)	Roll-to-roll compatibility
Contact printing:						
• Screen	0.1–5	35–50	30–50	0.1–100	10–15	Yes
• Gravure	0.1–1	25–45	10–50	0.1–1	10–1000	Yes
• Flexographic	0.05–0.5	10–30	20–100	0.005–1	100–500	Yes
Noncontact printing:						
• Inkjet	0.01–0.02	15–50	30–50	0.01–1	15–500	Partial
• Aerosol jet	0.001–0.16	–	10–120	0.01–3	Up to 12	Partial

324 Digital Manufacturing

increase the ink's viscosity and prevent the unintentional spreading of ink but can compromise the electrical conductivity of the printed patterns.

- Gravure printing, on the other hand, is an R2R contact printing technique that uses a rotating grooved cylinder containing a print pattern to transfer the ink onto the substrate. Unlike screen printing, this printing technique uses low viscosity ink, which allows for the printing of high conductivity traces.
- Flexographic printing techniques utilize a fountain roller, an anilox roller, and a printing plate cylinder. The anilox roller receives ink from a fountain roller and transfers ink to a printing plate cylinder. The anilox roller contains an array of cells, the number, size, and geometry of which can determine the amount of ink that will be delivered to the printing plate. The printing plate is normally made up of soft materials like polymer or rubber with the print pattern embossed on its surface. As such, the printing process is very similar to a stamping process. The soft print plate makes this technique suitable for printing conductive tracks on soft and compressible substrates, pressure-sensitive foils, and metalized films to be printed. Similar to screen printing and gravure printing, this technique is R2R compatible which allows for fast fabrication parts. However, its print resolution is relatively lower than screen printing and gravure printing.

Inkjet and aerosol jet printings are some examples of noncontact printing techniques. There are two kinds of inkjet printing techniques, continuous inkjet printing and drop-on-demand inkjet printing.

- Continuous inkjet printing ejects the ink droplets constantly. It controls the drop pattern by deflecting the ink using electrostatic force by charging up the droplets to prevent ink from reaching the substrates.
- Drop-on-demand inkjet printing, on the other hand, controls the deposition of ink using mechanisms such as thermal, piezoelectric, and electrostatic actuation. The aerosol jet printing technique is a noncontact droplet deposition technique that consists of multiple stages. First, the functional ink is atomized to form tiny droplets using ultrasonic or pneumatic methods. Then, the tiny droplets are then channeled into the deposition head via a carrier gas through a delivery tubing. At the deposition head the ink droplets are then aerodynamically focused by the sheath gas which encompasses the ink to form a tiny beam of aerosol jet.

The high exit velocity of the ink droplets in aerosol jetting allows the printing to happen at a large standoff distance, allowing noncontact printing on irregular surfaces. As the ejection of inks from the nozzle of aerosol jet printing is a continuous process, a mechanical shutter or pneumatic shutter is used to deflect the ink to allow the control of ink deposition for forming patterns. Together with the five-axis motion controller, aerosol jet printing allows the printing of conductive patterns on curved surfaces to realize conformal printing.

Most of these 3D electronics printers are using techniques such as liquid-dispensing printing, inkjet printing, aerosol jet printing or a combination of

The emerging frontiers in materials Chapter | 8   **325**

these techniques with a few exceptions using material extrusion technique. This allows the designers to have the flexibility in terms of choosing the substrate materials, ink materials, print resolution, and the subsequent postprocessing procedures. Apart from being more superior to the traditional counterparts in terms of their print resolutions, the fabrication of fine-feature design via 3D printing allows for lower fabrication cost compared to other conventional lithography techniques when used for prototyping. These direct-write techniques, such as liquid-dispensing printing, inkjet printing, and aerosol jet printing, not only reduce the cost by eliminating the need for tooling and mask but also considerably reduce the time for prototyping [186]. With the incorporation of various automations, nonplanar circuit designs and electrical components can be integrated into structures or even printed directly onto a surface of a part realizing multifunctional structures that not only save weight but also space. This is especially valuable and useful in the aviation industry where weight and space are of the utmost importance.

8.5.2 Materials for three-dimensional printing of electronics

Materials selection is an important aspect of the design stage of an electronic device as it determines the performance of the device. The fabrication of the main components of electronics such as conductors, semiconductors, and insulators has been successfully demonstrated using various printing techniques. Each of these components can be printed with various materials in different forms catering to specific applications. In general, these materials are made into inks that can be processed by these printing techniques, summarized in Table 8.6.

The inks typically consist of the main materials, stabilizing agents, solvents, and additives (Fig. 8.8). The main materials used in the inks are normally in the form of nanomaterials because nanomaterials can be easily made into homogeneous stable dispersion. They tend to have a lower melting temperature which is beneficial for the subsequent postprocessing steps. However, the nanomaterials tend to form aggregates, causing suspension instability. Therefore stabilizing agents are needed to prevent the aggregations of nanomaterials and to regard the oxidation of the nanomaterials for the case of metal nanoparticles (NPs). The stabilizing agents are normally organic compounds that are electrically insulating. Hence, postprocessing steps are required to remove the organic stabilizer layer by either rinsing or thermally decomposing them. The solvents are the liquid carriers that are used to disperse the nanomaterials for forming stable and printable ink suspensions. Typically, the nanomaterials can be dispersed in water, organic solvents like ethanol and xylene, etc. or a mixture of both. The solvents will determine the viscosity and the surface tension of the ink which will affect the choice of the printing techniques. Things to consider when choosing a

TABLE 8.6 Overview of additive manufacturing technologies used for processing materials for electronic printing.

Technology	Process	Material	Ref.
Material jetting	Inkjet printing	Conducting materials: • Silver nanoparticle • Gold nanoparticle • PEDOT: PSS	[187–189]
		Semiconducting materials: • Carbon nanotube • Graphene • Indium tin oxide	[190–192]
	Aerosol jet printing	Conducting materials: • Silver nanoparticle • Gold nanoparticle	[193,194]
		Semiconducting materials: • Carbon nanotube • Graphene	[195,196]
Material extrusion	Liquid dispensing	• Carbon • Silver composite	[197,198]
	Filament extrusion	• Carbon black, graphene, copper filler—thermoplastic composite	[199]
Powder bed fusion	Selective laser melting	• Copper/high-density polyethylene composite	[200]

FIGURE 8.8 Different components of printable inks.

solvent for nanomaterial inks are the volatility, toxicity, and compatibility with the substrates. The particle loading or the number of dispersed NPs in the solvent will affect the viscosity of the ink and determine the electrical property of the printed trace. Lastly, the additives such as dispersants, wetting agents, and binding agents are added to the nanomaterial inks to improve the suspension homogeneity and compatibility of the ink to the substrate. The final electrical property will depend on the final deposition pattern of the ink which is dictated by evaporation mode and particle—substrate interactions [1,12].

Conductors are materials that can conduct electricity when a potential difference is applied across the materials. Conductors are required for power transmission and signal transmission of electronic devices. Also, they can be used for developing passive electronic components such as resistors and inductors. The materials used for printing conductors are typically metal nanomaterials, metal oxide nanomaterials, carbon-based nanomaterials, and conducting polymer. Nanomaterials can be found in the forms of NP, nanowires, and nanoflakes. Among all the conducting materials, silver NPs inks are the most used materials because they have the best conductivity compared to the rest. Existing copper NPs inks are still not a viable low-cost substitute for the silver NPs ink due to the low conductivities. Gold NPs inks, despite being more expensive, are widely used for fabricating electrodes for electrochemical sensors, interdigitated electrodes, and microelectrodes arrays due to their inert nature. Palladium NPs inks are starting to receive attention for use as sensing electrodes due to their unique electrocatalytic properties. For electromagnetic applications, nickel NPs inks are suitable candidates due to their unique ferromagnetic properties. On the other hand, metal nanowires, indium tin oxide NPs and poly(3,4-ethylenedioxythiophene) polystyrene sulfonate (PEDOT: PSS) inks have been successfully attempted for the fabrication of transparent electronics. Lastly, graphene, carbon nanotubes (CNTs), and nanowires inks have been found suitable for flexible electronics due to the flexible nature of the materials and the elastic nature of the intertwined networks. Generally, most of the inks will require postprinting treatment such as thermal sintering to make the ink more conductive except CNT inks.

Transistors are the most important components of electronic devices. The fabrication of transistors requires the use of semiconducting material that allows current flow between the drain and source electrode when voltage is applied and block the current flow when there is no voltage applied. Semiconducting materials are critical building blocks for developing printed electronic devices and they must have high purities for high-transport properties. In general, they can be grouped into inorganic and organic semiconducting materials. Inorganic semiconductors are normally used for high-performance electronics devices due to their superior electrical properties and environmental stability. Organic semiconductors, on the other hand, are typically used for low-cost electronics devices.

Printable inorganic semiconductors are usually based on one-dimensional (1D) materials such as grown nanowires, nanoribbons, nanotubes, and 2D materials such as graphene and exfoliated semiconductors. Compared to NPs, these materials have fewer grain boundaries and allow charge to flow continuously thus allowing for higher charge carrier transport. A wide variety of semiconductors have been attempted such as ZnO, GaAs, InAs, AlGaAs nanowires, and CNTs for 1D materials, whereas graphene and MoS_2 for 2D materials.

Organic semiconductors are generally low cost and can be easily processed under a less controlled environment compared to inorganic semiconductors.

328 Digital Manufacturing

There are two main groups of organic semiconductors for printed electronics, oligomeric and polymer semiconductors.

- Oligomeric semiconductors are small molecules semiconductors, such as pentacene, Dihexyl-α-6T, and Buckminsterfullerene (C_{60}). They are generally more stable to electrochemical modification but tend to have low solubility [201]. They also tend to exhibit better electrical properties than solution-processed polymers due to the crystalline structures of the semiconductors.
- Polymer semiconductors, such as P3HT, can form good film-forming properties. However, it is very challenging to obtain pure polymer semiconducting materials because they cannot survive under normal purification processes, such as vacuum sublimation and chromatography. Also, polymers have defined grain boundaries, thus hindering them from being used as high-performance semiconducting materials.

In general, organic semiconducting materials have lower environmental stability such as sensitivity to humidity and prone to oxidations. Therefore they are mainly used for low-cost electronic devices that do not require stringent requirements on the devices' stability and performance. Typically, organic semiconducting materials are used for thin-film transistors and organic light-emitting diodes.

Dielectric is an insulating material that can be polarized by an applied electric field. It plays an important role in deciding the performance of organic transistors. Dielectrics can be grouped into inorganic and organic dielectrics. Inorganic dielectrics such as silicon dioxide and silicon nitride are normally processed by vapor deposition techniques, making them unsuitable for printed electronics. As such, organic dielectric materials are normally used because they can be solution-processed. Commonly used printable polymeric insulators are polyvinyl phenol, polystyrene, polyimide, sol$-$gel derived-siloxane-based hybrid polymers, polyester, polymethylmethacrylate, polyvinyl alcohol, and poly (vinylpyrrolidone). It is important to note that the device performance is dependent on the interfacial property of the dielectric with the conductors and semiconductors, such as the surface roughness.

Other than these materials, many nanocomposite materials are also widely explored for printed electronics [202,203]. For instance, the mixing of conductive materials such as CNTs and graphene with a matrix material such as silicone has been attempted to produce conductive stretchable materials with strain sensing capability.

8.5.3 Challenges and potential in three-dimensional printing electronics

AM of electronics has numerous benefits, such as faster prototyping process, shorter time-to-market, and the ability to increase part complexity and mass customization in a shorter time frame at a lower cost [185].

Fabrication of PCB through conventional methods typically takes a few weeks before the prototype can be tested. This makes iterative designs of circuit board inefficient and uneconomical. Often, this has become the bottleneck in the prototyping stage. With 3D printing techniques, the time-to-market of a product can be shortened significantly as the time for prototyping is drastically reduced. Apart from that, designers would be able to iterate the design of a product at a much lower cost as compared to conventional manufacturing methods.

The integration of the AM techniques for electronics with the existing 3D printing techniques allows for the fabrication of multifunctional devices with electronics printed on or embedded in the structures [204]. This changes the product development cycles from designing the structures and electronics separately to combining both components at the early design stage. The streamlining of product development not only opens the new possibility but also allows for a better-optimized structure in terms of weight, space, and cost. For instance, wires can be printed onto the structures directly, thereby eliminating the need for conventional wires that add weight and occupy space.

Other than that, 3D printing of electronics can help achieve better efficiency in supply chains, where inventories of preproduced electronics components can be substantially reduced, thus lowering logistic cost. With 3D printing, only the inks are required to be stored and electronics components can be printed on-demand. This also reduces the chance of fabricating and stocking up parts that will obsolete over time.

The ability of 3D printing to use a wide variety of substrate materials has allowed for innovation and new applications that are not achievable by conventional electronic manufacturing methods. This is especially true for flexible and stretchable applications where the substrates are temperature-sensitive materials. Some examples of printed electronics applications are flexible and stretchable electronics for wearables [205,206], optoelectronics devices, energy harvesting devices, electrochemical devices, and green and biodegradable electronics.

For printed electronics the key challenges mainly come from two aspects, the materials and the printing process.

8.5.3.1 Challenges involving the materials

From the material's perspective, challenges arise from the interplay between charge carrier mobility, complementary devices, material formulation, efficiency, and environmental stability. In most cases the charge carrier mobility of printed transistors is at least 2−3 order lesser in magnitude compared to the silicon counterpart, limiting their capability to produce high-performance electronics. Besides, p-type printable semiconductors are found to be more abundant while n-type is less common. Also, printable n-type semiconductors have much

330 Digital Manufacturing

lesser charge mobility than the *p*-type semiconductor. This limits the design of printable transistor to using a unipolar design that suffers in terms of gain, noise margin, voltage level shifting, etc. It is also imperative to ensure the material formulation can provide for high mobility and high uniformity. This is to ensure that the printed components can perform as per the design and, at the same time, reproducible in industrial quantities. As mentioned in the earlier section, most printable organic semiconductors are lacking in terms of environmental stability. They are less resistant to humidity and prone to oxidation. This causes them to degrade easily and shorten their operational lifetime.

8.5.3.2 Limitation on the printing process

Some considerations arising from the electronic printing process include the print resolution, alignment, uniformity, throughput, and cost. In terms of electronics fabrication, high print resolution is imperative for the fabrication of miniature devices because the parasitic is lesser, yielding higher speed transistors. The alignment of different print layers is important when printing devices with a multilayer design. This is especially important when dealing with high-resolution printing because the variability of the components will be high if the alignment is poor. Currently, the greatest challenge of printed electronic circuits is the large variation and poor matching between the thin film transistors. Despite the fact that some printing techniques, such as gravure printing and screen printing are compatible with the high-throughput R2R fabrication technique, they tend to have a lower print resolution. In contrast, the high-resolution digital printing technique, such as inkjet printing and aerosol jet printing, is not as scalable as their counterparts. Thus their throughput is limited. Although 3D printing is suitable for product customization, it is only economical when there is sufficient market demand to bring down the overall cost of fabrication.

8.5.3.3 Concerns with environmental and safety issues

The environmental issues comprise the use of novel materials during production and the recyclability of the printed electronic products. The use of novel materials, such as inorganic NPs and CNTs during the production, could cause an environmental issue if they were disposed of inappropriately. Although 3D printing is known for producing almost no waste, these materials can still be released to the environment during the cleaning and maintenance process if measures are not taken. Another environmental issue arises at the end-of-life phase of printed electronics. Recycling materials is difficult as the printed devices normally contain a mix of materials such as polymers and metals. At times, it is challenging to recycle the materials due to the homogeneity of the materials and the complexity of the products. Other than that, the toxicity of the nanomaterials is another important issue regarding the safety of the production workers. It is known that NPs have a large

The emerging frontiers in materials Chapter | 8 **331**

surface-to-volume ratio and high specific surface energy, making them thermodynamically unstable. Together with their small sizes, these nanomaterials can permeate cell membranes and interact with cells or tissues within the body. Besides, volatile organic compounds (VOCs) used as the solvents of inks can cause long-term health effects. The primary concern is the absorption of the airborne nanomaterials and VOC into the body through inhalation, especially the nanotubes. The nanomaterials can get airborne by various means such as atomization for the case of aerosol jet printing. Therefore workers should wear proper personal protective equipment during production and adhere to the safety regulation when handling nanomaterials.

8.6 Summary and moving forward

In this chapter, we examined ways to inject functionality into 3D printed parts. We started by framing various approaches in adding functionality and then dove into composites, bioprinting, ceramics for biomedical implants, 3D printing of ceramics for various applications, and 3D printed electronics. We highlighted the challenges and potential industrial applications. Moving forward, success of 3D AM printed parts in general, and 3D printed functional parts will require us to focus the research on academic institutes and industries in the following areas. Establishing standards throughout the different stages in 3D printing would increase the confidence of using 3D printed parts to replace conventional manufacturing processes. AM printed parts have not been widely used in the industry due to a lack of certification and qualification methods [207]. To alleviate such uncertainties in using 3D printed parts, there is a need for quality assurance and quality control of feedstock materials and the printing processes. Advancement in measurement science will enable better understanding and characterization of both feedstock and printed material for the certification of 3D printed parts.

Artificial intelligence (AI) is another key area that will aid 3D printing through facilitating automated production and reducing the risk of error [208]. AI in 3D printing can assist in planning for the print job, such as object placement or preparation, and guiding the development of materials and processing parameters [209,210]. The current fabrication processes in 3D printing are mainly deterministic, where the print path and print parameters are predetermined by the user prior to printing. With AI-aided 3D printing, printing parameters (e.g., planar vs complex surface print path in in situ printing, input parameters such as laser energy density and material flow) can be adjusted. Sensor technology that assists in in situ process monitoring will be imperative for open- and closed-loop adjustment of the print parameter.

Lastly, introducing and controlling functionality at each fundamental printing units (i.e., voxels in droplet-based printing and laser/optic-based processes, such as SLM, SLS, SLA, DLP, and 2PP; filaments/strands in

332 Digital Manufacturing

extrusion-based printing) will elevate the worth of 3D printed parts and serve as unique advantages. Noteworthy is that technology advancement and innovation in the areas highlighted previously revolves around material sciences and methods to process materials using 3D printing. These would be the main driving forces for higher adoption of 3D printing as the main manufacturing process and revolutionize the manufacturing industry.

References

[1] G.L. Goh, S. Agarwala, W.Y. Yeong, Aerosol-jet-printed preferentially aligned carbon nanotube twin-lines for printed electronics, ACS Applied Materials & Interfaces 11 (46) (2019) 43719–43730. 2019/11/20.

[2] Y. Jo, et al., 3D-printable, highly conductive hybrid composites employing chemically-reinforced, complex dimensional fillers and thermoplastic triblock copolymers, Nanoscale 9 (16) (2017) 5072–5084. Available from: https://doi.org/10.1039/C6NR09610G.

[3] A. Mostafaei, P. Rodriguez De Vecchis, E.L. Stevens, M. Chmielus, Sintering regimes and resulting microstructure and properties of binder jet 3D printed Ni-Mn-Ga magnetic shape memory alloys, Acta Materialia 154 (2018) 355–364. 2018/08/01.

[4] Y.Y.C. Choong, S. Maleksaeedi, H. Eng, J. Wei, P.-C. Su, 4D printing of high performance shape memory polymer using stereolithography, Materials & Design 126 (2017) 219–225. 2017/07/15.

[5] E.B. Joyee, A. Szmelter, D. Eddington, Y. Pan, Magnetic field-assisted stereolithography for productions of multimaterial hierarchical surface structures, ACS Applied Materials & Interfaces 12 (37) (2020) 42357–42368.

[6] J.M. Lee, W.Y. Yeong, Engineering macroscale cell alignment through coordinated toolpath design using support-assisted 3D bioprinting, Journal of the Royal Society Interface 17 (168) (2020) 20200294.

[7] W.L. Ng, J.T.Z. Qi, W.Y. Yeong, M.W. Naing, Proof-of-concept: 3D bioprinting of pigmented human skin constructs, Biofabrication 10 (2) (2018).

[8] M. Kesti, et al., A versatile bioink for three-dimensional printing of cellular scaffolds based on thermally and photo-triggered tandem gelation, Acta Biomaterialia 11 (2015) 162–172.

[9] S. Heid, A.R. Boccaccini, Advancing bioinks for 3D bioprinting using reactive fillers: a review, Acta Biomaterialia 113 (2020) 1–22.

[10] M. Mahmoudi, et al., On the printability and transformation behavior of nickel-titanium shape memory alloys fabricated using laser powder-bed fusion additive manufacturing, Journal of Manufacturing Processes 35 (2018) 672–680. 2018/10/01.

[11] J.M. Lee, W.L. Ng, W.Y. Yeong, Resolution and shape in bioprinting: strategizing towards complex tissue and organ printing, Applied Physics Reviews 6 (1) (2019) 011307.

[12] G.L. Goh, N. Saengchairat, S. Agarwala, W.Y. Yeong, T. Tran, Sessile droplets containing carbon nanotubes: a study of evaporation dynamics and CNT alignment for printed electronics, Nanoscale (2019). Available from: https://doi.org/10.1039/C9NR03261D.

[13] A.S. Gladman, E.A. Matsumoto, R.G. Nuzzo, L. Mahadevan, J.A. Lewis, Biomimetic 4D printing, Nature Materials 15 (4) (2016) 413.

[14] L. Friedrich, M. Begley, Printing direction dependent microstructures in direct ink writing, Additive Manufacturing 34 (2020) 101192.

[15] Y. Sriphutkiat, S. Kasetsirikul, D. Ketpun, Y. Zhou, Cell alignment and accumulation using acoustic nozzle for bioprinting, Scientific Reports 9 (1) (2019) 17774. 2019/11/28.

The emerging frontiers in materials Chapter | 8 **333**

[16] J.R. Raney, B.G. Compton, J. Mueller, T.J. Ober, K. Shea, J.A. Lewis, Rotational 3D printing of damage-tolerant composites with programmable mechanics, Proceedings of the National Academy of Sciences of the United States of America 115 (6) (2018) 1198−1203.

[17] D. Kokkinis, M. Schaffner, A.R. Studart, Multimaterial magnetically assisted 3D printing of composite materials, Nature Communications 6 (2015) 8643. online.

[18] X. Li, et al., Limpet tooth-inspired painless microneedles fabricated by magnetic field-assisted 3D printing, Advanced Functional Materials 31 (2021) 2003725.

[19] M. Luo, et al., Bi-scale interfacial bond behaviors of CCF/PEEK composites by plasma-laser cooperatively assisted 3D printing process, Composites Part A: Applied Science and Manufacturing 131 (2020) 105812.

[20] G. Weisgrab, A. Ovsianikov, P.F. Costa, Functional 3D printing for microfluidic chips, Advanced Materials Technologies 4 (10) (2019) 1900275.

[21] S.J. Keating, M.I. Gariboldi, W.G. Patrick, S. Sharma, D.S. Kong, N. Oxman, 3D printed multimaterial microfluidic valve, PLoS One 11 (8) (2016) e0160624.

[22] J. Kim, D.-J. Yoo, 3D printed compact heat exchangers with mathematically defined core structures, Journal of Computational Design and Engineering 7 (4) (2020) 527−550.

[23] L. Xiao, et al., Programmable 3D printed wheat awn-like system for high-performance fogdrop collection, Chemical Engineering Journal 399 (2020) 125139.

[24] H. Zhang, et al., The effects of ultrasonic nanocrystal surface modification on the fatigue performance of 3D-printed Ti64, International Journal of Fatigue 103 (2017) 136−146. 2017/10/01.

[25] A. Baux, et al., Synthesis and properties of macroporous SiC ceramics synthesized by 3D printing and chemical vapor infiltration/deposition, Journal of the European Ceramic Society 40 (8) (2020) 2834−2854.

[26] Y. Zhang, Post-printing surface modification and functionalization of 3D-printed biomedical device, International Journal of Bioprinting 3 (2) (2017) 7.

[27] G.D. Goh, Process-Structure-Properties of Additively Manufactured Continuous Carbon Fiber Reinforced Thermoplastic Composite,", PhD, School of Mechanical & Aerospace Engineering, Nanyang Technological University, Singapore, 2019.

[28] H. Milkert, Jelwek's 3D Printed, Wood Composite Watches Launch on Indiegogo, 2014. Available from: <https://3dprint.com/30947/jelwek-wood-3d-printed-watch/>.

[29] P. Waterman, Stratasys Now Shipping Dedicated Carbon Fiber Edition 3D Printer, 2018. Available from: <https://www.digitalengineering247.com/article/stratasys-now-shipping-dedicated-carbon-fiber-edition-3d-printer>.

[30] L. Blain, US$5,000 for the World's First 3D Carbon Fiber Printer, 2014. Available from: <https://newatlas.com/markforged-mark-one-carbon-fiber-3d-printer/30642/>.

[31] N. Hall, Envisiontec Unveils Slcom 1, 2016, 3/9. Available from: <https://3dprintingindustry.com/news/envisiontec-unveils-slcomm-1-79294/>.

[32] G.D. Goh, Y.L. Yap, S. Agarwala, W.Y. Yeong, Recent progress in additive manufacturing of fiber reinforced polymer composite, Advanced Materials Technologies 4 (1) (2019).

[33] S.M.F. Kabir, K. Mathur, A.-F.M. Seyam, A critical review on 3D printed continuous fiber-reinforced composites: history, mechanism, materials and properties, Composite Structures 232 (2020).

[34] Fortify, An Introduction to Digital Composite Manufacturing, 2019.

[35] A.N. Dickson, J.N. Barry, K.A. McDonnell, D.P. Dowling, Fabrication of continuous carbon, glass and Kevlar fibre reinforced polymer composites using additive manufacturing, Additive Manufacturing 16 (2017) 146−152. 2017/08/01.

[36] D.N. Saheb, J.P. Jog, Natural fiber polymer composites: a review, Advances in Polymer Technology 18 (4) (1999) 351−363.

334 Digital Manufacturing

[37] D. Filgueira, S. Holmen, J.K. Melbø, D. Moldes, A.T. Echtermeyer, G. Chinga-Carrasco, Enzymatic-assisted modification of thermomechanical pulp fibers to improve the interfacial adhesion with poly(lactic acid) for 3D printing, ACS Sustainable Chemistry & Engineering 5 (10) (2017) 9338−9346. 2017/10/02.

[38] M.L. Shofner, K. Lozano, F.J. Rodríguez-Macías, E.V. Barrera, Nanofiber-reinforced polymers prepared by fused deposition modeling, Journal of Applied Polymer Science 89 (11) (2003) 3081−3090.

[39] S. Dul, L. Fambri, A. Pegoretti, Fused deposition modelling with ABS−graphene nanocomposites, Composites Part A: Applied Science and Manufacturing 85 (2016) 181−191.

[40] H.L. Tekinalp, et al., Highly oriented carbon fiber−polymer composites via additive manufacturing, Composites Science and Technology 105 (2014) 144−150.

[41] F. Van Der Klift, Y. Koga, A. Todoroki, M. Ueda, Y. Hirano, R. Matsuzaki, 3D printing of continuous carbon fibre reinforced thermo-plastic (CFRTP) tensile test specimens, Open Journal of Composite Materials 6 (01) (2015) 18.

[42] B.G. Compton, J.A. Lewis, 3D-printing of lightweight cellular composites, Advanced Materials 26 (34) (2014) 5930−5935.

[43] J.H. Kim, et al., Three-dimensional printing of highly conductive carbon nanotube microarchitectures with fluid ink, ACS Nano 10 (9) (2016) 8879−8887.

[44] M. Gurr, D. Hofmann, M. Ehm, Y. Thomann, R. Kübler, R. Mülhaupt, Acrylic nanocomposite resins for use in stereolithography and structural light modulation based rapid prototyping and rapid manufacturing technologies, Advanced Functional Materials 18 (16) (2008) 2390−2397.

[45] T. Nakamoto, O. Kanehisa, Y. Sugawa, Whisker alignment in microparts using laser stereolithography with applied electric field, Journal of Advanced Mechanical Design, Systems, and Manufacturing 7 (6) (2013) 888−902.

[46] A. Sakly, S. Kenzari, D. Bonina, S. Corbel, V. Fournée, A novel quasicrystal-resin composite for stereolithography, Materials & Design 56 (2014) 280−285.

[47] R. Charan, T. Renault, A.A. Ogale, A. Bagchi, Automated fiber-reinforced composite prototypes, in: Fifth International Conference on Rapid Prototyping, 1994, pp. 91−97.

[48] A. Gupta, A.A. Ogale, Dual curing of carbon fiber reinforced photoresins for rapid prototyping, Polymer Composites 23 (6) (2002) 1162−1170.

[49] D. Klosterman, R. Chartoff, G. Graves, N. Osborne, B. Priore, Interfacial characteristics of composites fabricated by laminated object manufacturing, Composites Part A: Applied Science and Manufacturing 29 (9) (1998) 1165−1174.

[50] Impossible-Objects, Mechanical Properties of CBAM by Impossible Objects, 2017, 24 February. Available from: <http://impossible-objects.com/technology/#top>.

[51] S. Yuan, J. Bai, K.C. Chua, J. Wei, K. Zhou, Material evaluation and process optimization of CNT-coated polymer powders for selective laser sintering, Polymers 8 (10) (2016) 370.

[52] M.S. Wahab, K.W. Dalgarno, B. Cochrane, Synthesis of polymer nanocomposites for selective laser sintering (SLS), Journal of Mechanics Engineering and Automation 1 (2) (2011) 100−107.

[53] C. Browning, G. Husman, J. Whitney, Moisture effects in epoxy matrix composites, Composite Materials: Testing and Design (Fourth Conference), ASTM International, 1977.

[54] I. Hamerton, J. Kratz, The use of thermosets in modern aerospace applications, Thermosets, Elsevier, 2018, pp. 303−340.

[55] A. Murari, C. Vinante, M. Monari, Comparison of PEEK and VESPEL® SP1 characteristics as vacuum seals for fusion applications, Vacuum 65 (2) (2002) 137−145.

[56] G. Liao, et al., Properties of oriented carbon fiber/polyamide 12 composite parts fabricated by fused deposition modeling, Materials & Design 139 (2018) 283−292. 2018/02/05.

[57] C. Provin, S. Monneret, Complex ceramic-polymer composite microparts made by micro-stereolithography, IEEE Transactions on Electronics Packaging Manufacturing 25 (1) (2002) 59−63.

[58] S.R. Athreya, K. Kalaitzidou, S. Das, Mechanical and microstructural properties of Nylon-12/carbon black composites: selective laser sintering vs melt compounding and injection molding, Composites Science and Technology 71 (4) (2011) 506−510.

[59] N. Li, Y. Li, S. Liu, Rapid prototyping of continuous carbon fiber reinforced polylactic acid composites by 3D printing, Journal of Materials Processing Technology 238 (2016) 218−225.

[60] K.V. Wong, A. Hernandez, A review of additive manufacturing, ISRN Mechanical Engineering 2012 (2012). Article ID 208760.

[61] G.D. Goh, et al., Characterization of mechanical properties and fracture mode of additively manufactured carbon fiber and glass fiber reinforced thermoplastics, Materials & Design 137 (2018) 79−89. 2018/01/05.

[62] N. van de Werken, H. Tekinalp, P. Khanbolouki, S. Ozcan, A. Williams, M. Tehrani, Additively manufactured carbon fiber-reinforced composites: state of the art and perspective, Additive Manufacturing 31 (2020).

[63] A. M. Ltd, Composite 3D Printing: An Emerging Technology With a Bright Future, 2020. Available from: <https://amfg.ai/2020/02/25/composite-3d-printing-an-emerging-technology-with-a-bright-future/>.

[64] Z. Quan, et al., Additive manufacturing of multi-directional preforms for composites: opportunities and challenges, Materials Today 18 (9) (2015) 503−512.

[65] R.D. Farahani, M. Dube, D. Therriault, Three-dimensional printing of multifunctional nanocomposites: manufacturing techniques and applications, Advanced Materials 28 (28) (2016) 5794−5821.

[66] T.A. Campbell, O.S. Ivanova, 3D printing of multifunctional nanocomposites, Nano Today 8 (2) (2013) 119−120.

[67] U. Kalsoom, P.N. Nesterenko, B. Paull, Recent developments in 3D printable composite materials, RSC Advances 6 (65) (2016) 60355−60371.

[68] H. Wu, et al., Recent developments in polymers/polymer nanocomposites for additive manufacturing, Progress in Materials Science 111 (2020).

[69] C.E. Duty, et al., Structure and mechanical behavior of Big Area Additive Manufacturing (BAAM) materials, Rapid Prototyping Journal (2017).

[70] W.Y. Yeong, G.D. Goh, 3D printing of carbon fiber composite: the future of composite industry? Matter 2 (6) (2020) 1361−1363.

[71] X. Li, B. Liu, B. Pei, J. Chen, D. Zhou, J. Peng, X. Zhang, W. Jia, T. Xu, Inkjet bioprinting of biomaterials, Chemical Reviews 120 (19) (2020) 10793−10833.

[72] W.L. Ng, X. Huang, V. Shkolnikov, G.L. Goh, R. Suntornnond, W.Y. Yeong, Controlling droplet impact velocity and droplet volume are key factors to achieving high cell viability in sub-nanoliter droplet-based bioprinting, International Journal of Bioprinting 8 (1) (2022) 424.

[73] W.L. Ng, J.M. Lee, M.M. Zhou, Y.W. Chen, K.X.A. Lee, W.Y. Yeong, Y.F. Shen, Vat polymerization-based bioprinting − process, materials, applications and regulatory challenges, Biofabrication 12 (2) (2020) 022001.

[74] J. Zhang, Q. Hu, S. Wang, J. Tao, M. Gou, Digital light processing based three-dimensional printing for medical applications, International Journal of Bioprinting 6 (1) (2020) 242.

[75] Z. Fu, S. Naghieh, C. Xu, C. Wang, W. Sun, X. Chen, Printability in extrusion bioprinting, Biofabrication 13 (3) (2021) 033001.

336 Digital Manufacturing

[76] P. Zhuang, W.L. Ng, J. An, C.K. Chua, L.P. Tan, Layer-by-layer ultraviolet assisted extrusion-based (UAE) bioprinting of hydrogel constructs with high aspect ratio for soft tissue engineering applications, PloS One 14 (6) (2019) e0216776.

[77] J.M. Lee, S.L. Sing, M. Zhou, W.Y. Yeong, 3D bioprinting processes: a perspective on classification and terminology, International Journal of Bioprinting 4 (2) (2018).

[78] G. Jürgen, et al., Biofabrication: reappraising the definition of an evolving field, Biofabrication 8 (1) (2016) 013001.

[79] W.L. Ng, C.K. Chua, Y.F. Shen, Print me an organ! Why we are not there yet, Progress in Polymer Science 97 (2019) 101145.

[80] W.L. Ng, J.M. Lee, M. Zhou, W.Y. Yeong, Hydrogels for 3-D bioprinting-based tissue engineering, Rapid Prototyping of Biomaterials: Techniques in Additive Manufacturing, 2nd, Woodhead Publishing, 2020, pp. 183−204.

[81] J. Groll, J.A. Burdick, D.W. Cho, B. Derby, M. Gelinsky, S.C. Heilshorn, T. Jungst, J. Malda, V.A. Mironov, K. Nakayama, A. Ovsianikov, W. Sun, S. Takeuchi, J.J. Yoo, T.B. F. Woodfield, A definition of bioinks and their distinction from biomaterial inks, Biofabrication 11 (1) (2018) 013001.

[82] A. Al-Shamkhani, R. Duncan, Radioiodination of alginate via covalently-bound tyrosina-mide allows monitoring of its fate in vivo, Journal of Bioactive and Compatible Polymers 10 (1) (1995) 4−13.

[83] J.A. Rowley, G. Madlambayan, D.J. Mooney, Alginate hydrogels as synthetic extracellular matrix materials, Biomaterials 20 (1) (1999) 45−53.

[84] C. Xu, W. Chai, Y. Huang, R.R. Markwald, Scaffold-free inkjet printing of three-dimensional zigzag cellular tubes, Biotechnology and Bioengineering 109 (12) (2012) 3152−3160.

[85] T. Xu, C. Baicu, M. Aho, M. Zile, T. Boland, Fabrication and characterization of bio-engineered cardiac pseudo tissues, Biofabrication 1 (3) (2009) 035001. 1-035006.

[86] P.A. Smethurst, et al., Structural basis for the platelet-collagen interaction, Journal of Biological Chemistry 282 (2) (2007) 1296−1304.

[87] A.D. Konitsiotis, N. Raynal, D. Bihan, E. Hohenester, R.W. Farndale, B. Leitinger, Characterization of high affinity binding motifs for the discoidin domain receptor DDR2 in collagen, Journal of Biological Chemistry 283 (11) (2008) 6861−6868.

[88] E.O. Osidak, V.I. Kozhukhov, M.S. Osidak, S.P. Domogatsky, Collagen as bioink for bio-printing: A comprehensive review, International Journal of Bioprinting 6 (3) (2020) 270.

[89] J.M. Lee, S.K.Q. Suen, W.L. Ng, W.C. Ma, W.Y. Yeong, Bioprinting of collagen: consid-erations, potentials, and applications, Macromolecular Bioscience 21 (1) (2021) 2000280.

[90] C.M. Smith, et al., Three-dimensional bioassembly tool for generating viable tissue-engineered constructs, Tissue Engineering 10 (9−10) (2004) 1566−1576.

[91] S. Michael, et al., Tissue engineered skin substitutes created by laser-assisted bioprinting form skin-like structures in the dorsal skin fold chamber in mice, PLoS One 8 (3) (2013) e57741. 1−12.

[92] L. Koch, et al., Skin tissue generation by laser cell printing, Biotechnology and Bioengineering 109 (7) (2012) 1855−1863.

[93] M. Djabourov, et al., 3D analysis of gelatin gel networks from transmission electron microscopy imaging, Journal de Physique II 3 (5) (1993) 611−624.

[94] S. Sakai, K. Hirose, K. Taguchi, Y. Ogushi, K. Kawakami, An injectable, in situ enzymat-ically gellable, gelatin derivative for drug delivery and tissue engineering, Biomaterials 30 (20) (2009) 3371−3377. 2009/07/01.

[95] S.A. Irvine, et al., Printing cell-laden gelatin constructs by free-form fabrication and enzy-matic protein crosslinking, Biomedical Microdevices 17 (1) (2015) 1−8.

The emerging frontiers in materials **Chapter | 8 337**

[96] H. Babin, E. Dickinson, Influence of transglutaminase treatment on the thermoreversible gelation of gelatin, Food Hydrocolloids 15 (3) (2001) 271−276.

[97] M.K. Cowman, S. Matsuoka, The intrinsic viscosity of hyaluronan, in: J.F. Kennedy, G. O. Phillips, P.A. Williams (Eds.), Hyaluronan, Proceedings of an International Meeting, September 2000, Hyaluronan, Woodhead Publishing, 2002.

[98] M.N. Collins, C. Birkinshaw, Hyaluronic acid based scaffolds for tissue engineering—a review, Carbohydrate Polymers 92 (2) (2013) 1262−1279.

[99] A.I. Van Den Bulcke, B. Bogdanov, N. De Rooze, E.H. Schacht, M. Cornelissen, H. Berghmans, Structural and rheological properties of methacrylamide modified gelatin hydrogels, Biomacromolecules 1 (1) (2000) 31−38.

[100] J.W. Nichol, S.T. Koshy, H. Bae, C.M. Hwang, S. Yamanlar, A. Khademhosseini, Cell-laden microengineered gelatin methacrylate hydrogels, Biomaterials 31 (21) (2010) 5536−5544.

[101] P.A. Gunatillake, R. Adhikari, Biodegradable synthetic polymers for tissue engineering, European Cells & Materials 5 (1) (2003) 1−16.

[102] W.-W. Hu, et al., The use of reactive polymer coatings to facilitate gene delivery from poly(ε-caprolactone) scaffolds, Biomaterials 30 (29) (2009) 5785−5792.

[103] M.-C. Serrano, R. Pagani, M. Vallet-Regí, J. Peña, J.-V. Comas, M.-T. Portolés, Nitric oxide production by endothelial cells derived from blood progenitors cultured on NaOH-treated polycaprolactone films: a biofunctionality study, Acta Biomaterialia 5 (6) (2009) 2045−2053.

[104] F. Causa, E. Battista, R. Della Moglie, D. Guarnieri, M. Iannone, P.A. Netti, Surface investigation on biomimetic materials to control cell adhesion: the case of RGD conjugation on PCL, Langmuir 26 (12) (2010) 9875−9884.

[105] J. Zhu, Bioactive modification of poly(ethylene glycol) hydrogels for tissue engineering, Biomaterials 31 (17) (2010) 4639−4656.

[106] C.N. Salinas, K.S. Anseth, The enhancement of chondrogenic differentiation of human mesenchymal stem cells by enzymatically regulated RGD functionalities, Biomaterials 29 (15) (2008) 2370−2377.

[107] G. Dumortier, N. El Kateb, M. Sahli, S. Kedjar, A. Boulliat, J. Chaumeil, Development of a thermogelling ophthalmic formulation of cysteine, Drug Development and Industrial Pharmacy 32 (1) (2006) 63−72.

[108] M. Di Biase, P. de Leonardis, V. Castelletto, I.W. Hamley, B. Derby, N. Tirelli, Photopolymerization of Pluronic F127 diacrylate: a colloid-templated polymerization, Soft Matter 7 (10) (2011) 4928−4937.

[109] Y. Xu, Y. Hu, C. Liu, H. Yao, B. Liu, S. Mi, A novel strategy for creating tissue-engineered biomimetic blood vessels using 3D bioprinting technology, Materials 11 (9) (2018) 1581. 1−15.

[110] M. Hospodiuk, M. Dey, D. Sosnoski, I.T. Ozbolat, The bioink: a comprehensive review on bioprintable materials, Biotechnology Advances 35 (2) (2017) 217−239. 2017/03/01.

[111] J. Jia, et al., Engineering alginate as bioink for bioprinting, Acta Biomaterialia 10 (10) (2014) 4323−4331.

[112] X. Wang, et al., Generation of three-dimensional hepatocyte/gelatin structures with rapid prototyping system, Tissue Engineering 12 (1) (2006) 83−90.

[113] D. Petta, A. Armiento, D. Grijpma, M. Alini, D. Eglin, M. D'Este, 3D bioprinting of a hyaluronan bioink through enzymatic-and visible light-crosslinking, Biofabrication 10 (4) (2018) 044104. 1-044111.

[114] L.E. Bertassoni, et al., Direct-write bioprinting of cell-laden methacrylated gelatin hydrogels, Biofabrication 6 (2) (2014) 024105. 1-024111.

338 Digital Manufacturing

[115] L. Hockaday, et al., Rapid 3D printing of anatomically accurate and mechanically heterogeneous aortic valve hydrogel scaffolds, Biofabrication 4 (3) (2012) 035005. 1−12.

[116] J.-H. Shim, J.-S. Lee, J.Y. Kim, D.-W. Cho, Bioprinting of a mechanically enhanced three-dimensional dual cell-laden construct for osteochondral tissue engineering using a multi-head tissue/organ building system, Journal of Micromechanics and Microengineering 22 (8) (2012) 085014.

[117] H. Gudupati, M. Dey, I. Ozbolat, A comprehensive review on droplet-based bioprinting: past, present and future, Biomaterials 102 (2016) 20−42.

[118] E. Hoch, T. Hirth, G.E. Tovar, K. Borchers, Chemical tailoring of gelatin to adjust its chemical and physical properties for functional bioprinting, Journal of Materials Chemistry B 1 (41) (2013) 5675−5685.

[119] G. Gao, T. Yonezawa, K. Hubbell, G. Dai, X. Cui, Inkjet-bioprinted acrylated peptides and PEG hydrogel with human mesenchymal stem cells promote robust bone and cartilage formation with minimal printhead clogging, Biotechnology Journal 10 (10) (2015) 1568−1577.

[120] H. Gudupati, J. Yan, Y. Huang, D.B. Chrisey, Alginate gelation-induced cell death during laser-assisted cell printing, Biofabrication 6 (3) (2014) 035022. 1−11.

[121] N.R. Schiele, D.B. Chrisey, D.T. Corr, Gelatin-based laser direct-write technique for the precise spatial patterning of cells, Tissue Engineering Part C: Methods 17 (3) (2010) 289−298.

[122] L. Koch, et al., Laser bioprinting of human induced pluripotent stem cells—the effect of printing and biomaterials on cell survival, pluripotency, and differentiation, Biofabrication 10 (3) (2018) 035005. 1-035021.

[123] W. Lee, et al., Multi-layered culture of human skin fibroblasts and keratinocytes through three-dimensional freeform fabrication, Biomaterials 30 (8) (2009) 1587−1595. 2009/03/01.

[124] I. Henriksson, P. Gatenholm, D. Hägg, Increased lipid accumulation and adipogenic gene expression of adipocytes in 3D bioprinted nanocellulose scaffolds, Biofabrication 9 (1) (2017) 015022. 1-015029.

[125] P. Soman, P.H. Chung, A.P. Zhang, S. Chen, Digital microfabrication of user-defined 3D microstructures in cell-laden hydrogels, Biotechnology and Bioengineering 110 (11) (2013) 3038−3047.

[126] H. Lin, et al., Application of visible light-based projection stereolithography for live cell-scaffold fabrication with designed architecture, Biomaterials 34 (2) (2013) 331−339.

[127] C. Loebel, N. Broguiere, M. Alini, M. Zenobi-Wong, D. Eglin, Microfabrication of photo-cross-linked hyaluronan hydrogels by single-and two-photon tyramine oxidation, Biomacromolecules 16 (9) (2015) 2624−2630.

[128] N. Diamantides, et al., Correlating rheological properties and printability of collagen bioinks: the effects of riboflavin photocrosslinking and pH, Biofabrication 9 (3) (2017) 034102.

[129] M. Kesti, et al., Bioprinting complex cartilaginous structures with clinically compliant biomaterials, Advanced Functional Materials 25 (48) (2015) 7406−7417.

[130] W.L. Ng, W.Y. Yeong, M.W. Naing, Polyvinylpyrrolidone-based bio-ink improves cell viability and homogeneity during drop-on-demand printing, Materials 10 (2) (2017) 190.

[131] W.L. Ng, J.M. Lee, W.Y. Yeong, M.W. Naing, Microvalve-based bioprinting−process, bio-inks and applications, Biomaterials Science 5 (4) (2017) 689−699.

[132] Z.B. Sendekie, P. Bacchin, Colloidal jamming dynamics in microchannel bottlenecks, Langmuir 32 (6) (2016) 1478−1488.

[133] M. Lovett, K. Lee, A. Edwards, D.L. Kaplan, Vascularization strategies for tissue engineering, Tissue Engineering Part B 15 (3) (2009) 353−370.

The emerging frontiers in materials Chapter | 8 **339**

[134] A. Blaeser, D.F.D. Campos, U. Puster, W. Richtering, M.M. Stevens, H. Fischer, Controlling shear stress in 3D bioprinting is a key factor to balance printing resolution and stem cell integrity, Advanced Healthcare Materials 5 (3) (2016) 326–333.

[135] F. Chen, J.-M. Wu, H.-Q. Wu, Y. Chen, C.-H. Li, Y.-S. Shi, Microstructure and mechanical properties of 3Y-TZP dental ceramics fabricated by selective laser sintering combined with cold isostatic pressing, International Journal of Lightweight Materials and Manufacture 1 (4) (2018) 239–245.

[136] W. Wang, S. Ma, J.Y.H. Fuh, L. Lu, Y. Liu, Processing and characterization of laser-sintered $Al_2O_3/ZrO_2/SiO_2$, The International Journal of Advanced Manufacturing Technology 68 (9–12) (2013) 2565–2569.

[137] D. Grüner, Z. Shen, Ordered coalescence of nano-crystals during rapid solidification of ceramic melts, CrystEngComm 13 (17) (2011) 5303–5305.

[138] R.D. Goodridge, D.J. Wood, C. Ohtsuki, K.W. Dalgarno, Biological evaluation of an apatite-mullite glass-ceramic produced via selective laser sintering, Acta Biomaterialia 3 (2) (2007) 221–231.

[139] J. Liu, H. Hu, P. Li, C. Shuai, S. Peng, Fabrication and characterization of porous 45S5 glass scaffolds via direct selective laser sintering, Materials and Manufacturing Processes 28 (6) (2013) 610–615.

[140] J. Wilkes, Y.C. Hagedorn, W. Meiners, K. Wissenbach, Additive manufacturing of ZrO_2-Al_2O_3 ceramic components by selective laser melting, Rapid Prototyping Journal 19 (1) (2013) 51–57.

[141] Q. Lian, W. Sui, X. Wu, F. Yang, S. Yang, Additive manufacturing of ZrO_2 ceramic dental bridges by stereolithography, Rapid Prototyping Journal 24 (1) (2017) 114–119.

[142] P. Tesavibul, et al., Processing of 45S5 Bioglass® by lithography-based additive manufacturing, Materials Letters 74 (2012) 81–84.

[143] S.L. Sing, J. An, W.Y. Yeong, F.E. Wiria, Laser and electron-beam powder-bed additive manufacturing of metallic implants: a review on processes, materials and designs, Journal of Orthopaedic Research 34 (3) (2016) 369–385.

[144] Y. Li, Y. Hu, W. Cong, L. Zhi, Z. Guo, Additive manufacturing of alumina using laser engineered net shaping: effects of deposition variables, Ceramics International 3 (10) (2017) 7768–7775.

[145] C.J. Huang, et al., Advanced brass-based composites via cold-spray additive-manufacturing and its potential in component repairing, Surface and Coatings Technology 371 (2019) 211–223.

[146] H.-H. Tang, H.-C. Yen, Slurry-based additive manufacturing of ceramic parts by selective laser burn-out, Journal of the European Ceramic Society 35 (3) (2015) 981–987.

[147] U. Scheithauer, E. Schwarzer, H.-J. Richter, T. Moritz, Thermoplastic 3D printing—an additive manufacturing method for producing dense ceramics, International Journal of Applied Ceramic Technology 12 (1) (2015) 26–31.

[148] L. Rueschhoff, W. Costakis, M. Michie, J. Youngblood, T. Rodney, Additive manufacturing of dense ceramic parts via direct ink writing of aqueous alumina suspensions, International Journal of Applied Ceramic Technology 13 (5) (2016) 821–830.

[149] W. Li, A. Ghazanfari, D. McMillen, M.C. Leu, G.E. Hilmas, J. Watts, Characterization of zirconia specimens fabricated by ceramic on-demand extrusion, Ceramics International 44 (11) (2018) 12245–12252.

[150] M. Schwentenwein, J. Homa, Additive manufacturing of dense alumina ceramics, International Journal of Applied Ceramic Technology 12 (1) (2015) 1–7.

340 Digital Manufacturing

[151] W. Harrer, M. Schwentenwein, T. Lube, R. Danzer, Fractography of zirconia-specimens made using additive manufacturing (LCM) technology, Journal of the European Ceramic Society 37 (14) (2017) 4331−4338.

[152] G. Mitteramskogler, et al., Light curing strategies for lithography-based additive manufacturing of customized ceramics, Additive Manufacturing 1−4 (2014) 110−118.

[153] Z. Liu, et al., Additive manufacturing of hydroxyapatite bone scaffolds via digital light processing and in vitro compatibility, Ceramics International 45 (8) (2019) 11079−11086.

[154] C.-J. Bae, D. Kim, J.W. Halloran, Mechanical and kinetic studies on the refractory fused silica of integrally cored ceramic mold fabricated by additive manufacturing, Journal of the European Ceramic Society 39 (2−3) (2019) 618−623.

[155] T. Muhler, J. Heinrich, Slurry-based additive manufacturing of ceramics, International Journal of Applied Ceramic Technology 12 (1) (2015) 18−25.

[156] K. Liu, et al., Laser additive manufacturing and homogeneous densification of complicated shape SiC ceramic parts, Ceramics International 44 (17) (2018) 21067−21075.

[157] J.A. Gonzalez, J. Mireles, Y. Lin, R.B. Wicker, Characterization of ceramic components fabricated using binder jetting additive manufacturing technology, Ceramics International 42 (9) (2016) 10559−10564.

[158] S.M. Gaytan, et al., Fabrication of barium titanate by binder jetting additive manufacturing, Ceramics International 41 (5A) (2015) 6610−6619.

[159] D. Wu, et al., Al_2O_3-YAG eutectic ceramic prepared by laser additive manufacturing with water-cooled substrate, Ceramics International 45 (3) (2019) 4119−4122.

[160] M. Mamivand, M. Asle Zaeem, H. El Kadiri, Phase field modeling of stress-induced tetragonal-to-monoclinic transformation in zirconia and its effect on transformation toughening, Acta Materialia 64 (2014) 208−219.

[161] I. Denry, J.A. Holloway, Ceramics for dental applications: a review, Materials 3 (1) (2010) 351−368.

[162] L. Ferrage, G. Betand, P. Lenormand, Dense yttria-stabilized zirconia obtained by direct selective laser sintering, Additive Manufacturing 21 (2018) 472−478.

[163] T. Yu, Z. Zhang, Q. Liu, R. Kuliiev, N. Orlovskaya, D. Wu, Extrusion-based additive manufacturing of yttria-partially-stabilized zirconia ceramics, Ceramics International 46 (4) (2020) 5020−5027.

[164] Z. Fan, M. Lu, H. Huang, Selective laser melting of alumina: a single track study, Ceramics International 44 (8) (2018) 9484−9493.

[165] E.M. Fayed, A.S. Elmesalamy, M. Sobih, Y. Elshaer, Characterization of direct selective laser sintering of alumina, The International Journal of Advanced Manufacturing Technology 94 (5−8) (2018) 2333−2341.

[166] K. Zhang, T. Liu, W. Liao, C. Zhang, Y. Yan, D. Du, Influence of laser parameters on the surface morphology of slurry-based Al_2O_3 parts produced through selective laser melting, Rapid Prototyping Journal 24 (2) (2018) 333−341.

[167] H. Exner, et al., Laser micro sintering: a new method to generate metal and ceramic parts of high resolution with sub-micrometer powder, Virtual and Physical Prototyping 3 (1) (2008) 3−11.

[168] I. Shishkovsky, I. Yadroitsev, P. Bertrand, I. Smurov, Alumina−zirconium ceramics synthesis by selective laser sintering/melting, Applied Surface Science 254 (4) (2007) 966−970.

[169] P. Bertrand, F. Bayle, C. Combe, P. Goeuriot, I. Smurov, Ceramic components manufacturing by selective laser sintering, Applied Surface Science 254 (4) (2007) 989−992.

[170] H. Yves-Christian, W. Jan, M. Wilhelm, W. Konrad, P. Reinhart, Net shaped high performance oxide ceramic parts by selective laser melting, Physics Procedia 5 (2010) 587–594.

[171] H. Wu, et al., Fabrication of dense zirconia-toughened alumina ceramics through a stereolithography-based additive manufacturing, Ceramics International 43 (1B) (2017) 968–972.

[172] M.X. Gan, C.H. Wong, Properties of selective laser melted spodumene glass-ceramic, Journal of the European Ceramic Society 37 (13) (2017) 4147–4154.

[173] T. Minasyan, et al., A novel approach to fabricate Si_3N_4 by selective laser melting, Ceramics International 44 (12) (2018) 13689–13694.

[174] A. Danezan, et al., Selective laser sintering of porcelain, Journal of the European Ceramic Society 38 (2) (2018) 769–775.

[175] X. Wang, F. Schmidt, D. Hanaor, P.H. Kamm, S. Li, A. Gurlo, Additive manufacturing of ceramics from preceramic polymers: a versatile stereolithographic approach assisted by thiol-ene click chemistry, Additive Manufacturing 27 (2019) 80–90.

[176] C.F. Revelo, H.A. Colorado, 3D printing of kaolinite clay ceramics using the Direct Ink Writing (DIW) technique, Ceramics International 44 (5) (2018) 5673–5682.

[177] F.-H. Liu, Synthesis of bioceramic scaffolds for bone tissue engineering by rapid prototyping technique, Journal of Sol-Gel Science and Technology 64 (3) (2012) 704–710.

[178] T. Marcu, M. Todea, I. Gligor, P. Berce, C. Popa, Effect of surface conditioning on the flowability of Ti6Al7Nb powder for selective laser melting applications, Applied Surface Science 258 (7) (2012) 3276–3282.

[179] W.H. Yu, S.L. Sing, C.K. Chua, C.N. Kuo, X.L. Tian, Particle-reinforced metal matrix nanocomposites fabricated by selective laser melting: a state of the art review, Progress in Materials Science 104 (2019) 330–379.

[180] P. Mercelis, J.P. Kruth, Residual stresses in selective laser sintering and selective laser melting, Rapid Prototyping Journal 12 (5) (2006) 254–265.

[181] F. Wakai, Superplasticity of ceramics, Ceramics International 17 (3) (1991) 153–163.

[182] J.H. Ouyang, S. Nowotny, A. Richter, E. Beyer, Laser cladding of yttria partially stabilized ZrO_2 (YPSZ) ceramic coatings on aluminum alloys, Ceramics International 27 (1) (2001) 15–24.

[183] J.N. Burghartz, C. Harendt, T. Hoang, A. Kiss, M. Zimmermann, Ultra-thin chip fabrication for next-generation silicon processes, in: 2009 IEEE Bipolar/BiCMOS Circuits and Technology Meeting, IEEE, 2009, pp. 131–137.

[184] I. Hui, C. Li, H. Lau, Hierarchical environmental impact evaluation of a process in printed circuit board manufacturing, International Journal of Production Research 41 (6) (2003) 1149–1165.

[185] H.W. Tan, T. Tran, C.K. Chua, A review of printed passive electronic components through fully additive manufacturing methods, Virtual and Physical Prototyping 11 (4) (2016) 271–288. 2016/10/01.

[186] G.L. Goh, S. Agarwala, W.Y. Yeong, Directed and on-demand alignment of carbon nanotube: a review toward 3D printing of electronics, Advanced Materials Interfaces 6 (2019) 1801318.

[187] T.T. Nge, M. Nogi, K. Suganuma, Electrical functionality of inkjet-printed silver nanoparticle conductive tracks on nanostructured paper compared with those on plastic substrates, Journal of Materials Chemistry C 1 (34) (2013) 5235–5243.

[188] B. Bachmann, et al., All-inkjet-printed gold microelectrode arrays for extracellular recording of action potentials, Flexible and Printed Electronics 2 (3) (2017) 035003.

342 Digital Manufacturing

[189] S.H. Eom, et al., Polymer solar cells based on inkjet-printed PEDOT: PSS layer, Organic Electronics 10 (3) (2009) 536–542.

[190] K. Kordás, et al., Inkjet printing of electrically conductive patterns of carbon nanotubes, Small 2 (8–9) (2006) 1021–1025.

[191] A. Capasso, A.D.R. Castillo, H. Sun, A. Ansaldo, V. Pellegrini, F. Bonaccorso, Ink-jet printing of graphene for flexible electronics: an environmentally-friendly approach, Solid State Communications 224 (2015) 53–63.

[192] M.-S. Hwang, B.-Y. Jeong, J. Moon, S.-K. Chun, J. Kim, Inkjet-printing of indium tin oxide (ITO) films for transparent conducting electrodes, Materials Science and Engineering: B 176 (14) (2011) 1128–1131.

[193] H.W. Tan, N. Saengchairat, G.L. Goh, J. An, C.K. Chua, T. Tran, Induction sintering of silver nanoparticle inks on polyimide substrates, Advanced Materials Technologies (2019) 1900897.

[194] S. Khan, T.P. Nguyen, M. Lubej, L. Thiery, P. Vairac, D. Briand, Low-power printed micro-hotplates through aerosol jetting of gold on thin polyimide membranes, Microelectronic Engineering 194 (2018) 71–78.

[195] J.B. Andrews, C. Cao, M.A. Brooke, A.D. Franklin, Noninvasive material thickness detection by aerosol jet printed sensors enhanced through metallic carbon nanotube ink, IEEE Sensors Journal 17 (14) (2017) 4612–4618.

[196] E.B. Secor, S. Lim, H. Zhang, C.D. Frisbie, L.F. Francis, M.C. Hersam, Gravure printing of graphene for large-area flexible electronics, Advanced Materials 26 (26) (2014) 4533–4538.

[197] J.T. Muth, et al., Embedded 3D printing of strain sensors within highly stretchable elastomers, Advanced Materials 26 (36) (2014) 6307–6312.

[198] A.D. Valentine, et al., Hybrid 3D printing of soft electronics, Advanced Materials 29 (40) (2017) 1703817.

[199] P.F. Flowers, C. Reyes, S. Ye, M.J. Kim, B.J. Wiley, 3D printing electronic components and circuits with conductive thermoplastic filament, Additive Manufacturing 18 (2017) 156–163.

[200] S. Hou, S. Qi, D.A. Hutt, J.R. Tyrer, M. Mu, Z. Zhou, Three dimensional printed electronic devices realised by selective laser melting of copper/high-density-polyethylene powder mixtures, Journal of Materials Processing Technology 254 (2018) 310–324.

[201] J. Rogers, H. Katz, Printable organic and polymeric semiconducting materials and devices, Journal of Materials Chemistry 9 (9) (1999) 1895–1904.

[202] V. Dikshit, G.D. Goh, A.P. Nagalingam, G.L. Goh, W.Y. Yeong, Chapter 17 – Recent progress in 3D printing of fiber-reinforced composite and nanocomposites, in: B. Han, S. Sharma, T.A. Nguyen, L. Longbiao, K.S. Bhat (Eds.), Fiber-Reinforced Nanocomposites: Fundamentals and Applications, Elsevier, 2020, pp. 371–394.

[203] S. Agarwala, G.L. Goh, G.D. Goh, V. Dikshit, W.Y. Yeong, Chapter 10 – 3D and 4D printing of polymer/CNTs-based conductive composites, in: K.K. Sadasivuni, K. Deshmukh, M.A. Almaadeed (Eds.), 3D and 4D Printing of Polymer Nanocomposite Materials, Elsevier, 2020, pp. 297–324.

[204] G.L. Goh, et al., Additively manufactured multi-material free-form structure with printed electronics, The International Journal of Advanced Manufacturing Technology 94 (1–4) (2018) 1309–1316.

[205] G.L. Goh, S. Agarwala, Y.J. Tan, W.Y. Yeong, A low cost and flexible carbon nanotube pH sensor fabricated using aerosol jet technology for live cell applications, Sensors and Actuators B: Chemical 260 (2018) 227–235.

[206] S. Agarwala, et al., Wearable bandage based strain sensor for home healthcare: combining 3D aerosol jet printing and laser sintering, ACS Sensors (2018).

[207] C.K. Chua, C.H. Wong, W.Y. Yeong, Chapter Three — Measurement science roadmap for additive manufacturing, in: C.K. Chua, C.H. Wong, W.Y. Yeong (Eds.), Standards, Quality Control, and Measurement Sciences in 3D Printing and Additive Manufacturing, Academic Press, 2017, pp. 57–73.

[208] Z. Zhu, D.W.H. Ng, H.S. Park, M.C. McAlpine, 3D-printed multifunctional materials enabled by artificial-intelligence-assisted fabrication technologies, Nature Reviews Materials (2020) 1–21.

[209] J.-P. Correa-Baena, et al., Accelerating materials development via automation, machine learning, and high-performance computing, Joule 2 (8) (2018) 1410–1420. 2018/08/15.

[210] T.J. Wang, T.H. Kwok, C. Zhou, S. Vader, In-situ droplet inspection and closed-loop control system using machine learning for liquid metal jet printing, Journal of Manufacturing Systems 47 (2018) 83–92.

Chapter 9

Three-dimensional (3D) printing for building and construction

Mingyang Li[1], Xu Zhang[1], Yi Wei Daniel Tay[1], Guan Heng Andrew Ting[1], Bing Lu[1] and Ming Jen Tan[1,2]

[1]Singapore Centre for 3D Printing, School of Mechanical and Aerospace Engineering, Nanyang Technological University, Singapore, [2]HP-NTU Digital Manufacturing Corporate Lab, School of Mechanical and Aerospace Engineering, Nanyang Technological University, Singapore

List of abbreviations

3D	three-dimensional
3DP	three-dimensional printing
3DCP	three-dimensional concrete printing
AM	additive manufacturing
ASR	alkali-silica reaction
CAD	Computer-aided design
DCP	Digital construction platform
GGBFS	Ground granulated blast furnace slag
NTU	Nanyang Technological University
SCC	Self-compacting concrete
SCM	Supplementary cementitious materials

9.1 Introduction

9.1.1 Digital transformation and automation in building and construction

The world of construction lags behind manufacturing, agriculture, and wholesale/retail in terms of productivity over the last few centuries. Although the productivity of farming and production has increased 10−15 times in the past 80 years, construction remains stuck at the same pace as the 1950s, i.e., at about 1% per annum. The growth of automation in construction has been slow despite construction being one of the largest industries in the world economy, which contributes approximately 13% of GDP or US$10 trillion globally.

Digital Manufacturing. DOI: https://doi.org/10.1016/B978-0-323-95062-6.00004-8
© 2022 Elsevier Inc. All rights reserved.

Construction remains manual mainly due to the deeper cyclical nature of the industry, which makes long-term investments untenable and the flexible use of labor/migrant labor more attractive. In contrast, the use of automation, sensing, and digital technologies have been widely adopted in manufacturing and other industries.

The construction industry faces various significant challenges. These challenges include increasing labor efficiency and decreasing accident rates at construction sites. Because of some conditions on the building sites that workers face, including dangerous fixtures or inclement weather conditions, the construction industry has the highest rate of work-related injuries (59 per 1000 workers, according to the Australian Bureau of Statistics).

Another consideration is the production of traditional cement contributes about 5% to the carbon in the atmosphere. Moreover, a large shift in climate change worsening will require a significant reduction in these polluting substances, which new technology and sustainable materials are needed; moreover, modern buildings and housing are mainly built with concrete. More than 20 billion tons of concrete are consumed every year, second only to water in usage.

Three-dimensional (3D) printing can automatically produce complex shape geometries without any dies, fixtures, and tooling from a 3D computer-aided design (CAD) model. 3D printing (3DP) for building and construction is a potential solution to the construction industry's problems. The technique will also allow architects to produce more creative designs for new structures without considering the building procedure. Due to its significant advantages of fabricating functional products with minimum material wastage, less human intervention, and reasonable build time, 3DP today has already been applied to other industries. Furthermore, researchers are looking forward to more applications of 3DP in building and construction to improve traditional building strategies while reducing material waste, human resources, and high capital investments in the built environment. Research interest to employ 3DP for building and construction has significantly increased in the past few years [1].

In general, a 5–10 times productivity boost can be achieved by moving to a manufacturing-style system, such as automation and robots, and less dependence on labor and repetitive work, thus increasing productivity through automation and preventing worksite hazards/fatalities. Hence, a 24/7 round-the-clock production can be achieved.

One of the major economic drivers of 3D concrete printing (3DCP) is the potential to eliminate the cost of formwork. For concrete construction, materials such as formwork and timber account for approximately 60% of the total cost. Because these materials are discarded eventually, it is also a significant waste of resources. According to different studies, the construction industry produces waste between 40% and 80% of total waste worldwide.

In considering the difficulty and high cost of making bespoke formwork, the traditional formwork cast technique also limits the creativity of architects to build unique shapes. However, with free-form additive construction, architects could express their thoughts with more freedom. When the complex shape does not increase the cost of producing a structural component, the construction can have more diverse designs from the current building architecture.

The ability of digital transformation in this industry can not only enable mass customization, elimination of formwork but also simplifies supply chains, eliminates waste with just-in-time, just-in-place, just-enough concepts. It leads to more sustainable solutions and less environmental damages. This greater use of data can also be taken further by building/printing structures that can respond to harness light, heat, movement through sensors or the Internet of Things. As a result, enabling energy savings and improving comfort in buildings and homes.

9.1.2 Short history of construction three-dimensional printing

Automation in construction has grown slowly despite the fast advanced automation in manufacturing and other industries. Besides the nature of the cyclical business of construction (briefly mentioned in the previous section), the conventional methods of manufacturing automation are also different from large structures with internal features in the construction sector. This can be a reason for the slow adoption rate in construction automation.

In the past 20 years, pre-cast or pre-fabrication construction has grown rapidly in several developed and newly industrialized countries. It has replaced the traditional way of mixing and casting concrete on-site to a significant degree. However, further automation still can benefit the construction sector by reducing construction time and labor, reducing environmental impact, and improving quality.

Behrokh Khoshnevis (University of Southern California, United States) in 2004 [2] was the first to pioneer and demonstrate the great potential in automated construction through a technique he named "Contour Crafting." Contour crafting describes fabrication technology in a layered manner, which has laid the groundwork for 3DP for concrete/cementitious materials. Using this process, a single house, or several houses with different designs, may be automatically constructed in a single run with all the conduits for air-conditioning, electrical, and plumbing embedded in each house. It also addresses the application of Contour Crafting in building habitats on the Moon or on Mars, which are being targeted in the next 100 years for human colonization.

More recently, Winsun built their first 3D-printed structures and showed in 2018 to much publicity. Other significant milestones are a pedestrian/cyclist bridge printed in Eindhoven by Theo Salat in late 2017, and a

348 Digital Manufacturing

collection of houses due completed in 2021, and re-creation of an ancient bridge modeled on a 1,400-year-old stone arch bridge on the campus of the Hebei University of Technology in Tianjin in 2019, measuring over 28 m in length.

For about 10 years, significant effort has been made to address matters in buildability, workability, pumpability, and extrudability of concrete material by various researchers worldwide. These aspects are still the critical research and development areas for 3DCP, considering various material requirements in the printing process [3].

It is imperative to characterize the mechanical properties of 3DCP elements by standardized material tests so that engineering design and construction could learn from the established experience. The compressive and indirect tensile strength of 3DCP elements should be tested based on existing test methods for concrete and new tests to document other inherent differences, such as inhomogeneities across layers and time differences in layer solidification. Also, interlayer bond strength is crucial to structural integrity and must be characterized as appropriate design references.

Reinforcement is required to prevent quasi-brittle failure of concrete materials. Several studies in 3DCP research have reported potential solutions by manual placement of assembly of bars or steel reinforcement bars as inter-filament or inter-layer reinforcement. However, placement of reinforcement should ideally be automated through other devices or integrated as part of the 3DCP process. In manual and automated processes, the bond between the printed concrete and the reinforcement needs to be studied carefully to provide design guidelines for appropriate curtailment and anchorage of reinforcement in 3DCP structures and structural elements.

Standards for structural design, materials specification, manufacturing, and testing are required for acceptance of 3DCP as a construction technology by building and construction regulation authorities. Such standards and their practical implementation in design and construction must ensure that acceptable levels of reliability are achieved [3].

9.1.3 Technology trends and needs—why 3D printing?

3DP was first introduced 40 years ago as a method of rapid prototyping [1]. In contrast to most traditional manufacturing methods that subtract material, 3DP techniques fabricate parts by adding material layer by layer. It is also called additive manufacturing (AM). In the past few years, the AM technique has also been applied to building artificial organs (e.g., knee/hip/other joint, arms, legs, feet, hands) and enabling higher levels of customization in consumable products, toys, and jewelry [4]. Due to their unique capabilities for fabricating highly customized and geometrically complex products using various materials, 3DP is at the center of a design and fabrication shift. For the building and construction industry, this dramatic change also presents significant opportunities to design and build the next generation structures and habitats [4].

Typically, AM processes fabricate complex 3D designs in a layer-by-layer manner using CAD models as blueprints. Due to rapid developments and increasing interest in material science, 3DP has since been used in many other industrial applications such as the military, aerospace, automobile, and medical sectors [1]. This technology has recently expanded into building and construction applications with the development of 3D printable construction materials. As such, a number of 3D-printed construction projects have been introduced to demonstrate the successful implementation of this type of automated construction [2]. However, 3DP technology in building and construction is in its early stage. The difficulties of meeting the size requirement for building and construction applications need to be tackled to increase its application substantially. The development of this technology is also dictated by to the strict supervision of standards and regulatory bodies and conservatism in building and construction industries worldwide.

Japanese companies made initial attempts to incorporate automated construction in the 1980s, most of which were for two types of work [5]. In the first category, simple labor activities on construction sites were replaced by single-task robots. Robots conducted concrete floor finishing, spray painting, tile inspection, and materials handling. In the second category, high-rise steel buildings or steel-reinforced concrete buildings using prefabricated components were constructed by fully automated systems. An example of this approach was the world's first automated construction system for building a precisely defined concrete structure named Big-Canopy. It used a centralized information control system to schedule all tasks and used an overhead crane supported by four independent masts to deliver the components [6].

Because traditional concrete can be molded into various shapes, produced at low cost, and has relatively high durability and high thermal resistance, it is used worldwide as one of the major in situ and prefabricated construction materials. Typically, reinforced concrete construction is composed of three components: formwork, reinforcement, and concrete. Formwork may consume approximately 60% of the total construction time and account for up to half of the total construction cost [7]. However, with 3DCP as the new construction method, both the time and cost for construction can be significantly reduced. A considerable amount of building and demolition waste is generated by the construction industry. Only 10% is crushed for reuse, approximately 40% is used for land reclamation, and 50% goes to landfills. The implementation of 3DCP technology is anticipated to dramatically reduce construction waste, as the amount of concrete mixed and printed is accurately controlled by the automated system. Therefore, 3DCP structures could potentially limit the environmental impact by using less material and producing less waste, thus saving on construction cost and improving productivity.

350 Digital Manufacturing

In the near future, sustainability will be paramount as we move from a shareholder to a stakeholder economy. The need for just-in-time, just-in-place, and just-enough will change global supply chains, and the way we build houses and 3DP will be one big step in this endeavor.

9.2 Current concrete printing technologies

This section presents an overview of existing concrete printing systems and some printing process control strategies. From the perspective of system mechanisms, existing concrete printing systems can be categorized primarily as gantry systems and robotic arm systems.

9.2.1 Gantry-based systems

Since applying 3DP in construction was first proposed, gantry-based systems have been widely adopted in the current building and construction industry. Such a system uses a gantry to position the print head(s) in XYZ Cartesian coordinates, and the enclosed volume of the gantry determines its build.

Among the gantry-based systems, there are some dominant methods: contour crafting [2], concrete printing [8], and D-shape [9]. Extrusion and backfilling are two main printing processes in contour crafting. The boundaries of each layer are printed in the extrusion phase, and they are used as a mold for the target printed structure. There are two trowels attached to the extrusion nozzle, which are utilized to create a smooth surface finish of the printed specimen. After the boundaries are printed, the backfilling phase begins by injecting concrete to fill the internal volume in several batches. Concrete at the lower part of the form hardens to some extent before the next batch of material is poured. By doing so, it is expected to produce minimal lateral pressure on the form. However, this strategy produces weak bonding strength by making fresh concrete adjoin hardened concrete at the junctions [8].

Concrete Printing, which is similar to but distinct from contour crafting, is purely an extrusion-based method. Its printing approach resembles the prominent fused deposition modeling in AM. The method uses a smaller extrusion nozzle than contour crafting, thus achieving finer deposition resolution and better geometrical control. As there is only one extrusion nozzle adopted in the concrete printing system, it poses a limit in material deposition rate. Therefore, the aforementioned bond strength problem is likely to arise, especially for a target printing structure that has a large cross-sectional area.

D-shape, another gantry-based printing method, uses a different printing strategy. It adopts the binder jetting technique, in which binders are selectively deposited on a powder bed to create stone-like objects. When printing a layer, the powder material is first spread over the bed and flattened to the desired thickness. The printer then injects binders at the target locations that need to be solidified. The unbounded powder material is not removed from the bed until the whole structure is printed, which acts as temporary support. A D-shape printer has a line of nozzles that can toggle on and off to extrude binders selectively so that only a few traverses are required for each layer to improve the printing efficiency.

Fig. 9.1 illustrates the schematic system setups and the printed specimens of (A and B) contour crafting, (C and D) concrete printing, and (E and F) D-shape [10]. These three techniques differ in their printing mechanisms. contour crafting and concrete printing adopt an extrusion-based method in AM. The D-shape uses a binder jetting technique to selectively deposit binders on a powder bed made up of magnesium-based materials and sand. The main advantage of binder jetting method is that binder jetting uses powder as

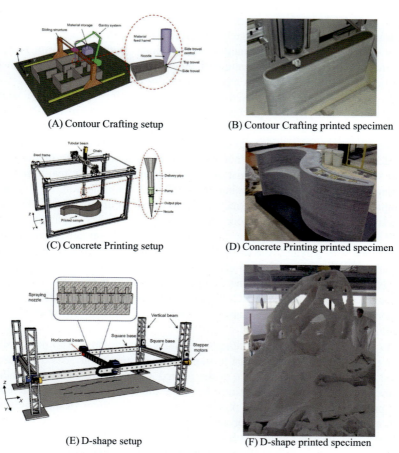

(A) Contour Crafting setup   (B) Contour Crafting printed specimen

(C) Concrete Printing setup   (D) Concrete Printing printed specimen

(E) D-shape setup   (F) D-shape printed specimen

FIGURE 9.1 Schematic setup and printed specimens for (A and B) contour crafting, (C and D) concrete printing, and (E and F) D-shape [10].

352 Digital Manufacturing

supports, which allows binder jetting method to print full 3D topology, while extrusion-based method is limited by the property of extruded materials to print limited inclined and overhanged structures. On the other hand, the main advantage of extrusion-based method is that extrusion-based method can print structures in a higher speed than the binder jetting method.

9.2.2 Arm-based systems

Compared to gantry-based printing systems, robotic arm-based printing systems are relatively new. They usually adopt a 6-axis robotic arm capable of achieving full 3D topologies instead of the 2.5D topologies generated by the vertical extrusion using conventional gantry-based systems with Cartesian kinematics. The robotic arm provides additional roll, pitch, and yaw controls to the end effector (print nozzle), which allows the printing system to accomplish more articulate print designs, such as printing with the tangential continuity method [11]. With this tangential continuity method, the print nozzle can maintain a continuous curvature change rate, produce a smoother transition between layers, and give a more esthetically pleasing look.

In addition to the fixed-based printing systems, recently, there are some mobile-based arm printing systems. Digital construction platform (DCP) developed at Massachusetts Institute of Technology [12] is one of them. DCP is constructed in a micro-macro manipulator configuration, where a robotic arm with a reach of more than 10 m is mounted on a track-driven mobile platform. A laser sensor is attached to the wrist of the robotic arm to provide real-time positioning feedback for fine positioning and oscillation compensation. DCP is also designed to be self-sufficient by recharging its electrical drive system with solar panels and is expected for outdoor on-site printing tasks. The capability of the DCP is demonstrated by 3DP a 14.6-m-diameter by 3.7-m-tall hemispherical open dome with polyurethane foam. One other mobile-based robotic arm system is Cybe RC 3Dp, which has a 6-axis robotic arm mounted on caterpillar tracks and is used in 3DP the R&Drone Laboratory in Dubai. It is worth noting that although both DCP and Cybe RC 3Dp are outfitted with mobile platforms, they demonstrate printing in a stationary style without any base movement during the printing process.

In a recent work [13], the printing-while-moving paradigm is proposed to alleviate the scalability issue common to most of the existing gantry-based and arm-based printing systems. The sizes of the printed structures are constrained either by the enclosed volume of the gantry or by the limited reach of the robotic arm. The proposed paradigm enables printing single-piece structures of arbitrary sizes with a single mobile robot printer and in a single take. The framework has been demonstrated in a hardware experiment, wherein the mobile printer shows the capability to print a structure larger than the reach of the robot arm.

Fig. 9.2 shows the system setups and printing illustrations of the above-mentioned arm-based systems.

Three-dimensional (3D) printing for building and construction Chapter | 9 **353**

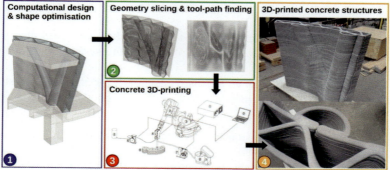

(A) 3D concrete printing process with a robotic arm

(B) Digital Construction Platform (DCP)

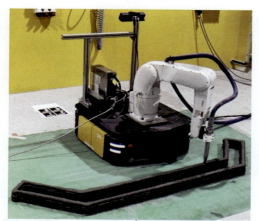

(C) Printing-while-moving paradigm at https://youtu.be/ZIbY00iTFwY

FIGURE 9.2 System setup and printing process for arm-based printing systems: (A) 3D concrete printing process with a robotic arm, (B) Digital Construction Platform (DCP), (C) Printing-while-moving paradigm at https://youtu.be/ZIbY00iTFwY [11–13].

9.2.3 Multirobot printing systems

Apart from conventional gantry-based systems and the above presented single robotic arm-based systems, there are other printing systems that involve more than one robot printer.

Minibuilders [14] presents an alternative approach for 3DCP. The system is comprised of three small mobile robots, each taking a unique role during the printing process.

1. The first robot builds the concrete foundation following an initial marked path.
2. The second robot grips the foundation with rollers and builds up the structure.
3. The last robot sucks on the wall and prints up the structure vertically for reinforcement.

Although a multiple agent system is introduced in Minibuilders, the robots require a hardened structure for climbing. Therefore, it has more specialized applications as it involves waiting for the printed concrete to grant sufficient strength before deployment.

Being motivated by the scalability and time-efficiency issues unveiled in most of the existing 3DP processes, the authors of [15] presented a 3DP system that employs multiple mobile robots that can print a large, single-piece structure concurrently. Such a system can potentially print single-piece structures of arbitrary sizes, depending on the number of deployed robots. The proposed strategy has been demonstrated by an actual printing experiment, where a single-piece concrete structure is printed by two mobile robot printers operating concurrently. Fig. 9.3 shows the concurrent printing scheme with multiple mobile robot printers [15].

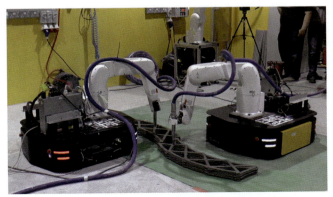

FIGURE 9.3 System setup and printing process of multirobot concurrent printing system. The full video of the experiment is at https://youtu.be/p_jcG25tUoo [15].

9.2.4 Printing process control

During actual printings, the print quality could be affected by a number of factors that are unlikely to be accurately and consistently controlled throughout the printing process. Consequently, it is desirable to introduce some feedback control schemes to monitor and regulate the printing process.

In [16], a 1D time-of-flight distance sensor is attached to the nozzle to measure the distance between the nozzle and the substrate. The distance is monitored and fed back to the system control unit in real-time. The nozzle position is adjusted accordingly to correct the deviation of the actual nozzle height from the intended one, which is typically caused by an imperfection in the hardware setup and the material deformation. Fig. 9.4 demonstrates tracking of the desired nozzle height with the proposed measurement and feedback system when printing over a non-flat surface that is not predefined.

Other than controlling the height of the nozzle to regulate the printing quality [16], Kazemian et al. developed a vision-based system that automatically adjusts the material extrusion rate based on the vision feedback [17]. A camera is attached to the nozzle to capture the top view of the printed filament. With the proposed vision algorithm, the filament shape is extracted, and the layer width is measured and compared with the desired one to detect over-extrusion or under-extrusion conditions. The material extrusion rate is thus adjusted accordingly to match the desired layer dimensions.

9.3 Fresh and harden properties of three-dimensional printable concrete

3D printable concrete use similar composites as conventional concrete. However, the unique layer-by-layer deposition process of 3DCP requires the

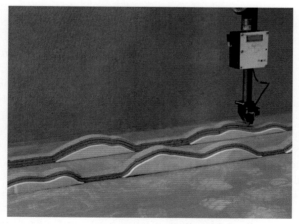

FIGURE 9.4 Printing over a non-flat surface whose geometry information is not predefined using the real-time nozzle height measurement and feedback control system [16].

356 Digital Manufacturing

concrete to have special fresh properties. With various printing parameters, 3DCP also produce parts with different hardened properties.

9.3.1 Different materials used and their effect on three-dimensional printing technology

The commonly used materials in the construction industry are cement, pozzolanic materials, sand, and chemical admixtures. The cementitious paste, which is formed by mixing cement, pozzolanic material, and water, binds the aggregate together. The current research at the Singapore Centre for 3D Printing has developed numerous printable construction materials such as geopolymer and fiber-reinforced mortar. Additionally, recycled glass, fly ash, and cenospheres are incorporated to improve the sustainability and lightweight properties of the material formulation [18].

9.3.1.1 The role of aggregates in three-dimensional concrete printing

The presence of particles in a Newtonian fluid induces a change to the flow characteristics. In the presence of a low volume fraction of suspended particles, the fluid is assumed to behave like a Newtonian fluid with a higher viscosity than the interstitial fluid. By increasing the volume fraction of suspended particles, the behavior of the system gradually changes from Newtonian to a shear-rate-dependent behavior [19].

Aggregates in concrete fulfill two main objectives - economy and strength. Aggregates are cheap and act as a good filler. The morphology of the aggregates has significant influences on the engineering properties of granular materials. In hardened concrete, aggregates with angular and irregular morphology with rough surface texture have a higher surface area of contact with the cement paste and better interlocking among the aggregate particle, thus have higher strength. Therefore, non-spherical and rough texture aggregates are preferred. Concrete made with angular and rough particles usually need to have higher mortar content to be pumpable. Aggregates with a spherical shape with smoother surface texture create a lower degree of particle interlock than the angular particles, which is generally good for workability in concrete. However, a material having good workability alone is not suitable for printing. Other than workability, the cementitious material needs high yield stress to retain its shape after extrusion,

Other factors, such as the maximum size and the gradation of particle sizes, can also affect the concrete's pumpability. The maximum size of the aggregates should be at least 4–5 times smaller than the hose diameter. As the hose gets smaller, more mortar paste is required. A 75 mm hose will require a mix containing 58% mortar, while a 50 mm hose will require 60% mortar [20]. In most studies, the maximum allowable size of aggregates is limited to 2 mm, as shown in Table 9.1.

TABLE 9.1 Printable mix design from other publications [11,21–25].

Institute/ company	University of Southern California	Yonsei University	TU Dresden	Loughborough University	University of Southern Brittany	XtreeE SAS	American University of Beirut
Main binder	• CEM II type cement (C/B: 1)	• CEM I type cement (C/B: 0.6) • Fly Ash (FA/B: 0.3) • Silica fume (SF/B: 0.1)	• CEM I type cement (C/B: 0.55) • Fly Ash (FA/B: 0.23) • Silica Fume (SF/B: 0.22)	• CEM I type cement (C/B: 0.7) • Fly Ash (FA/B: 0.2) • Silica fume (SF/B: 0.1)	• CEM I type cement (C/B: 0.5) • Limestone filler (LS/B: 0.25) • Kaolin clay (KC/B: 0.25)	• CEM I type cement (C/B: 0.6–0.8) • Limestone filler (LS/B: 0.1–0.2) • Silica Fume (SF/B: 0.1–0.2)	• Cement (Unknown type) (C/B: 1)
Fine aggregate/ sand	• S/B: 1.105	• Fine aggregate and silica sand (0.1–0.6 mm)	• S/B: 1.5897	• S/B: 1.5 (maximum particle size 2 mm)	• NIL	• S/B: 0.8–1 (crystalline silica)	• S/B: 1.92 (combination of fine aggregate and sand)
Water	• W/B: 0.505	• W/B: 0.35	• W/B: 0.23	• W/B: 0.26	• W/B: 0.205	• W/B: 0.2–0.233	• W/B: 0.39

(Continued)

TABLE 9.1 (Continued)

Institute/ company	University of Southern California	Yonsei University	TU Dresden	Loughborough University	University of Southern Brittany	XtreeE SAS	American University of Beirut
Plasticizer / Superplasticizer	• SP/B: 0.084	• NIL	• SP/B: 0.0128	• SP/B: 0.01 • Retarder/B: 0.005	• SP/B: 0.0015	• NIL	• Accelerator/ B: ∼0.008 • Retarder/B: ∼0.005
Other Materials / Remarks	• NIL	• Polysaccharide fibers (3 mm in length) • Thickening agent (TA/B: 0.001) • Styren-acrylic polymer resin (SAN/B: 0.001)	• NIL	• Polypropylene fibers (12 mm in length, 0.18 mm in diameter, 1.2 kg/m^3)	• Limestone filler (Particle size: 0.1–100 μm)	• NIL	• NIL
Average compressive strength (MPa)	• 18.9	• 52	• 80 (21 days)	• 110	• No harden test result	• 120 (Estimated from flexural strength 14.2 MPa @ 90 days)	• 42

9.3.1.2 The role of paste in three-dimensional concrete printing

A paste exhibits a continuous network of soft interactions throughout the interstitial fluid. Different from suspension, the interactions should be sufficiently strong to change the behavior of the mixture. According to Hoornahad [26], paste consistency is characterized by the yield stress of the paste. Pastes can show elastic and viscoelastic behaviors. When the shear stress remains below the yield stress of a material, it remains in an elastic behavior. As soon as the applied force is released, the initial configuration of the articles is restored. As the shear stress moves beyond the yield stress, the configuration of the particles is irreversibly broken, and the material starts flowing, displaying a viscoelastic behavior. The rheological properties of the cement paste are crucial to its flow behavior. They can be described by the Bingham plastic model with two defining parameters: yield stress and plastic viscosity. These parameters can affect the pumping characteristics of the fresh concrete. Furthermore, plastic viscosity is essential in estimating the resistance of concrete to flow in the pipe since this parameter is a measure of the change in shear stress with a change in shear rate.

According to Kennedy [27], there are two types of paste in a mixture: void paste and excess paste, as shown in Fig. 9.5. They are the fundamental theories on the close relation between paste and aggregate and the contribution to workability. The void paste is the paste that fills up the void space between the aggregates. It acts like glue to hold the aggregates together and keeps them in place. The excess paste is the layer of paste surrounding every particle in a mixture with a constant thickness. This excess paste helps in the flow of the aggregates and reduces internal friction as it pushes the aggregates apart. More excess paste layer will increase the flowability of the mixture. Deformability of fresh concrete is defined as the ease of being deformed by gravity and external forces. Yield stress is one of the governing parameters for deformability. The deformability of the mixture increases with the increasing volumetric fraction of the excess paste content and consistency of the paste.

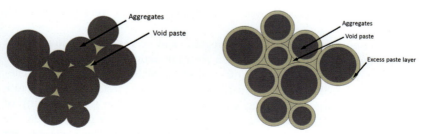

FIGURE 9.5 The mixture of aggregates and paste [27].

360 Digital Manufacturing

9.3.2 Fresh properties of three-dimensional printable concrete materials

The main challenge for the 3D printable concrete is the contradicting characteristic: it should be flowable enough to be pumped through the hose while having sufficient strength to support the subsequent layers during the layer-by-layer deposition. Therefore, the fresh properties of the 3D printable concrete need to be investigated and measured to satisfy this unique characteristic requirement.

9.3.2.1 Importance of yield stress and plastic viscosity in three-dimensional printing

It is common to characterize the concrete flowability by the Bingham model, as it allows the rheological properties such as the dynamic yield stress and the plastic viscosity to be obtained. The dynamic yield stress can be described as the shear stress required to maintain flow, while the plastic viscosity is defined as the resistance of material flow. During a low continuous shear rate, the shear stress of the material increases with time. As it reaches its yield point, the continuous shear breaks the microstructure of the material and causing the shear stress to decrease. This yield point, also known as the static yield stress, is described as the shear stress required to initiate flow. The presence of sand and gravel during the continuous shear further breaks down the microstructure for concrete and mortar mixtures.

In 3DCP, material parameters such as buildability and pumpability are crucial to the operation. The parameters are correlated to the material's plastic viscosity and yield stress. Pumpability can be defined as the ease of delivering the material through a hose to the extrusion nozzle. The study of pumping concrete material is complicated due to the time-varying behavior and dependency of the concrete properties on the pumping conditions. Pumping conditions can affect the bleeding and segregation of the concrete, and thus, it is difficult to predict the pumpability of the material. Deficient distribution of particle size or excessive water content can cause segregation of the cement paste from aggregates, resulting in clogging of the material delivery hose. Therefore, sufficient paste content should be present with excess to fill the gaps between the aggregates in a properly designed mixture.

The paste content coats the aggregates and serves a lubricating effect between the aggregate particles when shearing action is applied to the suspension during pumping. On the other hand, buildability, defined as a material property in 3DCP [25], is correlated to the yield stress and structure build-up rate of the concrete, as well as the cross-section shape stability of the layers. The static yield stress provides the material its rigidity during the initial stage of the printing, which can be correlated to the capability for shape retention upon extrusion. Shape stability is required before the structure build-up of the mixture, as it helps the material gain yield stress further. The structure build-up of the material is dependent on the flocculation rate of the binder particles due to early hydration. If the rate of stress application due to subsequent layers is higher than the material structure build-up rate, the structure collapses [25]. The control of the structure build-up rate of the concrete and the printing parameters is vital to achieving a highly buildable mixture. The addition of a set accelerating agent improves the material structure build-up rate. In addition, the shape retention capability of the material after extrusion can indicate the material's buildability. Due to the absence of formwork, the material is self-supported on the internal cohesiveness between the binder and the aggregates. For the filament to retain its intended shape upon extrusion and bear the load from that of the above layers, the material is required to have a higher static yield stress.

9.3.2.2 Printability studies with slump and slump flow test

Roussel [28] attempted to correlate the slump and flow diameter results to the yield stress of the cement paste. The results show that the fundamental correlation between yield stress and slump is independent of the material and the geometry of the slump cone mold. Conversely, the flow table test may not correlate with the yield stress of the material tested. Nevertheless, this test provided a cost-effective, simplified, and acceptably precise measurement of the material yield stress of self-compacting concrete (SCC). Marchment et al. [29] presented some work using the flow table. He tested the flow diameter at a different time interval and found that the flow of the material reduces with time. The main observation from the results shows that additives in the paste mixtures slow the stiffening process of the cement to maintain a malleable bond interface. Dhir et al. [30] used a flow table test to determine the consistency of the mortar. It is important as it allows the assessment of the degree of workability and the quantity of water required to obtain adequate plasticity in a practical manner. The result showed that an increase in the fine recycled aggregates would cause a reduction in the consistency. This decrease in consistency is mostly attributed to the higher water absorption capability of the recycled aggregates.

Existing research has proposed testing various mixtures using rheometers for measuring dynamic yield stress and the plastic viscosity in 3DCP [31]. However, these measurements with the rheometer are sensitive to the protocols and the data processing procedures if non-standardized geometries were used for the measurement. It can cause the results to vary. Tay et al. [32] introduced a simple test to characterize the material using the slump and flow table tests.

The slump test uses a cone-shaped mold, as shown in Fig. 9.6. The mold is lifted once the material is poured in. Material flow occurs if the shear stress inside the material generated by gravity exceeds the yield stress of the

362 Digital Manufacturing

FIGURE 9.6 Flow table test experimental setup: (A) before the mold is removed and (B) the spread of the material after dropping the table 25 times.

material. The slump is defined as height reduction in the concrete after the mold is lifted. The shear stress caused by gravity decreases with the thickness of the material as it flows. When the shear stress generated by gravity becomes lower than the yield stress, the material stops flowing. This slump of the concrete is therefore directly correlated to the material yield stress. The top surface area of the slumped material was determined as well. After dropping the flow table 25 times, the concrete is spread evenly, and the diameter of the spread concrete is measured and recorded as the flow values. The 25 drops cause the difference between the mixtures to break down the microstructure of the concrete. Thus, the slump value is closely relatable to the solid-like state of the microstructure and the static yield stress and, thus, the buildability of the concrete. The flow value is related to the dynamic yield stress and the pumpability. The slumped area is usually circular in shape, indicating a uniform distribution of spread from the deformation of the material. The area at the base is called the slump spread or flow diameter. Such a measurement technique is usually applied in the evaluation of SCC due to the low yield stress.

The result presented by Tay et al. [32] shows that printing a stiff material causes pores, while discontinuity can eventually cause the sample to lose structural integrity. Hence, producing low buildability. On the other hand, high slump and flow diameter value causes the filament at the bottom to be soft and is unable to support the weight of the subsequent layer. It can cause instability in the structure. Therefore, the middle region between the two extremes regions possesses a balance material behavior ideal for printing. This middle region should be between a slump of 4 and 8 mm and a flow diameter value of 150–190 mm.

In general, the result from Tay et al. [32] is limited to the material used and the printer's limitation. The investigation gave an insight into the performance of a printable material over a limited region. Such material performance is based on printer capability and material morphology.

9.3.3 Harden properties of three-dimensional printable materials

The unique layer-by-layer deposition process of 3DCP also produces unique performance of the printed structures. As well as many other 3DP processes, the strength of the printed parts is affected by the loading direction and the printing parameters.

9.3.3.1 Mechanical strength of three-dimensional concrete printed sample in different directions

Anisotropy is the main disadvantage of the mechanical properties of 3D-printed products. Furthermore, the printing parameters, including nozzle traveling speed, material flow rate, and the time interval between successive layers, significantly influence the final strength of the printed object. A study showed that the material is more compact in the horizontal direction than in the vertical direction [3]. Figs. 9.7 and 9.8 show the different loading directions of a 3D-printed specimen. It is further noted that the printing parameters, including the printing time gap between successive layers, layer contact area, and concrete viscosity, can affect the tensile strength of the printed object. The loading direction of the printed specimen was also reported to have influenced its mechanical properties [21].

Paul et al. [31] showed that the specimens extracted from the D3 direction have higher mechanical properties than samples tested in other directions. The higher values in the D3 direction agree with the results reported by Nerella et al. [33]. In the D3 specimens, both studies found that the strength increases about 15% compared to the cast specimens. Like the compressive strength, the trend was also observed in the flexural strength for

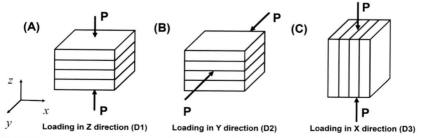

FIGURE 9.7 Application of compression load in printed objects related to the printing direction: (A) Z direction (B) Y direction (C) X direction [31].

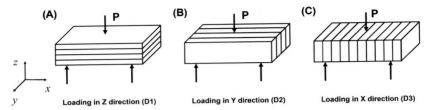

FIGURE 9.8 Application of flexural load in printed objects related to the printing direction: (A) Z direction (B) Y direction (C) X direction [31].

different directional tests. Paul et al. [31] and Nerella et al. [33] showed that the D2 direction has a higher flexural strength. Different results were also observed between the circular and rectangular/square-shaped nozzles in addition to the directional test. The circular-shaped nozzle has more voids in the printed products, which could be one of the main reasons for the weaker strength. For the rectangular or square-shaped nozzle, the issue on the voids created by the nozzle was not significant. A circular-shaped nozzle is advantageous for printing complex geometries since the nozzle can achieve a symmetrical cross-section at different rotational angles. However, this may not be achievable for the square or rectangular nozzle. Therefore, careful consideration of the printing parameters needs to be taken. Nozzle geometry, flow rate and travel speed have an important role in affecting the mechanical properties of the final print.

9.3.3.2 Time gap effect on the interlay bond strength

Research conducted by Roussel and Cussigh [34] used layered casting (casting of SCC) with a short amount of time for the initial layer to build its microstructure before pouring in successive layers. If the bottom layer of SCC rebuilds its structure such that the yield stress increases above a critical value, the intermixing between the two layers will decrease, resulting in a weak layer bond. Furthermore, the weak bond in the layers may create pores locally and increase its permeability to harmful substances. Hence, the time interval between the layer's deposition needs to be minimized to eliminate the creation of a weak interface.

Studies on the correlation between the deposition time gap on the fresh material's properties and the interlayer bond strength are interesting. Rheological results from Tay [35] gave insight into the possible reason for the weaker interlayer bond strength as the deposition time gap increases. It examined the time-dependent behavior of the paste under elevated shear rates, simulating the pumping action and the structural rebuilding of the paste, which simulates the material at rest after deposition.

Fig. 9.9 illustrates the results from a tensile test, with the tensile strength decreasing as the time gap increases. The tensile strength reduces in a

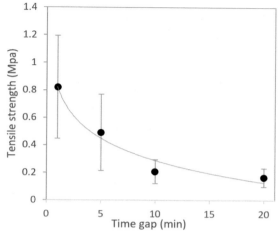

FIGURE 9.9 Tensile strength of samples at the different time gap [35].

logarithmic trend where the reduction was more notable between 1-minute and 5-minutes time gap. The decline was less when the time gap was beyond 10-minutes (See Fig. 9.9).

From Tay [35], it was found that the stiffening of the bottom layer is one of the causes of the weakening in the bond strength. As the layer interval time increases, the stiffening of the bottom layer increase, thus causing difficulty in the top layer to intermix with the bottom layer in the printing process. Fig. 9.10 shows the difference in the degree of the mixing at the layer interface with respect to the time gap. As the time gap increases, the top layer merely rests above the first layer, as shown in Fig. 9.10D, causing voids to appear.

A time interval between layers, which considerably influences the interlayer bond strength of the finished product, exists in 3D concrete extrusion-based printing due to the layering operation. The bottom layer's rheological properties impact the bonding process with the successive layer. The degree of adhesion between the layers holds the key to structural tensile integrity. The initial layer's modulus increases as the time gap become larger, while the subsequent layer's modulus is independent of the time gap. However, it can be explained that the higher modulus of the bottom layer did not permit good intermixing at the layer interface. As the interval time increases, more voids induced due to poor adhesion at the interface lead to the weakening of the bond strength logarithmically. Nonetheless, the high modulus is also required in the initial layer to withstand the weight of the layers above and maintain structural stability.

FIGURE 9.10 Samples printed at (A) 1-min time gap, (B) 5 min time gap, (C) 10 min time gap, (D) 20 min time gap [35].

9.3.4 Three-dimensional concrete printing parameters

Optimized printing parameters are essential in 3DCP process. While the printing quality would suffer from uncoupled printing parameters, it can also form a region with special function during the 3DCP process.

9.3.4.1 Effects of printing parameters on concrete filament

There is a significant effect on the printed filament by varying the printing parameters. Uncoordinated printing parameters can cause the printed filament to have void and cracks or an oversized filament. It is important to characterize the printing parameters such as material flow rate and nozzle travel speed to prevent abnormalities from appearing in the filament. Tay et al. [36] introduce an index to characterize the printing parameter. Fig. 9.11 shows the top surface area of a series of different filaments printed with different printing parameters.

Based on the filament characteristics such as voids, crack, and filament width, Tay et al. [36] differentiated them into the different filament categories (Fig. 9.12) based on $R_{m/i}$, which is defined as the ratio of the experimental surface area of the filament over the specified surface area of the filament. In Fig. 9.11, the filament surface area has increased steeply as the flow rate increases and print speed decreases, which also adversely results in longer print time. When the flow rate is high and print speed is low, an overflow of the material was observed, which reduces the print quality. On the other hand, the filament surface area decreases when the material flow rate decreases and printing speed increases. This can lead to the underflow of material, and extensive underflow can cause cracks and discontinuities development in the filaments. However, the time required for printing can be reduced by increasing the flow rate and the nozzle travel speed without any print quality change.

Three-dimensional (3D) printing for building and construction Chapter | 9 367

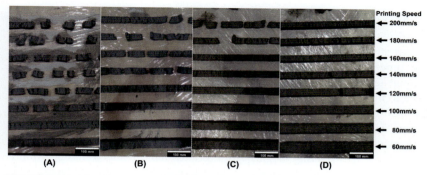

FIGURE 9.11 Filament printed with different flow rates. (A) flow rate: 37.9 mL/s, (B) flow rate: 45.2 mL/s, (C) flow rate: 48.0 mL/s, (D) flow rate: 51.3 mL/s.

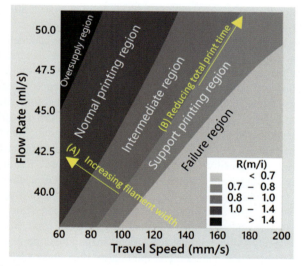

FIGURE 9.12 Contour plot of the ratio of measured surface area to the ideal surface area of the filament.

9.3.4.2 Printing of concrete supports materials

While extrusion-based 3D concrete, as demonstrated in existing literature, has been able to print complex structures, constructing overhanging structures continues to be a major hurdle for 3DCP. Powder-bed 3DP techniques allow the unused powder as support for the overhanging structures [9]. However, in extrusion-based printing, external physical support is required for the overhanging filament, which would otherwise collapse without a surface to rest on.

368 Digital Manufacturing

At the current state for 3D concrete extrusion-based printing, the common practice to resolve for the overhang structures is designing for no-support closed structure or using wooden planks as temporary supports in the printing process. A parabola-shaped window printed by Rudenko [37], is a support-less closed structure design has a maximum slope angle with reference to the printing direction that is printable without collapsing under gravitational forces. The maximum allowable slanting angle of the overhang design is dependent on the material. Although the design for the self-supporting overhanging structure is an ingenious method, this approach is limited by the slanting angle. Arup printed an on-site building with a portable robotic printer to showcase the capability of 3DCP by reducing inefficiency from the printing process. Temporary supports such as wooden planks and scaffoldings were used to hold the overhanging material in place. Existing solutions can mitigate the issue of overhanging structures, but they can also require further considerations such as human intervention and constraints in design, which need to be resolved.

Tay et al. [36] proposed a technique to vary the printing parameters and produce temporary support with concrete material to overcome the freeform issue. Employing such an approach can enable a complex structural design to be printed with overhanging components. A design with an overhanging component requiring supports was printed and used to verify the results obtained from previous sections, as shown in Fig. 9.13. Fig. 9.13A and B shows that the primary structure and removable supports were printed with the same material. Maintaining the flow rate in the printing process is easier due to the configuration of the printer. Thus, the nozzle travel speed was varied in the printing process. The applied parameters used for printing were selected using the experimental results obtained in Fig. 9.11. For the printing shown in Fig. 9.13, the flow rate used was 45.2 mL/s. The nozzle travel speeds used in the printing of the support structure and the main structure were selected at 160 mm/s ($R_{m/i} = 0.7$) and 80 mm/s ($R_{m/i} = 1.2$), respectively. After printing, the support material can be removed easily with the aid of a chisel and hammer, and the final product obtained (Fig. 9.13C).

In summary, the material retains its soft characteristic even after being extruded from the nozzle for 3D concrete extrusion printing. Temporary physical support is needed for the overhanging filament. Printing the support structure using the same material helps mitigate the need for a complex dual-nozzle system, which can increase the time gap between the layers due to the interchanging of the nozzles and, in turn, weaken the layer adhesiveness of the main structure [35].

9.4 Three-dimensional concrete printed applications and case study

The construction industry is also increasingly exploring 3DCP. In the past decade, several commercial companies and universities have developed

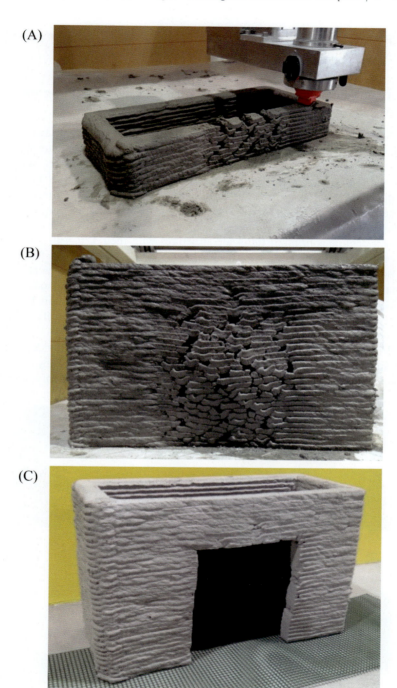

FIGURE 9.13 (A) Printing process with different printing parameters, (B) before post-processing and (C) after post-processing [36].

3DCP facilities to print houses, bridges, room units, and furniture to highlight its possibilities.

9.4.1 Applications of three-dimensional printing in building and construction

In 2015, a Chinese company Winsun built a multistorey apartment (Fig. 9.14A) paired with a villa (Fig. 9.14B) [38], which was the start of their showcases of 3D concrete printed buildings and structures. In 2016, Winsun built the first workable 3D-printed building in Dubai and used it as an office building (Fig. 9.14C) [38]. In 2020, Winsun dispatched 15 3D-printed quarantine rooms to Xianning Central Hospital to cover the lack of hospital beds during the Covid-19 pandemic. As one of the active 3DCP companies, Winsun continues printing full-scale structures with their $120 \times 40 \times 20$ feet 3D printer.

Besides Winsun, another commercial company Apis Cor also gained lots of publicity. In 2017, Apis Cor printed a round-shaped residential house with its iconic mobile printer in 24 hours (Fig. 9.15A) [39]. In 2020, Apis Cor 3D printed a two-storey administrative building for Dubai Municipality, which was considered the world's largest 3D printed building.

With other 3DCP applications are being implemented worldwide, more and more of them are put to real use. In 2017, a social housing named Yhnova was printed in Nantes, France (Fig. 9.15B) [40]. This house,

(A) A 3D printed 5-story building

(B) A 3D printed villa

(C) 3D printed Dubai Government Office

FIGURE 9.14 (A–C) Various 3D-printed buildings and constructures produced by Winsun.

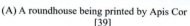

(A) A roundhouse being printed by Apis Cor [39]   (B) Picture of Yhnova™ without coating in Nantes, France [40]

FIGURE 9.15 3D-printed houses worldwide: (A) A roundhouse being printed by Apis Cor [39] (B) Yhnova without coating in Nantes, France [40].

considered the first 3D printed building given to a family, was printed with polymer foam and then filled with rebar and concrete. It has been occupied by a family since June 2018 [40]. On April 30, 2021, the first tenant of the first-ever Dutch 3D-printed concrete home received the key. The house in Eindhoven, which is the first of five within "Project Milestone," fully complies with all of the strict building requirements of The Netherlands [41]. In the United States, Mighty Buildings have developed a series of housing solutions based on 3DP technology, and expects to develop a community in California built by 3DP [42]. In August 2021, ICON and Texas Military Department 3D printed 3800-sq-ft training barracks, which plan to welcome their first inhabitants in fall 2021 [43].

Various other applications have adopted 3DCP technology too. In 2016, a bicycle bridge was 3D concrete printed by the Eindhoven University of Technology [44]. In order to guarantee this 3D printed structure is safe to be put in use, Finite Element Analysis was used to determine the structure design, a 1:2 scale destructive scale test was conducted to validate the computation. This 3D-printed bicycle bridge was opened to the public on October 17, 2017 (Fig. 9.16A) [38]. In 2019, a 3D-printed concrete bridge was unveiled in Tianjin, China. It's modeled on Zhaozhou Bridge, a 1400-year-old stone arch bridge, and printed by the Hebei University of Technology, China [45]. Around the same time, ETH Zurich printed "Concrete Choreography" (Fig. 9.16B), which was consisting of columns fabricated without formwork, and was inaugurated in Riom, Switzerland, as a stage for dancers [46]. In 2021, ETH Zurich 3D printed an unreinforced concrete bridge, which was built with concrete blocks that form an arch much like traditional masonry bridges. The concrete is not applied horizontally in the usual way but instead at specific angles such that they are orthogonal to the flow of compressive forces [47].

372 Digital Manufacturing

(A) Reinforced bicycle bridge 3D printed by the Eindhoven University of Technology at the opening in the Netherlands [38]

(B) "Concrete Choreography" printed by ETH Zurich [45]

FIGURE 9.16 Other 3D-printed structures worldwide: (A) Reinforced bicycle bridge 3D printed by the Eindhoven University of Technology at the opening in the Netherlands [38] (B) "Concrete Choreography" printed by ETH Zurich [46].

9.4.2 3D concrete printing technology developed by NTU Singapore

A large-scale concrete printer and the corresponding 3DCP process was designed, built, and tested in Nanyang Technological University, Singapore since 2016. Fig. 9.16 shows the positioning system used for this printing was a KUKA 6-axis robotic arm mounted on a fixed boost frame. This robot arm is a shelf-mounted type, and it is mounted on a boost frame to gain a printing capability of approximately 3000 mm in terms of the height of the printed part.

A modified MAI Pictor pump was used to deliver material to the nozzle through a hose. A digital controller was installed into the pump to replace the original analog knob to control the pump rotation speed and the pumping flow rate better (Fig. 9.17). The nozzle was designed to rotate around the Z-axis and used a rotational rectangular-shaped outlet to enhance the surface finishing of the printing.

Using the printer, a 1620 × 1500 × 2800 mm^3 (L × W × H) prefabricated bathroom unit with 100 mm wall thickness designed by Sembcorp Design & Construction was printed to test the developed 3DCP system and process. Printing was divided into two sections: a 2200 mm high bottom section with door openings and a 600 mm high top section. The entire printing process for the 600 mm top part was completed in 2 hours, as shown in Fig. 9.18. The entire printing process for this 2200 mm bottom section was completed in 7.5 hours, as shown in Fig. 9.19.

With the developed 3D concrete hardware and material, the 3DCP team in NTU was able to print for other industry applications, such as a concrete tree display for NTU Convocation 2019 and a reception table for Teambuild Construction Group shown in Figs. 9.20 and 9.21.

Three-dimensional (3D) printing for building and construction **Chapter | 9 373**

FIGURE 9.17 The positioning system using the KUKA robot arm installed on a boost frame.

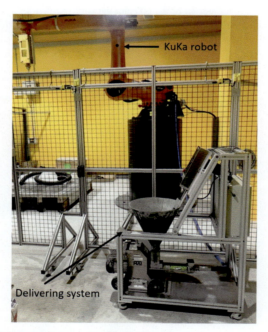

FIGURE 9.18 Delivery system on-site with KUKA robot.

9.5 Sustainable raw materials in concrete printing

3DCP can improve the sustainability of the construction industry by reducing material usage and construction waste. To further capitalize on the sustainability aspect of 3DCP, many recycled and sustainable construction materials have been adopted using this technology.

374 Digital Manufacturing

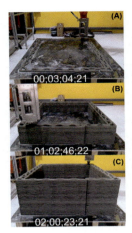

FIGURE 9.19 The printing process of the 600 mm top part: (A) first layer with the left-hand side standoff distance significantly larger than right-hand side, (B) middle of the printing process, (C) finishing up printing.

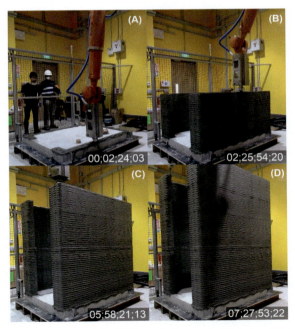

FIGURE 9.20 Printing process of the 2200 mm bottom part: (A) laying down the first layer, (B) printing the 59th layer, (C) printing the 118th layer, (D) nozzle was leaving the finishing point.

Three-dimensional (3D) printing for building and construction Chapter | 9 375

FIGURE 9.21 3D-concrete-printed tree "Growth" for NTU Convocation 2019 display.

9.5.1 Sustainable materials for cement replacement

Concrete is the second most utilized substance after water and the most broadly used human-made material worldwide. Raw materials of concrete for 3DP include cement and water, which make up the binder phase, sand as fine aggregates. Portland cement, the most widely used and essential material in concrete for the construction field, is manufactured from a mix of limestone and clay at precise proportions subjected to pyro-processing at an elevated temperature of 1450°C. Portland cement is then obtained by grinding the clinker from the kiln into powder form. Due to the decarbonization of limestone, the manufacturing of Portland cement results in large amounts of CO_2 emission. With the high demand for concrete materials, the CO_2 emission from cement production contributes about 5%−8% to the global human-made CO_2 emission. With global efforts to reduce greenhouse gas emissions, comprehensive research has been conducted in reducing the demand for Portland cement by introducing other types of cementing materials.

Blended cement has been developed by partially replacing the Portland cement with supplementary cementitious materials (SCM) [48]. SCM, also known as pozzolanic materials or pozzolans, are defined as materials that possess little or no cementitious value by themselves. However, when added to Portland cement, the SCM compliments the hydration of the cement paste through a secondary chemical reaction to cement hydration known as a pozzolanic reaction. SCM are common by-products of industrial manufacturing processes. Hence, the introduction of SCM into the concrete not only reduces the CO_2 emission but also provides a new channel for waste materials. The main characteristic of SCM is that the materials contain a large percentage of amorphous silica and chemically reacts with the Calcium Hydroxide, a

376 Digital Manufacturing

by-product from hydration reaction, to form compounds (Calcium Silicate Hydrates) that provide cementing properties to the concrete. SCM decreases the early-stage mechanical strength but increases the long-term mechanical properties of the concrete material. In 3DP for building and construction, SCM also contributes to improving the properties of the printable concrete materials in their fresh state, which is important in the 3DCP context.

Fly ash is derived from the pulverized coal burning in power generation plants as a by-product. The fine ash powder from the powdered coal combustion is transported through the flue gas and gathered in the electrostatic precipitators. ASTM C618 classifies fly ash into two different classes; class C and class F. The key difference between these two classes of fly ash is their chemical contents. Class F fly ash exhibits a smoother surface texture and more spherical morphology than class C, which permits higher flowability for the concrete [49]. As such, class F fly ash can be adopted as a constituent in the 3DCP materials since flowability plays an essential role in allowing ease of material delivery [50].

Silica fume is derived from the by-products from electric arc furnaces in the manufacturing process of elemental silicon or alloys containing silicon. Silicon dioxide vapor is emitted from the carbothermic reduction process from high-purity quartz into silicon materials at high temperatures of 200°C. The oxidization and condensation of the vapor at low temperatures then results in silica fume. Literature has shown that silica fume addition increases the yield stress and the plastic viscosity of the fresh concrete [49], which results in a more uniform and cohesive mixture that can be used for several applications. One such application is in the concrete mixture for 3DP. Silica fume addition into the printable mixture enhances the inter-layer cohesion of the print product. The inter-layer bonding of the materials often plays a crucial role in governing the structural integrity of the print since the layer interface acts as a weak link in the structure. Furthermore, silica fume addition to the 3DP concrete mixture results in better shape stability of the filament.

GGBFS is obtained from the smelting process in the iron and steel manufacturing industry as a by-product. The fine, granular non-crystalline powder is a result of quenching molten slag with water. Other than a sustainable solution to cement, the inclusion of slag into the 3DP mix increases the yield stress of the fresh material. Higher yield stress results in the better shape retention of the extrudate and allows multiple layers to be stacked above without excessive deformation. The addition of GGBFS also increases the late age strength of the printable concrete due to the creation of additional calcium silicate hydrate from the pozzolanic reaction.

9.5.2 Sustainable materials for natural sand replacement

Compared to conventional cement, the concrete material for 3DP contains about 50%–60% of fine aggregates [31]. The most common type of fine aggregates used concrete are construction sands, a depleting resource as the rate of construction sand usage is much higher than it can be replenished naturally [51]. Though it may seem that sand is an abundant resource due to the vast deserts, desert sands are unsuitable to use in concrete, where sand is mostly consumed, primarily due to the round grains from wind erosion. Sand used for construction is sourced from riverbeds and beaches. It is angularly shaped, which interlocks to provide mechanical resistance to the concrete. Nonetheless, there are alternatives to the use of sand in concrete that can be used.

Theoretically, glass can be recycled indefinitely without any degradation in quality. However, the end-of-life glass wastes are non-recyclable due to contamination and chemical incompatibility between different colored glass existing in the co-mingled waste stream. The sorting of these glass wastes is also unattractive as the classification process is difficult and costly, resulting in a low recycling rate of glass. Apart from having limited landfill space for glass waste disposal, glass is also not suitable for a landfill due to the material's non-biodegradable nature. Thus, recycled glass cullet acting as a replacement for natural river sand in the production of concrete has been widely studied for the past decades. Unlike natural sand, recycled glass sand has sharper edges, a higher aspect ratio, and smoother surfaces (Fig. 9.22).

The morphology and the surface texture of the fine aggregates affect the properties of the concrete material in both the fresh state and the hardened state. The smoother surface of the recycled glass aggregates improves the flowability of the concrete mixture as the inter-particle friction is reduced. The water absorption capacity of the recycled glass was lower than the natural sand, and more interstitial water present in the fresh mixture provides better flow properties. The strength reduction compared to that of the natural sand aggregates can also be attributed to the glass particles' smooth surfaces but sharp edges, which cause a weaker bond interface between the recycled glass particle and the binder matrix [51].

FIGURE 9.22 3D-concrete-printed reception table for Teambuild Construction Group (by NTU, 2019).

Alkali-silica reaction (ASR) is the primary concern in incorporating recycled glass in concrete. The silica within the glass can become reactive under a high alkalinity environment by the cement binder [51]. ASR results in a gel-like substance existing within the concrete pores, which absorbs moisture over time and expands, thus, creating stresses on the microstructure and leading to structural failure while the stress overcomes the strength of the material [52]. It is noted that ASR occurs over a long period, but the potential can be examined by accelerated methods proposed by standards organization. Nonetheless, the ASR effect can be reduced if the particle size distribution is more refined. In concrete 3DP, the aggregates used are generally limited to finer particles due to the material delivery system and extrusion nozzle opening in the 3D printer [24]. Segregation of large particles can occur at the nozzle and result in clogging of material. The reduction of particle size of the recycled glass sand used for 3DP not only prevents segregation but also mitigates the effect of ASR in the material [18].

9.5.3 Sustainable materials in spray-based three-dimensional printing

Spray-based 3DCP, or shotcrete 3DP [53], is a newly developed printing method to manufacture decorative profiles on vertical/overhead substrates [54] (Fig. 9.23). Unlike conventional spray concrete technology (shotcrete), spray-based 3DCP raises many strict requirements on uniform thickness distribution of spray material, which guarantees its dimensional accuracy. The even distribution of the sprayed material can be realized by specially designed mixtures, as reported in the literature [54,55].

9.5.3.1 Fly ash cenosphere

In the design of spray-based cementitious materials, fly ash cenosphere was one of the key raw ingredients. Fly ash cenosphere is a lightweight ingredient with spherical and hollow morphology characteristics (Fig. 9.24) [54].

FIGURE 9.23 The optical microscope images of natural sand and recycled glass.

Three-dimensional (3D) printing for building and construction Chapter | 9 379

FIGURE 9.24 NTU logo manufactured by spray-based 3D printing on an overhead substrate [54]. *Reproduced with permission from Elsevier.*

Similar to fly ash, it is one of the waste by-products in fire-power plants. It has very low chemical reactivity and is usually used as a lightweight filler due to its low density (0.4–0.8 g/cm^3) [56]. In spray-based 3DCP, fly ash cenosphere is applied to regulate the rheological property of the mixtures for better delivery and deposition performances.

In cement-based mixtures, the full substitution of silica sand by fly ash cenosphere leads to smaller dynamic yield stress as well as plastic viscosity. Hence, the introduction of fly ash cenosphere reduces required pumping pressure and contributes to better pumpability. On the other hand, although it results in a lower static yield stress, the reduction in fresh density is also needed when considering buildability. Besides, the full substitution of silica sand by fly ash cenosphere also leads to larger compressibility and contributes to more uniform thickness distribution. The improvement in uniform thickness distribution could be observed clearly in Fig. 9.25 (Fig. 9.26).

9.5.3.2 Magnesium oxide-based cement

Magnesium oxide (MgO) is a sustainable and green binder compared with cement. Calcination of MgO requires a much lower temperature (around 850°C) compared to cement (around 1450°C), thus, giving off less CO_2 emission [55]. Also, its precursor can be produced from the desalination of seawater [57], which is one of the key drinking water acquisition methods in island countries like Singapore. Carbonation of MgO-based materials also reduces the environmental impact of CO_2. Hence, it is promising to use MgO as an alternative binder in sustainable spray-based 3DP.

In reactive MgO-activated slag mixtures, MgO serves as an alkaline activator to activate slag, a sustainable raw ingredient. The mixtures can gain

380 Digital Manufacturing

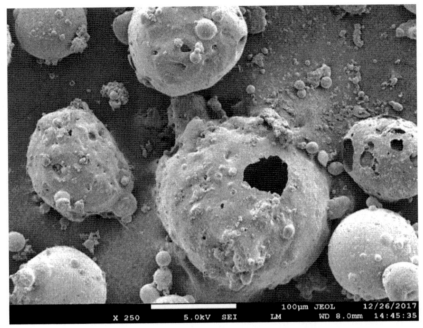

FIGURE 9.25 Image of fly ash cenosphere under the scanning electron microscope [54]. *Reproduced with permission from Elsevier.*

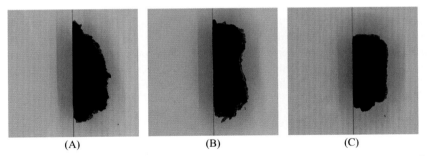

FIGURE 9.26 Thickness distribution of spray-printed filaments with different percentages of silica sand substitution by fly ash cenosphere: (A) 0% substitution, (B) 50% substitution, (C) 100% substitution [54]. *Reproduced with permission from Elsevier.*

strength with the formation of calcium silicate hydrate and hydrated magnesium hydrate. Lu et al. proposed to use fly ash cenosphere to regulate the rheological behavior of reactive-MgO slag mixtures [55]. The tailored mixture was confirmed to be feasible for spray-based 3DP.

9.6 Summary

3DP for building and construction has the potential to improve traditional building strategies by reducing material waste, human resources, and the high capital investments in the built environment. With its free-form ability, architects can express their thoughts with more freedom.

Research and applications in 3DP for building and construction have been developing rapidly in the past few years. Various printing techniques and systems, such as extrusion-based technique, powder-bed-based technique, gantry systems, robotic arm systems, and multirobots systems, have demonstrated good prospects. Traditional cementitious materials and sustainable materials like recycled aggregates and by-products from the energy and seawater desalination industries have also been explored. The effect of material components, material properties, and printing process parameters on the 3DP process and products have been investigated.

While 3DP for building and construction is still at an early development stage, many commercial companies and research institutes have implemented this technology into several novel applications. With the increasing need for a safer, more productive, and more sustainable construction industry, 3DP for building and construction will inevitably gain more attention and promise in the near future.

References

[1] C.K. Chua, K.F. Leong, 3D printing and additive manufacturing: principles and applications, World Scientific, Singapore, 2017.

[2] B. Khoshnevis, Automated construction by contour crafting—related robotics and information technologies, Automation in Construction 13 (2004) 5−19. Available from: https://doi.org/10.1016/j.autcon.2003.08.012.

[3] S.C. Paul, G.P.A.G. van Zijl, M.J. Tan, I. Gibson, A review of 3D concrete printing systems and materials properties: current status and future research prospects, Rapid Prototyping Journal 24 (2018) 784−798. Available from: https://doi.org/10.1108/RPJ-09-2016-0154.

[4] J. Gardiner, Exploring the emerging design territory of construction 3D printing - project led architectural research, Architecture and Design, RMIT University, 2011. Available from: http://researchbank.rmit.edu.au/view/rmit:160277. accessed May 9, 2020.

[5] M. Taylor, S. Wamuziri, I. Smith, Automated construction in Japan, ICE Proceedings Civil Engineering 156 (2003) 34−41.

[6] T. Wakisaka, N. Furuya, Y. Inoue, T. Shiokawa, Automated construction system for high-rise reinforced concrete buildings, Automation in Construction 9 (2000) 22.

[7] K.N. Jha, Formwork for Concrete Structures, McGraw Hill Education, New York, 2012.

[8] S. Lim, R.A. Buswell, T.T. Le, S.A. Austin, A.G.F. Gibb, T. Thorpe, Developments in construction-scale additive manufacturing processes, Automation in Construction 21 (2012) 262−268. Available from: https://doi.org/10.1016/j.autcon.2011.06.010.

382 Digital Manufacturing

[9] G. Cesaretti, E. Dini, X. De Kestelier, V. Colla, L. Pambaguian, Building components for an outpost on the lunar soil by means of a novel 3D printing technology, Acta Astronautica 93 (2014) 430−450. Available from: https://doi.org/10.1016/j.actaastro.2013.07.034.

[10] G. Ma, L. Wang, Y. Ju, State-of-the-art of 3D printing technology of cementitious material—An emerging technique for construction, Science China Technological Sciences 61 (2018) 475−495. Available from: https://doi.org/10.1007/s11431-016-9077-7.

[11] C. Gosselin, R. Duballet, Ph Roux, N. Gaudillière, J. Dirrenberger, Ph Morel, Large-scale 3D printing of ultra-high performance concrete − a new processing route for architects and builders, Materials & Design 100 (2016) 102−109. Available from: https://doi.org/10.1016/j.matdes.2016.03.097.

[12] A. Paolini, S. Kollmannsberger, E. Rank, Additive manufacturing in construction: a review on processes, applications, and digital planning methods, Additive Manufacturing 30 (2019) 100894. Available from: https://doi.org/10.1016/j.addma.2019.100894.

[13] M.E. Tiryaki, X. Zhang, Q.-C. Pham, Printing-while-moving: a new paradigm for large-scale robotic 3D Printing, in: 2019 IEEE/RSJ International Conference on Intelligent Robots and Systems (IROS), IEEE, Macau, China, 2019, pp. 2286−2291. https://doi.org/10.1109/IROS40897.2019.8967524.

[14] IAAC, Minibuilders, http://robots.iaac.net/ (accessed May 9, 2020).

[15] X. Zhang, M. Li, J.H. Lim, Y. Weng, Y.W.D. Tay, H. Pham, et al., Large-scale 3D printing by a team of mobile robots, Automation in Construction 95 (2018) 98−106. Available from: https://doi.org/10.1016/j.autcon.2018.08.004.

[16] R.J.M. Wolfs, F.P. Bos, E.C.F. van Strien, T.A.M. Salet, A real-time height measurement and feedback system for 3D concrete printing, High Tech Concrete: Where Technology and Engineering Meet, Springer, Cham, 2018, pp. 2474−2483. Available from: https://doi.org/10.1007/978-3-319-59471-2_282.

[17] A. Kazemian, X. Yuan, O. Davtalab, B. Khoshnevis, Computer vision for real-time extrusion quality monitoring and control in robotic construction, Automation in Construction 101 (2019) 92−98. Available from: https://doi.org/10.1016/j.autcon.2019.01.022.

[18] G.H.A. Ting, Y.W.D. Tay, Y. Qian, M.J. Tan, Utilization of recycled glass for 3D concrete printing: rheological and mechanical properties, Journal of Material Cycles and Waste Management 21 (2019) 994−1003. Available from: https://doi.org/10.1007/s10163-019-00857-x.

[19] D.A.R. Jones, B. Leary, D.V. Boger, The rheology of a concentrated colloidal suspension of hard spheres, Journal of Colloid and Interface Science 147 (1991) 479−495. Available from: https://doi.org/10.1016/0021-9797(91)90182-8.

[20] M. Thomas, Optimizing the Use of Fly Ash in Concrete, Portland Cement Association, Skokie, IL, United States, 2007.

[21] T.T. Le, S.A. Austin, S. Lim, R.A. Buswell, R. Law, A.G.F. Gibb, et al., Hardened properties of high-performance printing concrete, Cement and Concrete Research 42 (2012) 558−566. Available from: https://doi.org/10.1016/j.cemconres.2011.12.003.

[22] D. Hwang, B. Khoshnevis, Concrete wall fabrication by contour crafting, 2004 Proceedings of the 21st ISARC, ISARC, Jeju, Korea, 2004. Available from: https://doi.org/10.22260/ISARC2004/0057.

[23] K.-H. Jeon, M.-B. Park, M.-K. Kang, J.-H. Kim, Development of an automated freeform construction system and its construction materials, 2013 Proceedings of the 30th ISARC, ISARC, Montreal, Canada, 2013. Available from: https://doi.org/10.22260/ISARC2013/0153.

Three-dimensional (3D) printing for building and construction Chapter | 9 383

[24] Z. Malaeb, H. Hachem, A. Tourbah, T. Maalouf, N. Zarwi, F. Hamzeh, 3D concrete printing: machine and mix design, International Journal of Civil Engineering and Technology 6 (2015) 14–22.

[25] A. Perrot, D. Rangeard, A. Pierre, Structural built-up of cement-based materials used for 3D-printing extrusion techniques, Materials and Structures 49 (2016) 1213–1220. Available from: https://doi.org/10.1617/s11527-015-0571-0.

[26] H. Hoornahad, Toward development of self-compacting no-slump concrete mixtures, PhD Thesis, Delft University of Technology, 2014. http://resolver.tudelft.nl/uuid:431aa99f-f315–4030-838c-b550dff12ddf.

[27] C.T. Kennedy, The design of concrete mixes, Journal of the American Concrete Institute 36 (1940) 373–400. Available from: https://doi.org/10.14359/8528.

[28] N. Roussel, From industrial testing to rheological parameters for concrete, Understanding the Rheology of Concrete, Elsevier, 2012, pp. 83–95. Available from: https://doi.org/10.1533/9780857095282.1.83.

[29] T. Marchment, J.G. Sanjayan, B. Nematollahi, M. Xia, Interlayer strength of 3D printed concrete, 3D Concrete Printing Technology, Elsevier, 2019, pp. 241–264. Available from: https://doi.org/10.1016/B978-0-12–815481-6.00012-9.

[30] R.K. Dhir, J. de Brito, R.V. Silva, C.Q. Lye, Use of recycled aggregates in mortar, Sustainable Construction Materials, Elsevier, 2019, pp. 143–179. Available from: https://doi.org/10.1016/B978-0-08-100985-7.00006-6.

[31] S.C. Paul, Y.W.D. Tay, B. Panda, M.J. Tan, Fresh and hardened properties of 3D printable cementitious materials for building and construction, Archives of Civil and Mechanical Engineering 18 (2018) 311–319. Available from: https://doi.org/10.1016/j.acme.2017.02.008.

[32] Y.W.D. Tay, Y. Qian, M.J. Tan, Printability region for 3D concrete printing using slump and slump flow test, Composites Part B: Engineering 174 (2019) 106968. Available from: https://doi.org/10.1016/j.compositesb.2019.106968.

[33] V.N. Nerella, V. Mechtcherine, Studying the printability of fresh concrete for formwork-free concrete onsite 3D printing technology (CONPrint3D), Concrete Printing Technology, Elsevier, 2019, pp. 333–347. Available from: https://doi.org/10.1016/B978-0-12-815481-6.00016-6.

[34] N. Roussel, F. Cussigh, Distinct-layer casting of SCC: the mechanical consequences of thixotropy, Cement and Concrete Research 38 (2008) 624–632. Available from: https://doi.org/10.1016/j.cemconres.2007.09.023.

[35] Y.W.D. Tay, Large Scale 3D Concrete Printing: Process and Materials Properties, Nanyang Technological University, 2020, In press. Available from: https://hdl.handle.net/10356/137968.

[36] Y.W.D. Tay, M. Li, M.J. Tan, Effect of printing parameters in 3D concrete printing: Printing region and support structures, Journal of Materials Processing Technology 271 (2019) 261–270. Available from: https://doi.org/10.1016/j.jmatprotec.2019.04.007.

[37] Andrey Rudenko, 3D concrete house printer, 3D concrete print-design print medium-sized homes. (2014). http://totalkustom.com/video.html (accessed May 10, 2020).

[38] J. Zhang, J. Wang, S. Dong, X. Yu, B. Han, A review of the current progress and application of 3D printed concrete, Composites Part A: Applied Science and Manufacturing 125 (2019) 105533. Available from: https://doi.org/10.1016/j.compositesa.2019.105533.

[39] S.H. Ghaffar, J. Corker, M. Fan, Additive manufacturing technology and its implementation in construction as an eco-innovative solution, Automation in Construction 93 (2018) 1–11. Available from: https://doi.org/10.1016/j.autcon.2018.05.005.

384 Digital Manufacturing

[40] B. Furet, P. Poullain, S. Garnier, 3D printing for construction based on a complex wall of polymer-foam and concrete, Additive Manufacturing 28 (2019) 58–64. Available from: https://doi.org/10.1016/j.addma.2019.04.002.

[41] Eindhoven University of Technology, Printing our way out of the housing crisis: it is desperately needed, https://www.tue.nl/en/news-and-events/news-overview/30-04-2021-printing-our-way-out-of-the-housing-crisis-it-is-desperately-needed/ (accessed August 23, 2021).

[42] Mighty Buildings, Modern 3D printed prefab homes and ADUs, https://mightybuildings.com (accessed August 23, 2021).

[43] 3D Printing Media Network, ICON and Texas military department 3D print 3,800-sq-ft training barracks, https://www.3dprintingmedia.network/icon-texas-military-3d-print-training-barracks/, 2021 (accessed August 23, 2021).

[44] T.A.M. Salet, Z.Y. Ahmed, F.P. Bos, H.L.M. Laagland, Design of a 3D printed concrete bridge by testing, Virtual and Physical Prototyping 13 (2018) 222–236. Available from: https://doi.org/10.1080/17452759.2018.1476064.

[45] Xinhuanet, 3D-printed concrete bridge unveiled in North China, http://www.xinhuanet.com/english/2019-10/15/c_138474114.htm (accessed August 23, 2021).

[46] A. Anton, L. Reiter, T. Wangler, V. Frangez, R.J. Flatt, B. Dillenburger, A 3D concrete printing prefabrication platform for bespoke columns, Automation in Construction 122 (2021) 103467. Available from: https://doi.org/10.1016/j.autcon.2020.103467.

[47] ETH Zurich, First 3D printed and unreinforced concrete bridge, https://ethz.ch/en/news-and-events/eth-news/news/2021/07/3d-printed-and-unreinforced.html (accessed August 23, 2021).

[48] R.J. Flatt, N. Roussel, C.R. Cheeseman, Concrete: an eco material that needs to be improved, Journal of the European Ceramic Society 32 (2012) 2787–2798. Available from: https://doi.org/10.1016/j.jeurceramsoc.2011.11.012.

[49] D. Jiao, C. Shi, Q. Yuan, X. An, Y. Liu, H. Li, Effect of constituents on rheological properties of fresh concrete-A review, Cement and Concrete Composites 83 (2017) 146–159. Available from: https://doi.org/10.1016/j.cemconcomp.2017.07.016.

[50] B. Panda, M.J. Tan, Rheological behavior of high volume fly ash mixtures containing micro silica for digital construction application, Materials Letters 237 (2019) 348–351. Available from: https://doi.org/10.1016/j.matlet.2018.11.131.

[51] K.H. Tan, H. Du, Use of waste glass as sand in mortar: Part I – fresh, mechanical and durability properties, Cement and Concrete Composites 35 (2013) 109–117. Available from: https://doi.org/10.1016/j.cemconcomp.2012.08.028.

[52] H. Du, K.H. Tan, Use of waste glass as sand in mortar: Part II – alkali–silica reaction and mitigation methods, Cement and Concrete Composites 35 (2013) 118–126. Available from: https://doi.org/10.1016/j.cemconcomp.2012.08.029.

[53] H. Lindemann, R. Gerbers, S. Ibrahim, F. Dietrich, E. Herrmann, K. Dröder, et al., Development of a shotcrete 3D-printing (SC3DP) technology for additive manufacturing of reinforced freeform concrete structures, in: T. Wangler, R.J. Flatt (Eds.), First RILEM International Conference on Concrete and Digital Fabrication – Digital Concrete 2018, Springer International Publishing, Cham, 2019, pp. 287–298. https://doi.org/10.1007/978-3-319-99519-9_27.

[54] B. Lu, Y. Qian, M. Li, Y. Weng, K.F. Leong, M.J. Tan, et al., Designing spray-based 3D printable cementitious materials with fly ash cenosphere and air entraining agent, Construction and Building Materials 211 (2019) 1073–1084. Available from: https://doi.org/10.1016/j.conbuildmat.2019.03.186.

[55] B. Lu, W. Zhu, Y. Weng, Z. Liu, E.-H. Yang, K.F. Leong, et al., Study of MgO-activated slag as a cementless material for sustainable spray-based 3D printing, Journal of Cleaner Production 258 (2020) 120671. Available from: https://doi.org/10.1016/j.jclepro.2020.120671.

[56] A. Hanif, Z. Lu, Z. Li, Utilization of fly ash cenosphere as lightweight filler in cement-based composites — A review, Construction and Building Materials 144 (2017) 373—384. Available from: https://doi.org/10.1016/j.conbuildmat.2017.03.188.

[57] M. Seeger, W. Otto, W. Flick, F. Bickelhaupt, O.S. Akkerman, Magnesium compounds, Ullmann's Encyclopedia of Industrial Chemistry, Wiley-VCH Verlag GmbH & Co. KGaA, Weinheim, Germany, 2011. Available from: https://doi.org/10.1002/14356007.a15_595.pub2.

Chapter 10

Process monitoring and inspection

Tuan Tran[1] and Xuan Zhang[2]
[1]HP-NTU Digital Manufacturing Corporate Lab, School of Mechanical and Aerospace Engineering, Nanyang Technological University, Singapore, Singapore, [2]Singapore Centre for 3D Printing, School of Mechanical and Aerospace Engineering, Nanyang Technological University, Singapore, Singapore

Abbreviations

1D	one-dimensional
2D	two-dimensional
AE	acoustic emission
AM	additive manufacturing
ANN	Artificial Neural Network
ASTM	American Society for Testing and Materials
CCD	charge-coupled device
CMM	coordinate measuring machine
CMOS	complementary metal-oxide-semiconductor
CT	computed tomography
DED	directed energy deposition
DL	deep learning
DSLR	digital single-lens reflex
FDM	fused deposition modeling
FFF	fused filament fabrication
FPS	frames per second
LED	light-emitting diode
ML	machine learning
MPS	megapixels per second
NDT	nondestructive testing
NN	neural network
PBF	powder bed fusion
PID	proportional−integral−derivative
PV	process variable
RTD	resistance temperature detector
ROI	region of interest

Digital Manufacturing. DOI: https://doi.org/10.1016/B978-0-323-95062-6.00001-2
© 2022 Elsevier Inc. All rights reserved.

388 Digital Manufacturing

SLM selective laser melting
SP setpoint
UV ultraviolet

10.1 Introduction

There is a growing consensus among scientific researchers, industrial communities, and the public that additive manufacturing (AM) processes have many advantages over traditionally subtractive and formative manufacturing technologies, especially when dealing with parts having complex geometries and special functions. In recent decades, AM technologies have experienced rapid development, although their widespread adoption is still hampered by the lack of assurance of part quality and repeatability. Many researchers characterize process monitoring and inspections as an effective approach to overcome this challenge and identify the associated research with high priority in the development of AM ecosystem. This approach not only allows online monitoring of the build process and real-time detection of abnormalities or defects but also supports documentation and traceability of AM processes by storing recorded data. Furthermore, closed-up monitoring systems with feedback control can correct process parameters (e.g., power, scan velocity, and strategy) based on the deviation between printed parts and their designs. As a result, repeatability of AM processes is ensured, and the quality of the printed parts is improved.

This chapter will provide a comprehensive introduction to process monitoring and inspection techniques for AM processes. We start by presenting relevant techniques, sensors, and signals commonly used for process monitoring in Section 10.2. Subsequently, in Section 10.3, we conduct a detailed review of process monitoring applications in various AM processes. In Section 10.4, we discuss feedback controls used in typical monitoring systems. In Section 10.5, we outline several standards and toolkits for AM monitoring. In Section 10.6, we provide our insights into obstacles preventing widespread applications of process monitoring in AM processes and highlights milestones to be addressed by scientific and industrial communities. Finally, we summarize the chapter in Section 10.7.

10.2 Signals, sensors, and techniques for process monitoring

In AM processes, dynamic characteristics are reflected by the process signatures, which can be mainly classified into two categories: (1) observable and (2) derived. The former is measured and captured by various monitoring sensors during build process. The latter is derived from the former based on analytical models or simulations [1,2]. This section will discuss observable signatures (also known as signals), sensors and techniques used for monitoring AM processes. Subsequently, Section 10.3 will present derived signatures obtained from monitoring signals.

AM processes are complex processes both involving phase change and simultaneously releasing light, heat, sound, and vibration. Here, we focus on optical, thermal, X-ray and acoustic signals commonly used in practical monitoring systems. For each kind of signal, we introduce its corresponding sensors, practical use, advantages and disadvantages, and general applications in AM process monitoring.

10.2.1 Optical signals

Optical signals refer to light acquired in build processes including those in both visible (380−750 nm) and invisible (e.g., ultraviolet and infrared) spectra. While signals in the visible spectrum are typically used to obtain a more direct evaluation of build processes and part quality, those in the infrared spectrum (1−14 µm) are generally utilized to measure object temperature in build processes. For this purpose, acquired infrared signals, which will be introduced in Section 10.2.2, are often regarded as thermal signals. In addition, since the outcomes of using X-ray signals are optical images, we will include a discussion of the use of X-ray signals in Section 10.2.3.

10.2.1.1 Sensor types

Sensors used to capture optical signals are commonly categorized into four types according to their working principles, forms, and characteristics: photodiode, charge-coupled device (CCD), complementary metal-oxide-semiconductor (CMOS), and spectrometer.

10.2.1.1.1 Photodiodes

A photodiode, also known as a photodetector, photosensor or light detector, is a semiconductor device that converts a light signal into a current signal. When a photodiode absorbs photons having sufficient energy, it generates a current linearly proportional to the irradiance under a specific spectrum.

A photodiode is usually designed to operate in two working modes, photovoltaic mode and photoconductive mode. These two modes could be used for transformation of an optical signal to a digital signal and an "ON-OFF" switch in monitoring systems of AM and other industrial processes. The photovoltaic mode is widely applied to traditional solar cells for solar electric power generation. In practical use, a packaged photodiode, shown in Fig. 10.1A, always consists of a window (or an optical filter), a built-in lens and an optical fiber to connect the lens and sensor. The detected light is guided to the sensitive part of the sensor. The commonly used materials to fabricate photodiodes include silicon, germanium, and indium gallium

390 Digital Manufacturing

FIGURE 10.1 Images of different optical sensors. Photodiode (A), Nikon DSLR (B), and Photron high-speed camera (C). *Courtesy of the author, 2020.*

arsenide, with the corresponding detectable wavelength ranges being 190−1100, 400−1700, and 800−2600 nm, respectively [3]. The photoconductive mode is normally used in a photo-interrupter or gate, a transmission-type photosensor that integrates optical receiving and transmitting elements in a single package.

10.2.1.1.2 Charge-coupled device and complementary metal-oxide-semiconductor

Charge-coupled devices and complementary metal-oxide-semiconductor, known as CCD and CMOS, are based on metal-oxide-semiconductor (MOS) structures and popular optical sensors commonly used for digital imaging and video. Almost all modern CCD and CMOS sensors are integrated circuits incorporating arrays of pinned-photodiodes (PPD). Their main difference exists in the signal readout method. In a CCD sensor, all MOS capacitors representing pixels are coupled and covert projected light into electrons. Under the control of a circuit, the capacitors in each line transfer their contents one by one to the last capacitor. The number of electrons from each capacitor (pixel) is converted to a voltage by a couple of onboard transistors. In a CMOS sensor, the conversion of a light signal to a voltage is completed by a transistor and complex circuit installed at each pixel and thus easily operates at higher speeds. More detailed comparisons of CCD and CMOS sensors are given in Table 10.1.

Since CCD sensors easily produce high-quality images and with low noise and traditional CMOS sensors are more susceptible to noise, before 2000, CCD sensors tend to be used in professional, medical, and scientific applications that have special demands for high-quality images with high resolution and excellent light sensitivity. However, this difference has narrowed with the developments of the integrated circuit and other related technologies. CMOS sensors have gradually become an alternative to the users due to their lower price and higher frame rates. Since late 2010s, devices adopting CMOS sensors have dominated the consumer market, especially in applications with fewer requirements for high image quality.

TABLE 10.1 Comparisons of CCD and CMOS sensors.

Parameter	CCD	CMOS
Image quality	• High	• Lower
Noise	• Low	• Higher. Traditionally more susceptible to noise and need more light to create a low-noise image.
Sensitivity	• 60%–95%	• 75%–95%
Readout speed (MPS, megapixels per second)	• 1–40 MPS	• 100–400 MPS
Pixel Size	• 3–25 μm	• 2–9 μm
Resolution	• Higher	• Traditionally lower
Power consumption	• 100 times more than an equivalent CMOS sensor	• Low, longer battery life
Price	• Expensive	• Much more inexpensive • Fabricated on any standard silicon production line
Availability	• More mature	• Being improved rapidly

For monitoring of AM processes, commonly used CCD and CMOS sensors are digital single-lens reflex (DSLR) cameras (e.g., Canon, Nikon) and high-speed cameras (e.g., Photron, Phantom), as shown in Fig. 10.1B and C. The former usually records images or videos at relatively lower frame rates (below 50 frames per second (FPS)) and very high resolutions (typically 8000 pi × 6000 pi). The latter can capture images at much higher frame rates of over 250 FPS and relatively lower resolutions (typically 1024 pi × 1024 pi).

10.2.1.1.3 Spectrometers

As illustrated in Fig. 10.2, a spectrometer, also called a spectrophotometer, is a device that uses a grating or a prism to separate a beam of light into several spectral lines and measure their wavelengths and intensities. This working principle can be applied for a broad range of wavelengths, from grama ray to infrared light. The most utilized range is in the visible spectrum, especially in AM monitoring systems.

10.2.1.2 Practical use

At present, the most popular optical sensors used in monitoring systems are CCD and CMOS sensors, especially the DSLR and high-speed cameras.

392 Digital Manufacturing

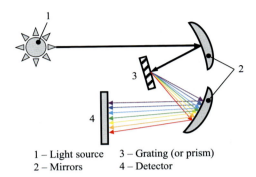

1 – Light source   3 – Grating (or prism)
2 – Mirrors   4 – Detector

FIGURE 10.2 Schematic of a grating spectrometer.

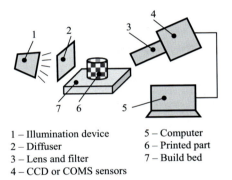

1 – Illumination device   5 – Computer
2 – Diffuser   6 – Printed part
3 – Lens and filter   7 – Build bed
4 – CCD or COMS sensors

FIGURE 10.3 Schematic of an optical monitoring system using CCD or CMOS sensors in AM process.

Fig. 10.3 depicts the schematic of an optical monitoring system using CCD or CMOS sensors in engineering applications. The monitoring system mainly contains an illumination device, a diffuser, a lens (including a filter), an optical sensor (or camera), and a controlling computer. In this type of systems, the following key points are usually considered in practical use. Some of these points may also be applicable in monitoring systems utilizing other optical sensors (e.g., photodiode, spectrometer) and will be subsequently introduced.

10.2.1.2.1 Sensor parameters

The sensor parameters are the most critical point in motoring system. Operations of CCD and CMOS sensors, including DSLR and high-speed cameras, are dependent on the resolutions, frame rates, exposure time, field of view, memory, interface. For photodiodes, spectral responsivity and response time are key parameters. The response time of a packaged photodiode usually increases with its surface area.

Process monitoring and inspection Chapter | 10 **393**

10.2.1.2.2 Installation method

Optical sensors could be installed co-axially or off-axis depending on the AM process, energy source and printing equipment. For example, the selective laser melting (SLM) process could adopt both co-axial and off-axis installation. Co-axial installation is not suitable for the directed energy deposition (DED) process using an electron beam as the energy source. During a build process, co-axially installed sensors can automatically trace the moving melt pool, while off-axis ones remain stationary [4].

10.2.1.2.3 Illumination devices and diffusers

To achieve uniform light and extract accurate visual information from images acquired from monitoring systems, supplemental illumination devices are required. These include but not limited to light-emitting diodes, laser light, halogen lamps, and UV arrays. Despite possible improvement of image accuracy and clarity, illumination may cause interferences to monitoring signals in a certain spectrum. Thus, when choosing illumination devices, the type of monitoring signal needs to be carefully considered. A diffuser may be used to render more uniform illumination. Under the same illumination condition, the quality of acquired images is also affected by the surface characteristics. For example, a rough surface may diffuse reflection of projected light, thus eliminate noises, and improve image quality and measurement accuracy [4].

10.2.1.2.4 Lenses and filters

For a monitoring system, the lens determines the working distance and magnification, while the filters are usually used to select a specific optical spectrum or angle of incidence and protect the lens from powder spattering or high temperature. A packaged photodiode typically has an integrated lens and a filter, which serves as a window. A stereo camera may use two or more lenses and filters to construct 3D pictures [4].

10.2.1.2.5 System testing and calibration

After completing the design and set-up of an optical monitoring system, calibration procedures, such as determination of optimal aperture stop, resolution, and frame rate, are required to optimize all system parameters used in the monitoring process. Calibration of a camera or an optical system involves estimating extrinsic, intrinsic parameters and distortion coefficients. The extrinsic parameters transform the world's 3D coordinate system into the camera's 3D coordinate system, while intrinsic parameters project the camera's 3D coordinate system onto the two-dimensional (2D) camera sensor. Distortion is different from other types of optical aberration as it may exist even if the image is in focus. In practice, distortion coefficients are introduced into the intrinsic parameters to model lens distortion. Determination of optical system parameters and distortion coefficients are critical to detect

394 Digital Manufacturing

and measure defective features accurately. This calibration procedure is also essential for post-processing processes, such as synthesizing images or matching features with other signal sources. For example, incorporating optical and thermal images obtained from optical and infrared cameras requires system calibration information when completing image registration. Generally, system calibration takes the following steps:

1. Choose calibration pattern
2. Prepare the calibration target
3. Collect calibration images
4. Evaluate calibration results
5. Improve calibration.

10.2.1.2.6 Data registration

In monitoring of AM processes, two or more sensors are usually utilized to acquire in-situ data. In that case, data registration, which transforms all acquired data to the same temporal and spatial scales, is necessary for further data processing and holistic analyses. For layer-wise optical and thermal images, a checkerboard is always employed to complete the image registration step.

10.2.1.3 Advantages and disadvantages

Optical monitoring system are low-cost and easy to develop. Monitoring optical signals within the visible spectrum is one of the most popular process monitoring techniques for AM processes. Acquired optical signals, typically in the form of images, can be intuitively interpreted or used to extract massive information of the monitored build process. However, the quality of optical signals is easily influenced by heat or light sources (e.g., a laser beam), spatter or scattering during acquisition. A typical monitoring system, due to the relatively large working distance between its lens and the melt pool, also has difficulty in monitoring the small melt pool.

10.2.1.4 Applications

Optical signals or images are applicable to online monitoring geometric features of the melt pool, track, slice/deposited layer, build plate/bed, feedstock, and plume [4]. Examples of these features include size, shape, or profile of the melt pool, and the deformation of the build layer. The signal from a photo-interrupter is used to trigger an action, and a spectrometer is used to analyze material composition.

10.2.2 Thermal signals

Thermal signals include temperature and any other information reflecting heat transfer in build processes. For many AM processes, the heat transfer is the driving force (e.g., melting and solidification of materials, cooing of printed parts), thus relevant to the build process and features of printed parts.

10.2.2.1 Sensor types
Common methods used for capturing thermal signals are mainly categorized as contact and noncontact. The former one usually uses thermometer as monitoring sensors, while the latter one uses pyrometers or infrared cameras.

10.2.2.1.1 Thermometers
Thermometers used in AM process monitoring systems include resistance temperature detectors (RTDs, e.g., Pt100) and thermocouples (e.g., Types K and T). The RTDs measure temperature by relating resistance and temperature changes, while the thermocouples transform thermal signals to voltages. Fig. 10.4A shows an image of a packaged Pt100 thermometer connected to a reader.

Since thermometers are always in contact with the measured objects, they yield more accurate temperature measurements than noncontact devices (e.g., infrared cameras), which typically are very sensitive to emissivity of the objects. Nonetheless, thermocouples should be used in their calibrated temperature ranges to avoid inaccurate measurements. For example, the temperature around the melt pool may either exceed the melting point of metals used to fabricate the sensors or influence their characteristics such as accuracy and response time.

10.2.2.1.2 Pyrometers
A pyrometer is a device used to remotely measure the high temperature of a heated object. Typically, a modern pyrometer consists of an optical system and a detector. Its working principle is that the intensity of light received by the detector is dependent on the distance from the detector to the observed heat source and the source's temperature. The relation between the detected signal (temperature T in-unit K of the observed object) and the thermal radiation or irradiance (E in unit W/m^2) obeys the Stefan–Boltzmann law:

$$E(\varepsilon, \quad T) = \varepsilon \sigma T^4 \quad (10.1)$$

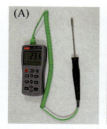

FIGURE 10.4 Images of different thermal sensors. Pt100 thermometer (A), sonel pyrometer (B), and InfraTEC infrared camera (C). *Courtesy of the author, 2020.*

396 Digital Manufacturing

where ε and σ respectively are the emissivity of the source object (unitless and ranging from 0 to 1) and Stefan–Boltzmann constant (5.670×10^{-8} W/m^2-K^4). Fig. 10.4B shows an image of a pyrometer.

10.2.2.1.3 Infrared cameras

An infrared camera, also called thermal camera or thermographic camera, is a remote-sensing instrument that produces thermal images based on infrared radiation from targeted objects, following the Stefan–Boltzmann law given in Eq. (10.1). Similar to optical cameras and high-speed cameras, which operate with light in the visible spectral range, that is, 400–700 nm, infrared cameras operate with wavelengths approximately from 1 to 14 μm. Infrared cameras also have wide ranges of frame rates and resolutions for different purposes. The commonly used brands include InfraTEC, FLIR and Fluke. Fig. 10.4C shows an image of a InfraTEC infrared camera.

10.2.2.2 Practical use

Infrared cameras are among the most popular thermal sensors used in AM monitoring systems. Monitoring systems employing infrared cameras mainly consist of similar components to those using optical cameras (Fig. 10.3), but without the illumination part. Thus, all key points except those for illumination devices involved in optical monitoring systems should also be considered when using infrared cameras for monitoring AM processes. Several special points only applicable to thermal monitoring systems are worth considering:

1. The range of temperatures to be measured is an important parameter when selecting a thermal sensor.
2. The emissivity of the observed object needs to be calibrated by a blackbody during system test and calibration since it directly affects the measurement accuracy and is related to the involving materials, surroundings, and working conditions.

10.2.2.3 Advantages and disadvantages

Thermal signals are obtained from the infrared spectrum and have the inherent ability to avoid interferences from strong visible light. Contact measurement methods, such as using thermocouples, although capable of acquiring accurate temperature data, are inconvenient for installation, especially in dynamic build processes. Noncontact measurement methods, such as using infrared cameras, are well capable of capturing dynamic temperatures. Their accuracy however is highly dependent on material emissivity, thus needs to be carefully calibrated before use.

10.2.2.4 Applications

Thermal signals in the form of images can be used to monitor dynamic thermal characteristics of the melt pool, track, slice/deposited layer, build plate/bed, nozzle. These characteristics include evolutions of temperature profile or thermal distribution of the melt pool [4]. Occasionally, thermal images could also be used to identify the size and shape of the melt pool after careful calibration.

A thermometer is usually installed at a fixed position and used to measure the substrate temperature. A pyrometer could be used to measure the temperature of a small region, such as the melt pool or a region of interest (ROI) of the build bed.

10.2.3 X-ray signals

X-ray is a form of electromagnetic radiation similar to visible light but having different wavelength (typically in the range 10 pm to 10 nm). X-rays with shorter wavelengths have higher frequencies and higher energy, allowing penetration of most objects. In AM processes, X-ray signals generated by users rather than the build process can pass through printed parts and monitoring their internal features by comparing the transmitted and received signals.

10.2.3.1 Working principle

Monitoring using X-rays is possible because X-ray attenuation is predominantly dependent on X-ray energy as well as physical and chemical characteristics (e.g., density, composition) of the material that X-rays pass through. For a monochromatic X-ray source, attenuation of an X-ray passing through a material is expressed by Beer's law,

$$I = I_0 e^{-cx} \tag{10.2}$$

where I_0 and I are the intensities of the emitted and finally X-ray, respectively; c is the linear attenuation coefficient (unit 1/length) of the material, and x is the distance the X-ray travels inside the material.

For a polychromatic X-ray source, the attenuation coefficient also changes with the X-ray energy (E) or wavelength. Thus, the final X-ray intensity is calculated by integrating Eq. (10.2) over the range of the X-ray energy spectrum, i.e.,

$$I = \int I_0(E) \sum_i e_i^{-c_i x_i} dE \tag{10.3}$$

In practical use, X-rays are usually reflected by a slice image where the local gray value varies with the X-ray attenuation and can be used to derive the proportion of absorbed or scattered X-rays.

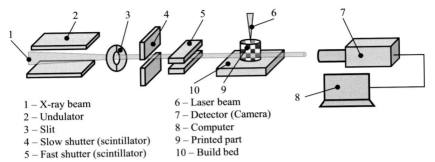

1 – X-ray beam
2 – Undulator
3 – Slit
4 – Slow shutter (scintillator)
5 – Fast shutter (scintillator)
6 – Laser beam
7 – Detector (Camera)
8 – Computer
9 – Printed part
10 – Build bed

FIGURE 10.5 Schematic of an monitoring system using X-ray signals in AM process [5].

10.2.3.2 Practical use

Fig. 10.5 shows the schematic of a monitoring system for AM processes using X-ray absorption or diffraction technology. The system consists of a synchrotron X-ray source, an undulator, a slit, a scintillator, a detector, and a computer [4]. The X-ray source can be monochromatic or polychromatic. The undulator is made of a periodic structure of dipole magnets, whose gap width controls the energy (or wavelength) of the emitted X-ray beam. The commonly used X-ray energy is 24.4 keV, corresponding to a wavelength of 0.508 Å. When passing through a pair of slits, the size of the X-ray beam is further adjusted. For different shooting speeds, the scintillator has both slow and fast shutters. The detector is generally an optical sensor (CCD or CMOS). When an X-ray beam penetrates ROI in a printed part, it will interact with the matter and be diffracted or scattered, inducing the X-ray attenuation. With the help of the scintillator, this attenuation is converted into visible light, generally recorded by a camera, and stored as a serial of images for further analysis [6].

10.2.3.3 Advantages and disadvantages

Although X-rays can intuitively monitor the morphology and location of internal defects in real-time, the cost of process monitoring techniques adopting X-ray signals is relatively high, and enhanced protection is necessary. It becomes much more expensive to use high-speed and high-resolution X-rays for process monitoring [7].

10.2.3.4 Applications

X-ray signals are widely used to detect internal defects, such as cracks, porosity, and holes. X-ray monitoring is generally used for validation of other monitoring methods [7]. A widely used technique, X-ray computed tomography (CT), uses X-ray for non-destructive testing (NDT), 3D measurement and reconstruction of internal structures of materials or objects.

10.2.4 Acoustic signals

In AM processes, formation of certain structures or defects may emit special sounds. These acoustic signals can be utilized to monitor build processes and evaluate quality of printed parts.

10.2.4.1 Working principle

Acoustic signals are used in various ways for monitoring of AM processes based on their working principles. As illustrated in Fig. 10.6, one way is to record acoustic emission directly and then analyze their characteristics such as frequency and amplitude. The other is to compare the difference between the emitted and received ultrasonic signals and derive the internal features using the delay time and the directional change between the two signals.

10.2.4.1.1 Acoustic emission

Acoustic emission (AE) is caused by transient elastic stress or pressure waves yielded by sudden release of energy stored in materials undergoing stress during fabrication [8]. As shown in Fig. 10.6A, an acoustic signal emitted from an AM process has main characteristics such as frequency, amplitude, energy/intensity, rising time, damping coefficient, and duration. The most common way to record this signal is by using a piezoelectric sensor, that is, a transducer working based on the piezoelectric effect. It is generally made from lead zirconate titanate (PZT) materials. The sensor converts the mechanical stress caused by an acoustic wave into an electric signal. Fig. 10.7 shows the general relationship between acoustic frequency and scale of detectable defect. Sensitivity of an ultrasonic testing method used in defect detection is higher with increasing acoustic frequency [9].

10.2.4.1.2 Ultrasonic waves

Ultrasonic waves typically refer to those with frequencies greater than the upper limit of human audible range (20 kHz). As shown in Fig. 10.6B, ultrasonic waves in conventional ultrasonic testing are emitted into a printed part

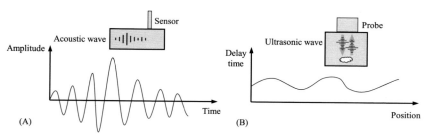

FIGURE 10.6 Working principles of acoustic emission (AE) method (A) [8] and ultrasonic testing (B) [9].

400 Digital Manufacturing

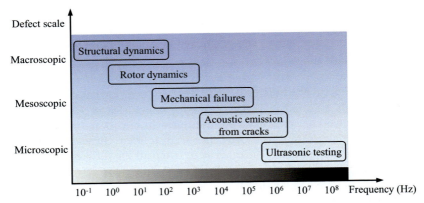

FIGURE 10.7 General relationship between the frequency spectrum of acoustic signal and defect scale [8].

and subsequently received by a probe after interacting with defects of the printed part via reflection, refraction, and scattering. The differences between the emitted and received ultrasonic signals, such as delay time, amplitude, and direction, are employed to evaluate the printing quality. Two categories of ultrasonic waves mainly used in testing are bulk waves and guided waves. The former includes longitudinal and shear waves, while the latter has many types, among which surface and plate waves are the most popular [9]. In conventional ultrasonic testing, a single piezoelectric transducer usually acts as both a transmitter and a receiver.

Over the past few decades, increasingly advanced ultrasonic testing techniques have been developed. Beside piezoelectric effect, other methods used to generate ultrasonic waves include electromagnetic acoustic transducers and laser ultrasonics [9]. Laser ultrasonics works on the principle that a laser beam projected a test surface causes the material to rapidly expand and generate ultrasound. In laser ultrasonic testing, a laser interferometer is commonly used as a receiver, although a piezoelectric element can also be considered [9]. As piezoelectric sensors become smaller, more piezoelectric elements can be incorporated into a single probe and even arranged in various patterns based on specific needs.

10.2.4.2 Practical use

As shown in Fig. 10.8, AM monitoring systems using acoustic sensors are straightforward. The most important component of each system is an acoustic sensor or probe used to record acoustic emission and ultrasonic signals. It should be noted that these signals are significantly affected by the sensors or probes' installation positions and directions.

Compared to ultrasonic signals, AE signals are more easily disturbed by environmental noise because acoustic waves can be generated by external

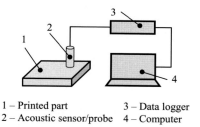

1 – Printed part　　　　3 – Data logger
2 – Acoustic sensor/probe　4 – Computer

FIGURE 10.8 Schematic of a monitoring system using acoustic sensors in AM process.

phenomena such as electromagnetic interferences, turbulent fluid flows, or mechanical vibrations. To eliminate the effect of noise signals, it is essential to identify the types of noise sources and extract the features of each kind of noise. Noise signals are then removed from the acquired acoustic signals by either analytical or numerical methods [8].

When performing ultrasonic testing, either contact mode and immersion mode could be adopted. In the former one, an ultrasonic probe directly touches the test sample, although in most practical contact tests, a thin layer of ultrasonic couplant (e.g., medical ultrasonic couplant) is usually used to better transmit ultrasonic waves to test samples. In the latter one, the probe is kept at a distance from the sample, and the space between them is filled with water. Besides, when guided waves are used in ultrasonic testing, a waveguide, i.e., a specific geometrical structure designed to direct waves in a preferred direction, is required to channel the waves and guide them through.

10.2.4.3 Advantages and disadvantages

Although acoustic monitoring systems are straightforward and inexpensive, recorded acoustic signals are easily disturbed by environmental noise. Relative positions and angles of acoustic signal sensors have a significant influence on recorded signals, and this needs to be considered during installation of the sensors. Greater efforts are required for post-processing and analyses of acquired signals to establish the corresponding relationship between the signal features and defects. Besides, ultrasonic signals are more sensitive to defects such as cracks and delamination due to lack of fusion but less sensitive to pores, slag inclusions, and defects near the surface. Compared with optical, thermal and X-Ray signals, using acoustic signals is less intuitive and easily results in omission of defects.

10.2.4.4 Application cases

Similar to X-ray signals, acoustic signals can be employed to diagnose internal defects such as crack, pore, lack of fusion [8]. Acoustic signals can also be applied to monitor the machine status such as powder mass flow rate and

402 Digital Manufacturing

mechanical vibrations from printers [10]. At present, acoustic signals are mainly used to monitor process status and defects of single-track scanning in SLM processes.

10.2.5 Other signals

Other signals can also be used to monitor AM processes, including but not limited to displacement signal, vibration signal, and current signal. These signals will be briefly introduced in the following subsections.

10.2.5.1 Displacement signals

Displacement signals are usually collected by displacement sensors, also known as distance or proximity sensors. A displacement sensor, typically consisting of a transmitter and a receiver, detects monitored objects and measures the distance between the sensor and an object. In an optical displacement sensor, the transmitter emits a laser signal, which is reflected after hitting an object's surface and finally collected by the receiver. The time difference between the sending and receiving moments is used to determine the distance between the sensor and the monitored surface. Displacement signals are commonly used to monitor the height of slices or deposited layers during build processes, especially those by the directed energy deposition (DED) technology [11]. Using displacement signals is a non-contact measurement method with high precision and short response time. However, it generally only provides measurements at a single point [12].

10.2.5.2 Vibration signals

External excitations in AM processes induce vibrations, and this kind of signals can be collected by piezoelectric vibration sensors or accelerometers. Vibration signals are usually used in fused filament fabrication (FFF) to monitor motions of build platforms and extruder heads [13,14].

10.2.5.3 Current signals

Current signals can be detected by current sensors working on the Hall effect. In a DED process, a supplied current combined with voltage signals can be used to calculate the total resistance of the wire and melt pool to monitor of build process [4]. In a fused deposition modeling (FDM) process, motor current of the filament feed pump is related to extrusion pressure on the FDM nozzle [15].

Table 10.2 summarizes the above-mentioned signals and sensors in this section. They can be used to monitor not only the defects during the build process but also the machines' state. In addition, some other sensors, such as interferometer, pressure sensor, anemometer, humidity sensor, rheometer, encoder, and magnetometer, are also used for monitoring [16].

TABLE 10.2 Summary of monitoring signals and sensors.

Signal	Typical sensor or machine	Advantages	Disadvantages	Applications
Optical signal (photography)	• Photodiode • CCD and CMOS • Spectrometer	• Easily available • Inexpensive • Intuitive • Provide massive information	• Easily influenced by other lights • Limitation of working distance and resolution	• Geometric characteristics • Triger • Composition.
Thermal signal (thermography)	• Thermometer • Pyrometer • Infrared camera	• Easily avoid the interference of strong visible light	• Difficult to track dynamics process for contact method. • Highly dependent on emissivity for contactless method • Careful calibration before use	• Temperature characteristics • Geometric characteristics
X-Ray signal (radiography)	• CT machine	• Ability to detect internal defects	• Expensive • Protection needed	• Internal defects • Validation of other methods • Nondestructive testing
Acoustic signals	• AE sensor • Ultrasonic probe	• Simple and cheap • Ability to detect internal defects	• Easily disturbed by the environmental noise and installation • Less intuitive	• Internal defects • Machine status
Others	• Displacement sensor • Vibration sensor/ Accelerometer • Current sensor	—	—	—

404 Digital Manufacturing

10.3 Applications in additive manufacturing processes

Based on the standard ISO/ASTM 52900:2015, all AM processes are classified into seven categories: [17,18]

1. Powder bed fusion (PBF): AM process in which thermal energy selectively fuses regions of a powder bed.
2. Directed energy deposition (DED): AM process in which focused thermal energy is used to fuse materials by melting as they are being deposited.
3. Material extrusion (ME): AM process in which material is selectively dispensed through a nozzle or orifice.
4. Vat photopolymerization (VM): AM process in which liquid photopolymer in a vat is selectively cured by light-activated polymerization.
5. Material jetting (MJ): AM process in which droplets of build material are selectively deposited.
6. Binder jetting (BJ): AM process in which a liquid bonding agent is selectively deposited to join powder materials.
7. Sheet lamination (SL): AM process in which sheets of material are bonded to form a part.

Table 10.3 lists the energy sources, technological names, machine brands (or players) and commonly used materials for each of the AM processes.

10.3.1 PBF processes

In a PBF process, powder is fused by laser (SLM, SLS, DMLS, DMLM, DMP, LPBF, LMF.), electron beam (EBM, EPBF), another heat source (SHS), or with agent and energy (MJF). When the thermal energy source is laser or electron beam, the commonly used materials include stainless steel, titanium, aluminum, cobalt chrome, steel, nickel alloy, and copper. For other heat sources, nylon (monochrome white thermoplastic powder) is the most popular material.

10.3.1.1 Monitoring systems

Optical or thermal signals captured by optical or thermal sensors are usually adopted to monitor PBF processes. Fig. 10.9 presents the schematics of coaxial and off-axis monitoring systems using optical and thermal sensors. Detailed configurations of a monitoring system depend on characteristics of a specific PBF process including the type of heat source, the printer layout, and the monitoring targets.

10.3.1.2 Monitored signatures

Signals obtained from a monitoring system can be used to derive signatures, which are classified into five different categories listed in Table 10.4, including the melt pool, track (scan path), slice (build layer scanned by the heat

TABLE 10.3 Seven categories of AM processes defined by ASTM/ISO standard [4,9,19].

AM process	Energy source	Technological name	Machine brand (or player)	Commonly used material
Powder bed fusion (PBF)	Laser	Selective laser melting (SLM)	• SLM solutions • Renishaw • Matsuura	Metals, plastics, ceramic powders, and sand
		Selective laser sintering (SLS)	• EOS • 3D Systems • Formlabs	
		Direct metal laser sintering (DMLS)	• SLM Solutions • EOS • DMG Mori • Sharebot	
		Direct metal laser melting (DMLM)	• Concept Laser (GE Additive)	
		Direct metal printing/production (DMP)	• 3D Systems	
		Laser metal fusion (LMF)	• Sisma • Trumpf	
		LaserCUSING	• Concept Laser (GE Additive)	
		Laser powder bed fusion (LPBF)	—	
	Electron beam	Electron beam melting (EBM)	• Arcam EBM (GE Additive)	
		Electron beam powder bed fusion (EPBF)	—	

(Continued)

TABLE 10.3 (Continued)

AM process	Energy source	Technological name	Machine brand (or player)	Commonly used material
Directed energy deposition (DED)	Other heat source	Selective heat sintering (SHS)	• Blue Printer	Metal powders, metal wires, and ceramic powders
	Agent and energy	Multi jet fusion (MJF)	• HP	
	Laser	Laser engineering net shape (LENS)	• Optomec	
		Laser metal deposition (LMD)	• Trumpf • FormAlloy • Additec	
		Direct metal deposition (DMD)	• Trumpf • DM3D technology • Sculpteo	
		Laser deposition technology (LDT)	• RPM Innovations Inc.	
		Laser deposition welding (LDW) and hybrid manufacturing	• MG MORI	
		Direct metal tooling (DMT)	• InssTek	
		Laser powder deposition (LPD) Laser consolidation Laser deposition	—	
	Electron beam	Electron beam direct manufacturing (EBDM)	• Sciaky Inc.	
		Electron beam additive manufacturing (EBAM)	• Sciaky Inc.	

	Plasma or electric arc	Wire arc additive manufacturing (WAAM)	• AML3D • Ramlab • MX3D	
		3D metal print (3DMP)	• Gefertec	
		Supersonic particle deposition (SPD) or cold spray (CS)	• RUAG	
		Rapid plasma deposition (RPD)	• Norsk Titanium	
		Shape metal deposition (SMD)	—	
Material extrusion	—	Fused filament fabrication (FFF)	• Kora 3D • Omni3D • RepRap	Thermoplastic polymer filaments
		Fused deposition modeling (FDM)	• Stratasys Inc. • 3D Systems • Discovery 3D Printers • Creatz3D	
		Robocasting (robotic material extrusion)	—	Ceramic slurries and clays
Vat photopolymerization	—	Stereolithographic apparatus (SLA)	• 3D Systems • Formlabs	Photopolymer resin
		Digital light processing (DLP)	• ADMATEC	
		Digital light processing (DLP) Continuous digital Light manufacturing (CDLM)	• EnvisionTEC	
		Scan, spin, and selectively photocure (3SP)	• EnvisionTEC • Realize Inc.	
		Continuous liquid interface production (CLIP)	• Carbon 3D	

(*Continued*)

TABLE 10.3 (Continued)

AM process	Energy source	Technological name	Machine brand (or player)	Commonly used material
Material jetting	—	Smooth curvatures printing (SCP)	• Solidscape (Stratasys Inc.)	Photopolymers, waxes and composites
		Multijet printing (MJP) Multijet modeling (MJM)	• 3D Systems	
		PolyJet/polyjet modeling (PJM)	• Objet (Stratasys Inc.)	
		Drop on demand (DoD)	• Solidscape (Stratasys Inc.)	
		Nanoparticle jetting (NPJ)	• XJet	
		Liquid metal jet printing (LMJP)	• Vader systems (Xerox)	
		UV cured material jetting	—	
Binder jetting	—	Metal jet technology	• HP	Plastics, metals, glass, sand and ceramic powders
		Binder jetting 3d printing (BJ3DP)	• ExOne	
		Universal binder jetting	• voxeljet	
		High-speed sintering (HSS)	• voxeljet	
		Digital metal	• Digital Metal	
		ColorJet printing (CPJ)	• ZCorp (3D Systems)	
		Drop-on-powder (DoP) Binder jet additive Manufacturing (BJAM)	—	

Sheet lamination	—	Laminated object manufacturing (LOM)	• Helisys Inc. (now Cubic Technologies)	Plastics, paper and metal sheets
		Selective deposition lamination (SDL)	• Mcor Technologies (now CleanGreen3D)	
		Selective lamination composite object manufacturing (SLCOM)	• EnvisionTEC	
		Ultrasonic additive manufacturing (UAM) Ultrasonic consolidation (UC)	• Solidica (now Fabrisonic)	
		Composite-based additive manufacturing (CBAM)	• Impossible Objects	
		Plastic sheet lamination (PSL)	• Solido 3D	
		Computer-aided manufacturing of laminated engineering materials (CAM-LEM)	• CAM-LEM, Inc.	

410 Digital Manufacturing

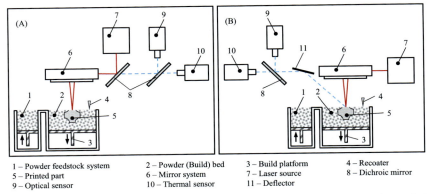

1 – Powder feedstock system 2 – Powder (Build) bed 3 – Build platform 4 – Recoater
5 – Printed part 6 – Mirror system 7 – Laser source 8 – Dichroic mirror
9 – Optical sensor 10 – Thermal sensor 11 – Deflector

FIGURE 10.9 Schematics of (A) co-axial and (B) off-axial monitoring systems using optical and thermal sensors for laser-based PBF processes [2].

TABLE 10.4 Monitored signatures derived from various process signals at different observation levels in PBF processes [2].

Monitored signature		Process signal	Sensor/Monitoring devices
Melt pool	Geometry (size, profile, shape)	Optical signal	• Optical camera
		Thermal signal	• Thermal camera
	Temperature intensity	Thermal signal	• Pyrometry • Thermal camera
	Temperature profile	Thermal signal	• Optical camera • Thermal camera
Track (Scan path)	Geometry	Optical signal	• Optical camera
		Others	• Interferometer
	Temperature intensity	Thermal signal	• Pyrometry • Thermal camera
	Temperature profile	Optical signal	• Optical camera
		Thermal signal	• Thermal camera
	Ejected material	Optical signal	• Optical camera
		Thermal signal	• Pyrometry • Thermal camera
Slice (Build layer)	Surface pattern	Optical signal	• Optical camera • Photodiode • Spectrometer
		Others	• Interferometer

(Continued)

Process monitoring and inspection Chapter | 10 **411**

TABLE 10.4 (Continued)

Monitored signature		Process signal	Sensor/ Monitoring devices
	Geometry	Optical signal Thermal signal	• Optical camera • Thermal camera
	Thickness profile	Optical signal	• Optical camera
	Temperature intensity	Thermal signal	• Pyrometry • Thermal camera
	Temperature profile	Optical signal	• Optical camera
		Thermal signal	• Thermal camera
Powder bed (Build bed)	Homogeneity	Optical signal	• Optical camera
		Others	• Interferometer
	Temperature intensity	Thermal signal	• Thermocouple • Pyrometry • Thermal camera
	Temperature profile	Optical signal	• Optical camera
		Thermal signal	• Thermal camera
Part	Geometry/Distortion	Optical signal	• Optical camera
		Thermal signal	• Thermal camera
	Void/porosity/hole/ crack	X-ray signal	• X-ray scanner
		Ultrasonic signal	• Ultrasonic probe
Others	Powder recoating system vibration	Vibration signal	• Accelerometer
	Environment	Gas flow	• Anemometer

source), powder bed (build bed), and (printed) parts. Several process signals may contribute to the same monitored signature. For example, both optical signal and thermal signal can generate geometrical information.

10.3.1.3 Process defects

After processing monitoring data and deriving monitored signatures, some process defects may be further acquired by developing analytical models. Since the build of each layer in a PBF process contains both deposition and printing processes, Table 10.5 presents defects commonly found in these processes with illustrating photos. These include powder bed discontinuity,

TABLE 10.5 Typical defects in deposition and printing of PBF processes [2,19].

Process	Defect	Illustration	Description
Deposition	Powder bed discontinuity	[20]	• Mainly caused by coating mechanism.
Printing	Spatter	[7]	• Caused by strong jet due to metal vaporization under laser beam. • Removes powder particles in scan path and induces powder shortages near the melt pool.
	• Porosity / • Voids / • Gas pores	[21]	• Intralayer defect. • Caused by entrapped gas pores inside the bulk of material. • Depends on the material types.
	Elongated pores	[21]	• Interlayer defect. • Caused by a lack of energy input to melt the feed material (powder or wire) between layers.

Balling	[22]	• Molten material solidifies into small spheres instead of bulk solid layers. • Depends on the process parameters (e.g., heat power, scan speed) and material properties (e.g., surface tension).
Surface defects	[23]	• Surface roughness and characteristics.
Geometric defects and dimensional inaccuracy	[24]	• Shrinkage, oversizing, warping, curling, and geometrical distortion.
Residual stresses, crack, and delamination	[25]	• Residual stresses are induced by large thermal gradients during the cooling stages of printed layers. • Cracks result from large tensile stress exceeding the material limit. • Delamination, also known as interlayer cracks, occurs between adjacent build layers.
Inhomogeneity and impurity	[26]	• Microstructural inhomogeneities are associated with the grain size characteristics and crystallographic textures. • Impurities involve inclusions and contaminations from other materials, oxide formations, and unfused powders.

414 Digital Manufacturing

spatter, voids, porosity or gas pores, elongated pores, balling, surface defects, geometric defects, crank, delamination and residual stresses, microstructural inhomogeneity and unfused powder, impurity, or contamination. All these defects influence the quality of the printed parts.

10.3.2 DED processes

Similar to PBF, DED processes are classified into the following groups by the energy source used to melt material:

1. Laser-based DED (e.g., LENS, LPD, LMD)
2. Electron beam-based DED (e.g., EBDM)
3. Plasma or electric arc-based DED (e.g., WAAM).

Table 10.6 provides advantages and disadvantages of DED with different energy sources. DED processes can use both powder (e.g., LENS, LPD, LMD) and wire (e.g., WAAM) as feed materials, while PBF processes only use powder. Although both powder-based and wire-based DED systems feed materials through nozzles, the former use laser or electron beams, while the latter use laser, electron beams, or plasma arcs to create the melt pool.

The most frequently used materials in DED processes are titanium, aluminum, stainless steel, or copper. DED processes can be used to produce components with composition gradients or hybrid structures consisting of multiple materials of varying compositions and structures.

10.3.2.1 Monitoring system

The most popular signals employed to monitor the DED processes are optical or thermal signals. Fig. 10.10 presents the schematics of co-axial and off-axis monitoring systems using optical and thermal sensors for laser-based DED processes. It is worth noting that detailed designs of monitoring

TABLE 10.6 Advantages and disadvantages of DED processes with different energy sources [4].

Heat source	Build volume	Detail resolution	Deposition rate	Coupling efficiency	Potential for contamination
Laser	☆	☆ ☆	☆ ☆	☆	☆ ☆ ☆
Electron beam	☆ ☆ ☆ ☆	☆	☆ ☆ ☆	☆ ☆ ☆ ☆	☆ ☆ ☆ ☆
Plasma or electric arc	☆	☆	☆	☆ ☆ ☆ ☆	☆ ☆

*A higher number of stars indicates better characteristics. This table is intended as a general guide. Variations in individual systems and process advancements may affect the characteristics of each process.

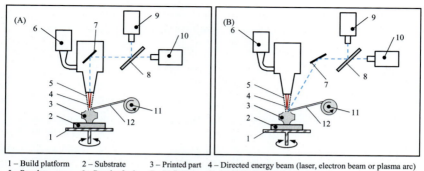

1 – Build platform 2 – Substrate 3 – Printed part 4 – Directed energy beam (laser, electron beam or plasma arc)
5 – Powder 6 – Powder feeder 7 – Deflector 8 – Dichroic mirror 9 – Optical sensor
10 – Thermal sensor 11 – Wire (Filament) coil 11 – Wire (Filament)

FIGURE 10.10 Schematics of (A) co-axial and (B) off-axial monitoring systems using optical and thermal sensors for laser-based DED processes [4].

systems also depend on characteristics of the intended DED process, such as the type of heat source, printer layout, and monitoring targets.

10.3.2.2 Monitored signatures

In DED processes, monitored signatures obtained from process monitoring signals comprise of characteristics of the melt pool, deposited layer (build layer), substrate, and feedstock. Table 10.7 lists monitored signatures derived from various process signals.

10.3.2.3 Process defects

As summarized in Table 10.5, the most common defects arising in DED processes are porosity, voids, or gas pores, surface defects, geometric defects, delamination, microstructural inhomogeneity, change in chemical compositions [4,27].

10.3.3 Material extrusion processes

Major techniques using material extrusion AM processes include Fused Filament Fabrication (FFF), Fused Deposition Modeling (FDM), and robocasting. The first two are more popular and generally use thermoplastic polymer filaments as feed materials, while the third one uses ceramic slurries and clays.

10.3.3.1 Monitoring systems

A variety of signals can be utilized to monitor material extrusion processes. Fig. 10.11 shows a schematic of a monitoring system using optical, thermal, acoustic and vibration signals for FFF processes. The system has an optical

416 Digital Manufacturing

TABLE 10.7 Monitored signatures derived from various process signals in DED process [4].

Monitored signature		Process signal	Sensor/ Monitoring devices
Melt pool	Geometry (size, profile/shape)	Optical signal	• Optical camera • Spectrometer
		Thermal signal	• Thermal camera
	Temperature intensity	Thermal signal	• Pyrometry • Thermal camera
	Temperature profile	Optical signal	• Optical camera
		Thermal signal	• Thermal camera
	Flow velocity	Optical signal	• Optical camera
Deposited layer (build layer)	Geometry	Optical signal	• Optical camera
		Thermal signal	• Thermal camera • 3D camera system
		Displacement signal	• Displacement sensor
	Temperature intensity	Thermal signal	• Pyrometry • Thermal camera
	Temperature profile	Optical signal	• Optical camera
		Thermal signal	• Thermal camera
Substrate	Geometry	Displacement signal	• Displacement sensor
		Optical signal	• Optical camera
		Thermal signal	• Thermal camera • 3D camera system
	Temperature intensity	Thermal signal	• Thermocouple • Pyrometry • Thermal camera
	Temperature profile	Optical signal	• Optical camera
		Thermal signal	• Thermal camera
Feedstock	Powder feed system	Optical signal	• Optical camera
		Thermal signal	• Thermocouple • Thermal camera

(Continued)

TABLE 10.7 (Continued)

Monitored signature		Process signal	Sensor/Monitoring devices
Others	Powder flow rate	Acoustic signal Optical signal	• AE sensor • Photodiode
	Plume	Optical signal	• Optical camera
	Distance between tool and substrate	Resistance signal	• Hall effect current sensor
	Elemental composition	Optical signal	• Spectrometer

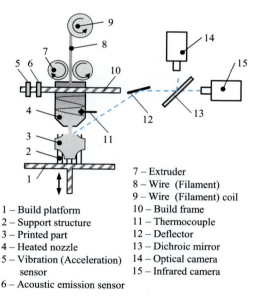

1 – Build platform
2 – Support structure
3 – Printed part
4 – Heated nozzle
5 – Vibration (Acceleration) sensor
6 – Acoustic emission sensor
7 – Extruder
8 – Wire (Filament)
9 – Wire (Filament) coil
10 – Build frame
11 – Thermocouple
12 – Deflector
13 – Dichroic mirror
14 – Optical camera
15 – Infrared camera

FIGURE 10.11 Schematic of a monitoring system using optical, thermal, acoustic and vibration signals for FFF processes [28].

camera, an infrared camera, a thermocouple, an AE sensor, a vibration sensor (or an accelerometer).

10.3.3.2 Monitored signatures

Monitored signatures of material extrusion processes can be divided into two categories, one about the machine state, and the other one about the part

418 Digital Manufacturing

quality. The machine state is obtained by monitoring the status of filament, extruder, nozzle, and build platform. The part quality is evaluated by monitoring the features of the melt pool, deposited layer (build layer) and (printed) parts. Table 10.8 summarizes all signatures derived from various process signals in FFF processes.

TABLE 10.8 Monitored signatures derived from process monitoring signals in FFF processes [16].

Category	Monitored signature		Process signal	Sensor/ Monitoring devices
Machine state	Filament	Breakage	Acoustic signal	• AE sensor • Photodiode
			Optical signal	• Optical camera
		Runout	Acoustic signal	• AE sensor
			Optical signal	• Optical camera
		Material classification (e.g., type and color)	Optical signal	• Optical camera
		Diameter	Optical signal	• Optical camera • Optical comparator
	Extruder	Filament loading state (transport slippage)	Vibration signal	• Vibration sensor • Accelerometer
			Acoustic signal	• AE sensor
			Optical signal	• Optical camera
		Filament feed rate	Optical signal	• Optical camera
		Blockage	Optical signal	• Optical camera
	Nozzle	Temperature intensity	Thermal signal	• Thermocouple • Thermal camera
		Vibration	Vibration signal	• Vibration sensor • Accelerometer
			Acoustic signal	• AE sensor
		Distortion	Optical signal	• Optical camera
		Blockage	Vibration signal	• Vibration sensor • Accelerometer

(Continued)

Process monitoring and inspection Chapter | 10   **419**

TABLE 10.8 (Continued)

Category	Monitored signature		Process signal	Sensor/ Monitoring devices
Part quality			Acoustic signal	• AE sensor
			Optical signal	• Optical camera
		Pressure, filament viscosity, filament flow rate	Pressure signal, viscosity signal, thermal signal	• Pressure sensor, rheometer, thermocouple
		Force between nozzle and build frame	Force signal	• Force sensor
	Build platform	Temperature intensity	Thermal signal	• Thermocouple • Thermal camera
		Vibration	Vibration signal	• Vibration sensor • Accelerometer
			Acoustic signal	• AE sensor
	Melt pool	Geometry	Optical signal	• Optical camera • Spectrometer
			Thermal signal	• Thermal camera
		Temperature intensity	Thermal signal	• Pyrometry • Thermal camera
		Temperature profile	Optical signal	• Optical camera
			Thermal signal	• Thermal camera
	Deposited layer (build layer, including infill)	Geometry	Displacement signal	• Displacement sensor
			Optical signal	• Optical camera
			Thermal signal	• Thermal camera • 3D camera system
		Temperature intensity	Thermal signal	• Pyrometry • Thermal camera
		Temperature profile	Optical signal	• Optical camera
			Thermal signal	• Thermal camera

(*Continued*)

420 Digital Manufacturing

TABLE 10.8 (Continued)

Category	Monitored signature		Process signal	Sensor/Monitoring devices
Part	Surface pattern		Optical signal	• Optical camera • Photodiode • Spectrometer
			Others	• Interferometer
	Geometry		Displacement signal	• Displacement sensor
			Optical signal	• Optical camera
			Thermal signal	• Thermal camera • 3D camera system
	Temperature intensity		Thermal signal	• Thermocouple • Pyrometry • Thermal camera
	Temperature profile		Optical signal	• Optical camera
			Thermal signal	• Thermal camera

10.3.3.3 Process defects

Typical defects in material extrusion processes are porosity, voids, or gas pores, surface defects, geometric defects, delamination, and microstructural inhomogeneity. Besides, material extrusion processes are constantly faced with printing failures caused by machine states such as filament slippage, breakage and runout, and blockage of extruders or nozzles.

10.3.4 Other additive manufacturing processes

Except for PBF, DED and material extrusion, the other four AM processes are vat photopolymerization (e.g., SLA, DLP, CDLP, 3SP, CLIP), material jetting (e.g., SCP, MJM, DOD, NPJ), binder jetting (e.g., BJAM, HSS) and sheet lamination (e.g., LOM, SDL, SCOM, UAM, PSL). As summarized in Table 10.3, different AM processes are applicable to different materials.

10.3.4.1 Monitoring systems

Development of monitoring systems for the remaining four AM processes is still on-going, as opposed to that for PBF, DED and material extrusion. Fig. 10.12 shows schematics of two monitoring systems, one using an ultrasonic transducer for an UAM process, and the other using optical and thermal cameras for a LMJP process. It is noted that monitoring systems for different AM processes have distinct components and configurations.

10.3.4.2 Monitored signatures and process defects

In principle, monitored signatures for AM processes other than PBF, DED, and material extrusion can also be derived from acquired process monitoring signals. However, at present, only few monitored signatures and process defects in these AM processes are reported, as listed in Table 10.9.

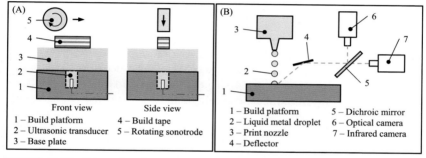

FIGURE 10.12 Schematics of monitoring systems, (A) using an ultrasonic transducer for an UAM process [29,30], and (B) using optical and thermal cameras for an LMJP process [31].

TABLE 10.9 Monitored signatures derived from various process signals in other AM processes.

AM process	Technical name	Monitored signature and process defects	Process signal	Sensor/Monitoring devices
Vat photo-polymerization	—	Conversion degree	Thermal signal	• Fourier transform infrared (FTIR) spectroscopy
Material jetting	LMJP	Droplet size, timing, and motion	Optical signal	• Optical camera
			Millimeter-wave signal	• Piezoelectric sensor (PZT)
Binder jetting	—	Edge detection	Optical signal	• Optical camera
Sheet lamination	UAM	Delamination and bond quality	Acoustic signal	• Ultrasonic transducer • AE sensor

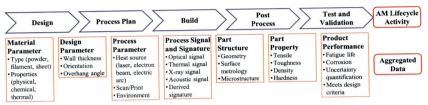

FIGURE 10.13 AM lifecycle and associated data [32].

10.4 Quality and feedback control

Generally, an AM lifecycle includes five major activities, i.e., design, process plan, build, post-process, test, and validation, as shown in Fig. 10.13 [32]. The first four activities directly influence the quality of printed parts and performance of the final production. Among them, the most important one is the build process, which is affected by various process parameters and is also the focus this section [33,34]. To enhance the part quality and repeatability, feedback or closed-loop control for the build process is necessary because it enables detection of deviations between the monitored and planned processes in real-time, as well as automatic adjustment of process parameters to correct the build process.

10.4.1 Process parameters

The build process of a designed part involves four categories of process parameters,

1. Material
2. Heat source
3. Scan (print)
4. Environment/machine

Table 10.10 summarizes all the process parameters involved in AM processes. Although all of them are responsible for the quality of printed parts, only the adjustable ones can be used in feedback or closed-loop control of the build process. The most commonly used parameter in feedback control is laser power.

10.4.2 Signal processing and feedback control

Fig. 10.14 presents the schematic of an AM process with a feedback control system. The control system acquires monitoring signals by sensors, extracts monitored signatures by processing the acquired signal data, and finally adjusts the process parameters online by comparing the monitored with the targeted signatures. Three crucial components of this system are monitoring sensors, data processing unit, and feedback control model, all of which close the AM process. A conventionally open-loop AM process lacks either a

TABLE 10.10 Process parameters in AM processes [33–35].

Category	Type	Parameter
Material (Powder, wire/filament sheet)	Uncontrollable	• Physical properties (size, shape and morphology, distribution, density, surface roughness) • Thermal properties (thermal conductivity, heat capacity, diffusivity, emissivity, absorptivity, latent heat of fusion, melting and boiling temperature, vapor pressure, coefficient of thermal expansion) • Flow properties (melt pool viscosity, flowability and fluidization, surface free energy) • Chemical properties (element composition, surface chemistry, crystal structure)
	Controllable	• Deposition/Feed rate
Heat source (Laser electron beam, electric arc, or plasma)	Uncontrollable	• Laser mode and wavelength, pulse width, frequency, polarization, intensity profile
	Controllable	• Power, spot size, offset, print temperature
Scan/Print	Uncontrollable	• Nozzle type and characteristics
	Controllable	• Scan velocity, scan pattern or strategy (hatch, zigzag, spiral), scan spacing, pulse distance, layer thickness, scaling factor
Environment and machine	Uncontrollable	• Inert gas type and properties
	Controllable	• Oxygen level, environmental pressure and temperature, gas flow rate, preheating temperature, recoater/extrusion velocity

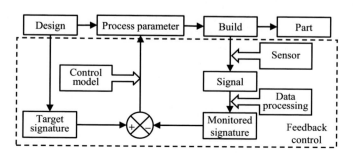

FIGURE 10.14 Schematic of an AM process with a feedback control system [36].

424 Digital Manufacturing

monitoring system or a feedback control one. As Section 10.2 has already exhaustively introduced the commonly used sensors and signals in the monitoring of AM processes, this section will focus on the data processing and feedback control methods.

10.4.2.1 Conventional methods

Traditional data processing and analysis methods used to establish the relationship between monitoring signals and process signatures mainly includes time-frequency domain transform, Fourier transform, and feature extraction (average, minimum and maximum value). These are adequate when dealing with a small amount of one-dimensional (1D) or two-dimensional (2D) monitoring data. However, they are stretched when facing with a large amount of 1D, 2D, or 3D monitoring data. More efficient methods are therefore needed to process massive signal data produced in monitoring AM processes.

Conventional control models usually adopt proportional-integral-derivative (PID) controllers, which introduce feedbacks into the closed-up control system to correct the system process parameters. The feedback signals are derived from the differences between the target setpoints (SP) and the process variables (PV) measured by online sensing devices. Corrections are applied based on proportional, integral, and derivative terms in the control model. Many industrial applications employ this type of controllers to achieve continuous feedback controls in their control systems.

10.4.2.2 Machine learning approach

Machine learning (ML) is an intelligent and fast data analyzing method capable of automatically building analytical models that continuously assess data and learn from the results. Thus, hidden insights into massive data could be revealed. Machine learning has been widely used in prediction, optimization, detection, classification, and regression. For example, email filtering and computer vision are two typical applications of ML in our daily life, where conventional methods usually fail to perform to required works reasonably fast.

Recent development of machine learning makes it easier to process and analyze large amounts of monitoring data and assess the print quality rapidly. After comparing these results with the targeted ones, ML models can make decisions and instruct feedback control systems how to adjust process parameters to achieve better print quality.

10.4.2.2.1 Learning methods

Traditional ML methods are classified into three main categories based on features of input and output data:

1. Supervised learning aims to obtain a general relation between the inputs and outputs. Both sample inputs and needed outputs are available to the learning system.

2. Unsupervised learning aims to discover the hidden features or structures of input and output data without being supplied with labels of inputs and outputs.
3. Reinforcement learning performs specific actions or tasks with feedbacks provided by a dynamic environment and aims to maximize the right ones (or the cumulative rewards).

Other ML methods have also been developed, including semi-supervised learning, self-supervised learning, feature learning, and deep learning (DL), but they do not strictly belong to either of the above three categories. For example, the DL method, which dominates the field of machine learning since 2020, can be regarded as either supervised, unsupervised, or semi-supervised. DL method contains multiple hidden layers in an artificial neural network (ANN) with representation learning. It seeks to learn the way the human brain converts light (or sound) into vision (hearing), which has been successfully applied in computer vision and speech recognition [37].

10.4.2.2.2 Learning algorithms

Each ML method includes various learning algorithms, as listed in Table 10.11. In supervised learning, the most widely used algorithms are linear discriminant analysis, linear regression, logistic regression, naive Bayes, support vector machines (SVM), k-nearest neighbor algorithm (k-NN), decision trees, neural networks (NN), and similarity learning. In unsupervised learning, the most popular ones are k-means clustering, hierarchical clustering and principal component analysis (PCA). Several ML algorithms are similar to the traditional statistical method and are primarily used to perform predictions by computers. More algorithms used in ML are provided in Table 10.11.

10.4.2.2.3 Learning steps

Generally, seven steps are recommended to build a machine learning model in practice. They are listed as follows:

1. Gathering data: Gather raw data for training. This step is of great importance because the quality and quantity of the acquired data directly determine the model accuracy.
2. Preparing data: Randomize data ordering; Visualize data to detect relevant relationships between variables or data imbalances; Clean data (e.g., duplicate removal, error correction, normalization, data type conversion); Split the acquired data into two sets, one for training and the other for evaluation.
3. Choosing a model: Choose a model according to data features and task types.

TABLE 10.11 ML methods and algorithms.

Method category	Sub-category	Algorithm	Application
Supervised learning	Classification	• Linear classifiers (Linear Discriminant Analysis (LDA), Logistic regression, Naive Bayes classifier, Perceptron, Support vector machines (SVM)) • K-nearest neighbors (k-NN) • Decision trees • Random forest network (RFN) • Learning vector quantization • Genetic algorithm • Neural networks (NN) (Artificial neural networks, Back propagation neural network, Feed forward neural network) • Bayesian networks • Relevance vector machine (RVM) • Similarity learning • Ensemble learning (Bagging, Boosting)	• Image classification • Identity fraud detection • Customer retention • Diagnostics
	Regression	• Linear regression • Logistic regression • Polynomial regression • Gaussian process regression	• Advertising popularity prediction • Weather forecasting • Market forecasting • Estimating life expectancy • Population growth prediction

Unsupervised learning	Clustering	• K-means clustering (centroid-based clustering) • Hierarchical clustering (connectivity-based clustering) • Self-consistent clustering • Expectation−maximization • Density-based clustering • Distribution-based clustering	• Recommender system • Targeted marketing • Customer segmentation
	Dimensionality reduction	• Principal component analysis (PCA) • Factor analysis • Independent component analysis	• Meaningful compression • Big data visualization • Structure discovery • Feature extraction
Reinforcement learning		• Markov decision process (MDP)	• Real-time decisions • Robot navigation • Learning tasks • Skill acquisition • Game AI

428   Digital Manufacturing

4. Training: Repeat to update parameters involved in the model using the training data set. Each updating iteration or cycle is called one training step.
5. Evaluation: Test the model against previously unused data and evaluate the model performance.
6. Hyperparameter tuning: Improve the trained model by adjusting or tuning hyperparameters (e.g., training steps, learning rate, initialization values, and distribution).
7. Prediction: Use the model to make predictions.

10.4.3 Applications of machine learning in additive manufacturing process and process monitoring

As shown in Fig. 10.13, machine learning can be applied in any lifecycle activity of AM processes. Table 10.12 lists ML applications in different lifecycle activities and corresponding ML methods [37,38]. When monitoring a build process, machine learning can be used for data pre-processing, process monitoring, defect detection and classification, monitoring of machine condition, environment monitoring, failure detection, closed-loop feedbacks, and quality control. Table 10.13 provides examples of ML applications in process monitoring of AM processes, including associated signatures, signals, ML algorithms, control, and adjustment of process parameters.

TABLE 10.12 Overview of ML applications in AM processes.

AM Lifecycle Activity	Category	ML method
Design	Design recommendation	• Unsupervised learning—clustering • Supervised learning—classification
	Topology optimization	• Supervised learning—classification
	Tolerancing and manufacturability assessment	• Supervised learning—classification
	Material classification and selection	• Supervised learning—classification
	Cost estimation	• Unsupervised learning—clustering • Supervised learning—regression
	Build-time prediction	• Supervised learning—classification

(Continued)

Process monitoring and inspection Chapter | 10   **429**

TABLE 10.12 (Continued)

AM Lifecycle Activity	Category	ML method
Process plan	Data pre-processing	• Unsupervised learning—dimensionality reduction
	Process parameter optimization	• Supervised learning—classification
	Property/quality/ performance prediction and optimization	• Supervised learning—classification • Supervised learning—regression • Unsupervised learning—clustering
Build	Data pre-processing	• Unsupervised learning—dimensionality Reduction
	Process monitoring (melt pool, build layer)	• Supervised learning—classification • Supervised learning—regression • Unsupervised learning—clustering
	Defect/anomaly detection and classification	• Supervised learning—classification • Unsupervised learning—clustering
	Machine condition monitoring (Feedstock and material status)	• Supervised learning—classification
	Environment monitoring	• Supervised learning—classification
	Fault/failure detection	• Supervised learning—classification • Unsupervised learning—clustering
	Feedback/closed-loop/real-time and quality control	• Supervised learning—classification • Supervised learning—regression • Reinforcement learning
Post process, Test and Validation	Data pre-processing	• Unsupervised learning—dimensionality Reduction
	Process end point identification	• Supervised learning—classification
	Defect/anomaly detection and classification	• Supervised learning—classification • Unsupervised learning—clustering
	Quality assessment (surface metrology, part property, product performance)	• Supervised learning—classification

TABLE 10.13 ML applications in monitoring of AM processes [32].

Application	Monitoring object/ signature	Process signal	ML algorithm	Control/ Detection object	Adjusted process parameter
Data pre-processing	• All data	• All signals	• PCA	• Data compression	—
Defect/anomaly detection and classification	• Melt pool • Slice/Deposited layer • Part • Substrate/Powder bed	• Optical signal • Thermal signal • X-ray-signal • Acoustic signal	• SVM • k-NN • K-means clustering	• Porosity • Anomaly	—
Process monitoring (Melt pool, build layer)	• Melt pool • Slice/deposited layer • Part • Substrate/powder bed	• Optical signal • Thermal signal • X-ray-signal • Acoustic signal	• SVM • Linear regression • K-means clustering	• Temperature Geometry • Surface pattern	—
Machine condition monitoring (feedstock and material status)	• Recoating system • Feed system	• Optical signal • Acoustic signal • Vibration signal	• SVM • k-NN • K-means clustering	• Material classification • Runout • Breakage • Blockage	—
Environment monitoring	• Protective gas	• Thermal signal • Humidity signal • Velocity signal	• SVM • k-NN	• Temperature • Humidity • Velocity	—
Fault/Failure detection	• All signatures	• All signals	• SVM • k-NN • K-means clustering	—	—
Feedback/Closed-loop/ Real-time and quality control	• All signatures	• All signals	• SVM • Linear regression • MDP	• Geometric deviation • Temperature uniformity • Part property and quality	• Heat source power • scan velocity • layer thickness

10.5 Standards and toolkits

Since AM has been widely used in many industrial sectors, numerous standards and toolkits for AM monitoring have gradually been developed.

10.5.1 Standards

Standards for AM processes are mainly developed by ASTM and ISO, as shown in Table 10.14. At present, there are no published standards for monitoring of AM processes. However, related standards have been established by ASTM committee E07 on Nondestructive Testing and by ASTM committee F42 on Additive Manufacturing Technologies, involving the use of X-ray, AE and digital images. ASTM work items WK73978 and WK74390 are currently developing guidelines for registration and conversion of in-process monitoring data. These guidelines are particularly prepared for AM processes with a focus on PBF. The requirements and specifications developed in these standards and guidelines are more rigorous than those used for general purposes [9].

TABLE 10.14 Related standards and work items developed by ASTM [32,38].

Committee	Subcommittee	Standards and work items
ASTM E07 Nondestructive Testing	ASTM E07.01 Radiology (X and Gamma) Method	• E1441−19 Standard Guide for Computed Tomography (CT) • E1570−19 Standard Practice for Fan Beam Computed Tomographic (CT) Examination • E1672−12(2020) Standard Guide for Computed Tomography (CT) System Selection • E1695−20 Standard Test Method for Measurement of Computed Tomography (CT) System Performance • E1931−16 Standard Guide for Non-computed X-Ray Compton Scatter Tomography • E1935−97(2019) Standard Test Method for Calibrating and Measuring CT Density • WK61161 Volumetric Computed Tomographic (CT) Examination • WK71550 Practice for Computed Tomographic Examination of Additive Manufactured Parts

(Continued)

432 Digital Manufacturing

TABLE 10.14 (Continued)

Committee	Subcommittee	Standards and work items
	ASTM E07.04 Acoustic Emission Method	• E569/E569M-20 Standard Practice for Acoustic Emission Monitoring of Structures During Controlled Stimulation • E650/E650M-17 Standard Guide for Mounting Piezoelectric Acoustic Emission Sensors • E749/E749M-17 Standard Practice for Acoustic Emission Monitoring During Continuous Welding • E2374–16 Standard Guide for Acoustic Emission System Performance Verification • E976–15 Standard Guide for Determining the Reproducibility of Acoustic Emission Sensor Response • E1139/E1139M-17 Standard Practice for Continuous Monitoring of Acoustic Emission from Metal Pressure Boundaries
	ASTM E07.06 Ultrasonic Method	• E1001–16 Standard Practice for Detection and Evaluation of Discontinuities by the Immersed Pulse-Echo Ultrasonic Method Using Longitudinal Waves • E2001–18 Standard Guide for Resonant Ultrasound Spectroscopy for Defect Detection in Both Metallic and Non-metallic Parts • E2580–17 Standard Practice for Ultrasonic Testing of Flat Panel Composites and Sandwich Core Materials Used in Aerospace Applications • E2700–20 Standard Practice for Contact Ultrasonic Testing of Welds Using Phased Arrays • E2904–17 Standard Practice for Characterization and Verification of Phased Array Probes • E2985/E2985M-14(2019) Standard Practice for Determination of Metal Purity Based on Elastic Constant Measurements Derived from Resonant Ultrasound Spectroscopy
	ASTM E07.09 Non-destructive Testing Agencies ASTM E07.10 Specialized NDT Methods	• E543–15 Standard Specification for Agencies Performing Non-destructive Testing • E1212–17 Standard Practice for Establishing Quality Management Systems for Non-destructive Testing Agencies • E1359–17 Standard Guide for Auditing and Evaluating Capabilities of Non-destructive Testing Agencies

(Continued)

Process monitoring and inspection Chapter | 10 **433**

TABLE 10.14 (Continued)

Committee	Subcommittee	Standards and work items
		• E1213−14(2018) Standard Practice for Minimum Resolvable Temperature Difference for Thermal Imaging Systems • E1311−14(2018) Standard Practice for Minimum Detectable Temperature Difference for Thermal Imaging Systems • E1543−14(2018) Standard Practice for Noise Equivalent Temperature Difference of Thermal Imaging Systems • E3166−20e1 Standard Guide for Non-destructive Examination of Metal Additively Manufactured Aerospace Parts After Build • WK62181 New Guide for Standard Guide for In-Situ Monitoring (IPM) of Metal Additively Manufactured Aerospace Parts • WK73289 New Guide for In-Situ Monitoring of Metal Additively Manufactured Aerospace Parts
	ASTM E07.11 Digital Imaging and Communication in Non-destructive Evaluation (DICONDE)	• E2339−15 Standard Practice for Digital Imaging and Communication in Non-destructive Evaluation (DICONDE) • E2663−14(2018) Standard Practice for Digital Imaging and Communication in Non-destructive Evaluation (DICONDE) for Ultrasonic Test Methods • E2767−13(2018) Standard Practice for Digital Imaging and Communication in Non-destructive Evaluation (DICONDE) for X-ray Computed Tomography (CT) Test Methods • E3147−18 Standard Practice for Evaluating DICONDE Interoperability of Non-destructive Testing and Inspection Systems • E3169−18 Standard Guide for Digital Imaging and Communication in Non-destructive Evaluation (DICONDE)
	ASTM E07.92 Editorial Review	• E1316−21 Standard Terminology for Non-destructive Examinations

(*Continued*)

434 Digital Manufacturing

TABLE 10.14 (Continued)

Committee	Subcommittee	Standards and work items
ASTM F42 Additive Manufacturing Technologies	ASTM F42.01 Test Methods	• F2971−13 Standard Practice for Reporting Data for Test Specimens Prepared by Additive Manufacturing • F3122−14 Standard Guide for Evaluating Mechanical Properties of Metal Materials Made via Additive Manufacturing Processes • ISO/ASTM52902−19 Additive Manufacturing - Test Artifacts - Geometric Capability Assessment of Additive Manufacturing Systems • ISO/ASTM52921−13(2019) Standard Terminology for Additive Manufacturing - Coordinate Systems and Test Methodologies • ISO/ASTM52907−19 Additive Manufacturing - Feedstock Materials - Methods to Characterize Metallic Powders • WK69731 Additive Manufacturing - Non-Destructive Testing (NDT) for Use in Directed Energy Deposition (DED) Additive Manufacturing Processes • WK66682 Evaluating Post-processing and Characterization Techniques for AM Part Surfaces • WK71395 Additive Manufacturing - Accelerated Quality Inspection of Build Health for Laser Beam Powder Bed Fusion Process • WK76038 Additive Manufacturing of Metals - Non-destructive testing and evaluation - Porosity Measurement with X-ray CT
	ASTM F42.04 Design	• ISO/ASTM52915−20 Specification for additive manufacturing file format (AMF) Version 1.2 • ISO/ASTM52950−21 additive manufacturing - general principles - overview of data processing • WK74006 additive manufacturing - data formats - specification for optimized medical image data

(Continued)

TABLE 10.14 (Continued)

Committee	Subcommittee	Standards and work items
	ASTM F42.05 Materials and Processes	• F3049−14 standard guide for characterizing properties of metal powders used for additive manufacturing processes • ISO/ASTM52903−1−20 additive manufacturing - material extrusion-based additive manufacturing of plastic materials - Part 1: feedstock materials
	ASTM F42.07 Applications	• WK75329 nondestructive testing (NDT), part quality, and acceptability levels of additively manufactured laser-based powder bed fusion aerospace components
	ASTM F42.08 Data	• WK73978 Additive Manufacturing - Data Registration • WK74390 Additive Manufacturing of Metals - Data - File Structure for In-Process Monitoring of Powder Bed Fusion
	ASTM F42.91 Terminology	• ISO/ASTM52900−15 Standard Terminology for Additive Manufacturing - General Principles - Terminology
Others	—	• ISO/IEC 9834−8:2014 Information technology - Procedures for the Operation of Object Identifier Registration Authorities - Part 8: Generation of Universally Unique Identifiers (UUIDs) and Their Use in Object Identifiers • ISO/IEC 11179−4:2004 Information Technology - Metadata Registries (MDR) - Part 4: Formulation of Data Definitions • ISO/IEC 20005:2013 Information Technology - Sensor Networks - Services and Interfaces Supporting Collaborative Information Processing in Intelligent Sensor Networks

10.5.2 Toolkits

To achieve the monitoring and feedback control of AM processes, users could develop their codes or tools. Many general monitoring systems and toolkits have been provided by commercial companies, especially for PBF and DED processes, as listed in Table 10.15. The table also summarizes

TABLE 10.15 Toolkits for AM process monitoring [2,4,19,33].

AM process	Sub-category	Toolkit name	Developer	Sensor	Monitoring object	Monitored signature	Adjusted process parameter
Powder bed fusion (PBF)	SLM	QM meltpool 3D concept	Concept laser	Co-axial camera and photodiode High-speed CMOS-camera	Melt pool	Area and intensity	Laser power
		QM coating	Concept Laser	Off-axial camera	Powder bed	—	—
		EOSTATE meltPool	EOS	Co-axial and off-axial sensors	Melt pool	—	—
		EOSTATE PowderBed	EOS	Off-axial camera	Powder bed	—	—
		PrintRite3D	B6 Sigma, Inc.	Set of co-axial and off-axial sensors available Thermo-couple and high-speed camera	Different monitoring possibility	—	—
		Melt pool monitoring (MPM) system	SLM Solutions	Co-axial pyrometer	Melt pool	—	—
		Layer control system (LCS)	SLM Solutions	Off-axial camera	Powder bed	—	—
	EBM	LayerQam	Arcam	Off-axial camera	Slice	Pattern and geometry	—

Directed energy deposition (DED)	Laser based	DMD closed loop control system; integrated DMD vision system	DM3D technology	Dual-color pyrometer and three high-speed CCD cameras	Melt pool	Height and temperature	Laser power
		ThermaViz system	Stratonics	Infrared cameras with one wavelength; infrared camera with two wavelengths Two-wavelength imaging pyrometer	Melt pool Deposited layer	Temperature	Laser power
		—	Optomec	Pyrometer	—	Thermal imaging	Closed loop process control
		Laser welding monitor	Precitec	CMOS camera	Laser welding	Temperature, back reflection	—
		—	3D-Hybrid solution	Dual-color pyrometer	Melt pool	Temperature	—
		—	FormAlloy	Thermal and optical sensors	Melt pool	Temperature Size	—
		—	DMG MORI	Dual-color pyrometer	Melt pool	Temperature	Laser power
		—	Hybrid Manufacturing Technologies	Thermal and optical sensing devices	—	—	—
		—	RPM Innovations Inc.	Cameras	—	—	—

(*Continued*)

TABLE 10.15 (Continued)

AM process	Sub-category	Toolkit name	Developer	Sensor	Monitoring object	Monitored signature	Adjusted process parameter
		—	BeAM	—	Melt pool	—	—
		—	Raycham	Pyrometer	Melt pool	Temperature	Laser power
		DMT close loop feedback control system	InssTek	—	—	—	—
		—	Additec	—	Deposited layer	Height	Nozzle to part distance
	Electron beam based	IRISS	Sciaky	—	—	—	Power; wire feed rate, CNC motion profiles, etc.
		LCC 100	DEMCON	Camera	Melt pool	—	Laser power
		LD-600	Laser depth	Inline coherent imaging	—	Depth measurement	Laser power
		Welding Monitor PD 2000	Promotec	CMOS camera	Melt pool	—	—
	Plasma based	Plasma Monitor PM 7000	Promotec	IR and UV photo detector	Melt pool	Emission	—

sensors, monitoring objects, monitored signatures, and adjusted process parameters for feedback control used in these systems.

10.6 Insights and future outlook

Although numerous works on AM process monitoring have been conducted, challenges still exist and hinder widespread applications of process monitoring technologies.

The biggest obstacle is to improve adoption of these technologies. Since additive manufacturing is currently in rapid development stage, AM machines are developed with both characteristics and integrated monitoring systems that vary widely between machine manufacturers. Monitoring systems usually employ different sensors and generate massive data of various types, including point-wise data (e.g., temperature sequence at a certain point), 1D data (e.g., temperature sequence at a certain line), 2D data (e.g., optical image, infrared image), and even 3D data (e.g., data from a CMM). Data acquired by different types of sensors have disparate time and space coordinates. It is therefore of great importance to develop suitable and widely accepted methods to share data among a wide array of monitoring systems or toolkits for calibration and validation. To overcome this obstacle, adequate sensors need to be selected and tested in various process monitoring systems. Based on the acquired results from these sensors, the standards for the AM process monitoring could be developed, including but not limited to system components, system calibration, data registration and file exchange.

Another challenge is to ensure reliability of process monitoring technologies. Most reported works are still at the level of scientific research. There is a long way to achieving reliability requirements in industrial applications. For example, thermal signals recorded by infrared cameras are sensitive to material emissivity, and utilization of X-ray signals is dependent on material composition and structures. Extra sensors are usually introduced for further validation when using these uncertain signals in the monitoring system. For example, an infrared camera is accompanied by a dichromatic pyrometer or a thermocouple. The development of ML also provides avenues to improve reliability of process monitoring and quality of printed parts.

Most available online monitoring systems acquire only a few process signals. Building a process monitoring system incorporating all signals released in build processes will undoubtedly expedite the development of process monitoring and quality control.

10.7 Summary

This chapter reviews applications and techniques of process monitoring and quality control of AM processes. Optical, thermal, X-ray and acoustic signals are commonly used in AM process monitoring, and each type of these

440 Digital Manufacturing

signals has its own advantages and disadvantages. They could be acquired by various sensors that are integratable process monitoring systems. Using these signals in AM processes enables monitoring of geometrical, temperature and property signatures of the melt pool, track, layer, powder bed and overall quality of printed parts. It also allows detection of process defects including porosity, gas pores, balling, geometrical abnormalities and microstructural inhomogeneity. Subsequently, quality control of printed parts can be implemented by introducing feedback control in process monitoring systems and adjusting process parameters based on monitoring results. In addition, machine learning is introduced as a highly capable tool that offers rapid data processing and real-time feedback control of AM processes. To promote widespread applications of AM process monitoring, it is meaningful to ensure universality and reliability of relevant technologies and methods aiming towards standard development and quality assurance of AM products.

References

[1] M. Mani, S. Feng, B. Lane, A. Donmez, S. Moylan, R. Fesperman, Measurement science needs for real-time control of additive manufacturing powder bed fusion processes, National Institute of Standards and Technology (2015).

[2] M. Grasso, B.M. Colosimo, Process defects and in situ monitoring methods in metal powder bed fusion: A review, Measurement Science & Technology 28 (4) (2017) 44005.

[3] G. Held, Introduction to light emitting diode technology and applications, CRC Press, Boca Raton, FL, 2008, p. 192.

[4] Z. Tang, W. Liu, Y. Wang, K.M. Saleheen, Z. Liu, S. Peng, et al., A review on in situ monitoring technology for directed energy deposition of metals, International Journal of Advanced Manufacturing Technology 108 (11−12) (2020) 3437−3463.

[5] C. Zhao, K. Fezzaa, R.W. Cunningham, H. Wen, F. De Carlo, L. Chen, et al., Real-time monitoring of laser powder bed fusion process using high-speed X-ray imaging and diffraction, Scientific Reports 7 (1) (2017) 3602.

[6] S.J. Wolff, H. Wu, N. Parab, C. Zhao, K.F. Ehmann, T. Sun, et al., In-situ high-speed X-ray imaging of piezo-driven directed energy deposition additive manufacturing, Scientific Reports 9 (1) (2019) 962.

[7] C.L.A. Leung, S. Marussi, R.C. Atwood, M. Towrie, P.J. Withers, P.D. Lee, In situ X-ray imaging of defect and molten pool dynamics in laser additive manufacturing, Nature Communications 9 (1) (2018) 1355.

[8] M.S. Hossain, H. Taheri, In situ process monitoring for additive manufacturing through acoustic techniques, Journal of Materials Engineering and Performance 29 (10) (2020) 6249−6262.

[9] F. Honarvar, A. Varvani-Farahani, A review of ultrasonic testing applications in additive manufacturing: Defect evaluation, material characterization, and process control, Ultrasonics 108 (106227) (2020) 106227.

[10] J. Whiting, A. Springer, F. Sciammarella, Real-time acoustic emission monitoring of powder mass flow rate for directed energy deposition, Additive Manufacturing 23 (2018) 312−318.

[11] G. Tapia, A. Elwany, A review on process monitoring and control in metal-based additive manufacturing, Journal of Manufacturing Science and Engineering 136 (6) (2014).

[12] W. He, W. Shi, J. Li, H. Xie, In-situ monitoring and deformation characterization by optical techniques; Part i: laser-aided direct metal deposition for additive manufacturing, Optics and Laser in Engineering 122 (2019) 74−88.

[13] Y. Li, W. Zhao, Q. Li, T. Wang, G. Wang, In-situ monitoring and diagnosing for fused filament fabrication process based on vibration sensors, Sensors 19 (11) (2019) 2589.

[14] J. Guo, A. Huang, R. Hu, H. Xu, G. Yang, S. Pang, An in-situ monitoring system for electron beam wire-feed additive manufacturing, Sensors and Actuators A: Physical 307 (111983) (2020) 111983.

[15] C. Kim, D. Espalin, A. Cuaron, M.A. Perez, E. MacDonald, R.B. Wicker, A study to detect a material deposition status in fused deposition modeling technology, IEEE (2015) 779−783.

[16] Y. Fu, A. Downey, L. Yuan, A. Pratt, Y. Balogun, In situ monitoring for fused filament fabrication process: a review, Additive Manufacturing 38 (2021) 101749.

[17] ASTM, ISO, Standard terminology for additive manufacturing — general principles — Terminology, ISO/ASTM 52900 2015 (E) (2015) 9.

[18] BS, ISO, Additive manufacturing — General principles — Part 2: Overview of process categories and feedstock, BS EN ISO 17296−2:2016, 2016, pp.

[19] S.K. Everton, M. Hirsch, P. Stravroulakis, R.K. Leach, A.T. Clare, Review of in-situ process monitoring and in-situ metrology for metal additive manufacturing, Materials & Design 95 (2016) 431−445.

[20] T. Craeghs, S. Clijsters, E. Yasa, J. Kruth, Online quality control of selective laser melting, Austin, TX, US, 2011, pp. 212−226.

[21] S. Tammas-Williams, H. Zhao, F. Léonard, F. Derguti, I. Todd, P.B. Prangnell, XCT analysis of the influence of melt strategies on defect population in Ti−6AL−4V components manufactured by selective electron beam melting, Materials Characterization 102 (2015) 47−61.

[22] R. Li, J. Liu, Y. Shi, L. Wang, W. Jiang, Balling behavior of stainless steel and nickel powder during selective laser melting process, International Journal of Advanced Manufacturing Technology 59 (9) (2012) 1025−1035.

[23] C. Guo, S. Li, S. Shi, X. Li, X. Hu, Q. Zhu, et al., Effect of processing parameters on surface roughness, porosity and cracking of as-built in738lc parts fabricated by laser powder bed fusion, Journal of Materials Processing Technology 285 (2020) 116788.

[24] M. Grasso, V. Laguzza, Q. Semeraro, B.M. Colosimo, In-process monitoring of selective laser melting: spatial detection of defects via image data analysis, Journal of Manufacturing Science and Engineering 139 (5) (2017).

[25] W.J. Sames, F.A. List, S. Pannala, R.R. Dehoff, S.S. Babu, The metallurgy and processing science of metal additive manufacturing, International Materials Reviews 61 (5) (2016) 315−360.

[26] R. Casati, J. Lemke, M. Vedani, Microstructure and fracture behavior of 316l austenitic stainless steel produced by selective laser melting, Journal of Materials Science & Technology 32 (8) (2016) 738−744.

[27] A. Dass, A. Moridi, State of the art in directed energy deposition: from additive manufacturing to materials design, Coatings 9 (7) (2019).

[28] P.K. Rao, J.P. Liu, D. Roberson, Z.J. Kong, C. Williams, Online real-time quality monitoring in additive manufacturing processes using heterogeneous sensors, Journal of Manufacturing Science and Engineering 137 (6) (2015).

[29] V.K. Nadimpalli, G.M. Karthik, G.D. Janakiram, P.B. Nagy, Monitoring and repair of defects in ultrasonic additive manufacturing, International Journal of Advanced Manufacturing Technology 108 (5) (2020) 1793−1810.

442 Digital Manufacturing

[30] V.K. Nadimpalli, L. Yang, P.B. Nagy, In-situ interfacial quality assessment of ultrasonic additive manufacturing components using ultrasonic NDE, NDT&E International 93 (2018) 117−130.

[31] D. Saenz-Castillo, M.I. Martín, S. Calvo, A. Güemes, Real-time monitoring of thermal history of thermoplastic automatic lamination with fbg sensors and process modelling validation, Smart M aterials and Structures 29 (11) (2020) 115004.

[32] S.S. Razvi, S. Feng, A. Narayanan, Y.T. Lee, P. Witherell, A review of machine learning applications in additive manufacturing, Anaheim, California, USA, 2019.

[33] T.G. Spears, S.A. Gold, In-process sensing in selective laser melting (SLM) additive manufacturing, Integrating Materials and Manufacturing Innovation 5 (1) (2016) 16−40.

[34] E.A. Papon, A. Haque, Review on process model, structure-property relationship of composites and future needs in fused filament fabrication, Journal of Reinforced Plastics and Composites 39 (19−20) (2020) 758−789.

[35] Z.Y. Chua, I.H. Ahn, S.K. Moon, Process monitoring and inspection systems in metal additive manufacturing: status and applications, International Journal of Precision Engineering and Manufacturing 4 (2) (2017) 235−245.

[36] T. Wang, T. Kwok, C. Zhou, S. Vader, In-situ droplet inspection and closed-loop control system using machine learning for liquid metal jet printing, Journal of Manufacturing Systems 47 (2018) 83−92.

[37] J. Wang, Y. Ma, L. Zhang, R.X. Gao, D. Wu, Deep learning for smart manufacturing: methods and applications, Journal of Manufacturing Systems 48 (2018) 144−156.

[38] L. Meng, B. McWilliams, W. Jarosinski, H. Park, Y. Jung, J. Lee, et al., Machine learning in additive manufacturing: a review, JOM 72 (6) (2020) 2363−2377.

Index

Note: Page numbers followed by "*f*" and "*t*" refer to figures and tables, respectively.

A

Abstract geometry features modeling, 71–72
Acoustic signals, 399–402
  advantages and disadvantages, 401
  application cases, 401–402
  other signals, 402–403
    current signals, 402–403
    displacement signals, 402
    monitoring signals and sensors, 403*t*
    vibration signals, 402
  practical use, 400–401
  working principle, 399–400, 399*f*
    acoustic emission, 399
    ultrasonic waves, 399–400
AD. *See* Axiomatic design (AD)
Addition polymers, 221–222
Additive manufacturing (AM), 41–42, 50, 78–79, 111, 146–151, 155, 176–177, 185, 221, 248, 303, 316, 348–349, 388
  advantages and process parameters of, 124*t*
  building block of digital manufacturing, 147
  conventional manufacturing *vs.*, 151
  design, optimization, and simulation, 154–155
    conversion to printable file format, 154–155
    design generation, 154
    optimization and simulation, 154
  design freedoms afforded by, 114–115, 115*t*
  design methodologies for, 117–140
    design stages of general DFMA framework, 119–136
    notable design for additive manufacturing research works, 117–119
  for digital transformation
    additive manufacturing market, 147–150

case studies during COVID-19 pandemic, 176–180
industry drivers for additive manufacturing adoption, 150–151
process chain, 152–157
technologies and processes, 158–176, 159*t*
and terminologies, 146–147, 147*t*
implementation level, 152–153
  direct part replacement, 152
  new product designs, 153
  parts consolidation, 153
  rapid prototyping and tooling, 152
industry drivers for additive manufacturing adoption, 150–151
level of additive manufacturing implementation, 152–153
macroscale modeling, 186–187
manufacturing, 155
  classifications of additive manufacturing processes, 157*t*
market, 147–150
  additive manufacturing systems market, 149
  additive manufacturing–building block of digital manufacturing, 147
  expiration of key patents and industrial trends, 149–150
  global, 148–149
  rise of additive manufacturing, 148
material selection, 155
  materials available in additive manufacturing market, 156*t*
mesoscale modeling, 187
microscale modeling, 187
ML applications in additive manufacturing process and process monitoring, 428–430, 428*t*, 430*t*
objectives, 188
optimization, 78–79

443

444 Index

Additive manufacturing (AM) (*Continued*)
  parameters optimization, 188
  postprocessing, 155−156
  process, 158−176
    chain, 152−157
    monitoring and validation, 157
    review of models employing in, 188−207
      crystallization/microstructure simulation, 198−206
      heat transfer and melt pool dynamics, 191−195
      light source simulation, 195−198
      powder interaction, 189−191
  systems market, 149
    key patents and industrial trends in, 150*t*
  topology optimization, 207−216
Additive printing techniques, 322−324
Additively manufactured polymer, 241
Adhesive energy, 239−240
Adhesive powders removal in postprocessing stage, 281−282
Advanced CAE systems, 59
Advanced robotics, 42
Advanced ultrasonic testing techniques, 400
Acoustic emission (AE), 399
AEM. *See* Assemblability evaluation method (AEM)
Aerodynamic drag coefficient, 86−89
Aerodynamic lifting coefficient, 86−89
Aerodynamic tests, 86
Aerosol jet printing, 324−325
Aerospace
  and automotive industries, DFMA application in, 109−110
    Beijing Automotive Technology Center, 109−110
    composites materials for, 302−310
    McDonnell Douglas Corporation, 109
  metal additive manufacturing application in, 285−288
    damage component repair, 287−288
    direct component manufacturing, 285−287
Aggregates role in three-dimensional concrete printing, 356−358
  printable mix design from other publications, 357*t*
Agrarian society, 3, 26−27
Agricultural revolution, 3−4
AI. *See* Artificial intelligence (AI)
Alginate, 311
  hydrogels, 311

Aligners, 163
Alkali−silica reaction (ASR), 377
Alleviating printing process, 251
Altair HyperWorks OptiStruct software, 214−215
Alumina (Al_2O_3), 316−318
Aluminum foil, 169
AM. *See* Additive manufacturing (AM)
AM File Format (AMF), 155
American Society for Testing and Materials (ASTM), 146, 249, 404
  ASTM E466−15, 268
  ASTM E2368−10, 268
  ASTM E2714−13, 268
  ASTM E2760, 268
  ASTM F2792, 158, 168, 174−175
American system of manufacturing, 7
Analytical approach, 82
Angle of repose (AOR), 235
Anisotropy, 241−242, 363
ANSYS CFX (CFD software), 84−85
ANSYS Discovery AIM software, 215
Application-driven child model, 77−78
Applications of metal additive manufacturing, 284−293, 284*f*
Applied science, 6
APT. *See* Automatically Programed Tool (APT)
Arbitrary Lagrange Eulerian method, 187
Architectural concept design, 66
Arginylglycylaspartic acid (RGD), 311
Arm-based systems, 352−353
Artificial intelligence (AI), 42, 47, 50, 58, 186, 331
  AI-aided CFD analysis, 85
  AI-based design algorithms, 59
ASF. *See* Assembly Sequence Flowchart (ASF)
ASR. *See* Alkali−silica reaction (ASR)
Assemblability, 102
  Hitachi assemblability evaluation method, 106−107
Assemblability evaluation method (AEM), 40, 101−102, 106−107
Assembly
  associations, 70
  cost, 104
  efficiency percentage, 104
  operation time, 106
Assembly Sequence Flowchart (ASF), 107
Association, 74−76
Associative assembly feature, 73

Index **445**

Associative feature concept, 76
Associative relations, 76
ASTM. *See* American Society for Testing and Materials (ASTM)
Asymptotic homogenization, 210
Atactic polymers, 223
Atomic force microscope (AFM), 280–281
Autoacceleration, 225
AutoCAD software, 154
Autodesk generative design software, 154, 215
Automated assembly machines, 11
Automated evaluation process, 105
Automated flexible manufacturing system, 20–23
    human operators, 22–23
    supporting hardware, 21–22
    workstation, 21
Automated guided vehicles, 21
Automated machines, 33
Automated manufacturing system, 11
Automated milling system, 20
Automated roller conveyor, 21
Automated system, 11
Automatically Programed Tool (APT), 13–14
Automation, 9–12, 47
    in building and construction, 345–347
Automotive industry, 150–151, 288–290
    commercial vehicles, 289–290
    racing vehicles, 290
Autonomous machines, 50
Autonomous robots, 47, 50
Avid Product Development, 178
Axiomatic design (AD), 71, 117–118
2,2'-azobis[2-methyl-n-(2-hydroxyethyl) propionamide], 312

B

B-Reps. *See* Boundary representations (B-Reps)
B-Spline Curve, 63–64
Batch production, 8–9
Beijing Automotive Technology Center (BATC), 109–110
Bellus3D Inc., 178
Benzylidene cycloketone-based PIs G2CK and P2CK, 312
Besides solving complicated manufacturing processes, 186
Bezier Curve, 63–64
Bidirectional evolutionary structural optimization (BESO), 211

Bifunctional monomers, 224
Big Canopy, 349
Big data analytics, 50–51
Binder jetting (BJ), 158, 173–174, 239–240, 249, 254–255, 350–351, 404.
    See also Material jetting (MJ)
    applications, 174
    benefits, 173–174
    drawbacks, 174
    range of materials available, 174, 174*t*
    working principles, 173
Bioactive ceramic, 315
Bioinert ceramic, 315
Bioinks for bioprinting, 311–313
    natural bioinks, 311
    synthetic bioinks, 312–313
Biological metabolism, 45
Biological nutrients, 45
Biomaterials for bioprinting, 310–315
    bioinks for bioprinting, 311–313
    bioprinting, 310
    challenges and potential in, 313–315, 314*t*
        added components, added complexity, 313
        limitation in print resolution, 314–315
        natural material has relatively weak strength, 313
Biomedical implants
    ceramic materials by three-dimensional printing for, 316–319
    ceramics for biomedical implants, 315–322
    SLS, 320*f*
Biomedical models, 167
Biomimicry, 44
Bioprinting process, 310
    bioinks for, 311–313
Bioresorbable ceramic, 315
Biosynthetic ink, 314–315
Blended cement, 375
Boolean Difference, 27–28
Boothroyd-Dewhurst Design for Manufacturing and Assembly, 102–106
Bottom-up configuration process, 162
Bottom-up process, 301
Boundary representations (B-Reps), 64
Brittle nature, 305
Broad selection of process–machine–material combinations, 123–125
Brownian motion, 261
Budmen Industries, 178
Buildability, 360–361

446 Index

Business Council for Sustainable Development, 43
Business proprietors, 9

C

Calibration, 393–394
CAD. *See* Computer-aided design (CAD)
CAD/CAM. *See* Computer-aided design and manufacturing (CAD/CAM)
CAE. *See* Computer-aided engineering (CAE)
CALPHAD method, 154
CAM. *See* Computer-aided manufacturing (CAM)
Cam-follower mechanism, 81
Capacitors, 16
Capitalism, 36
Capitalist system, 6
CAPP. *See* Computer-aided process planning (CAPP)
Carbon fiber, 303–305
Carbon nanotubes (CNTs), 327
"Card-a-matic" milling machine, 12–15
CASA/SME. *See* Computer and Automated Systems Association of the Society of Manufacturing Engineer (CASA/SME)
Casting patterns, 163
CATIA software, 154
CBAM technique. *See* Composite-based additive manufacturing technique (CBAM technique)
CCD. *See* Charge-coupled device (CCD)
Cellular manufacturing, 19–20
Cenospheres, 356
Central processing unit (CPU), 24
Ceramic–polymer composites, 162
Ceramics for biomedical implants, 315–322
   ceramic materials by three-dimensional printing for biomedical implants, 316–319
   challenges and potential in ceramics for three-dimensional printing, 319–322
   dimensional accuracy, 321–322
   flowability of, 320
   thermal and residual stresses, 320–321
   three-dimensional printed ceramic implants, 315–316
CFD. *See* Computational fluid dynamics (CFD)
"Chain reaction", 92
Charge-coupled device (CCD), 390–391
Chemical composition, 262–263

CIM. *See* Computer-integrated manufacturing (CIM)
Circular economic model, 44
Class C fly ash, 375–376
Class F fly ash, 375–376
Closed-loop control systems, 50
Cloud computing, 50–51
Cloud processing algorithms, 93–94
Cloud servers, 50–51
CM. *See* Conventional manufacturing (CM)
CMOS. *See* Complementary metal–oxide–semiconductor (CMOS)
CNC machines. *See* Computer numerical control machines, (CNC machines)
CNTs. *See* Carbon nanotubes (CNTs)
CO_2 laser, 171
Coal, 4
   burning of, 4–5
Cobots, 42, 50
Coding process, 13–14
Coefficient of thermal expansion (CTE), 232–233
Cold cracking, 273
Cold isostatic pressing, 319
Collaboration process, 40–41
Collaborative engineering
   approach, 40–41
   environment, 92
Collaborative product development, 112
Collagen, 313
   precursor, 311
   solution, 311
Combustion engines, 6
Commercial software for TO, 214–216
   Altair HyperWorks OptiStruct, 214–215
   ANSYS Discovery AIM, 215
   Autodesk generative design, 215
   COMSOL optimization module, 215
   Dassault Systèmes SE Tosca Structure, 215
   nTopology Element Pro, 215
   PTC generative topology optimization extension, 216
   Siemens solid edge generative design, 216
Communication network system, 21–22
Complementary metal–oxide–semiconductor (CMOS), 390–391
Complex modulus, 231–232
Complex viscosity, 231–232
Composite 3D printing, 308
Composite industry, 303–304
Composite-based additive manufacturing technique (CBAM technique), 303–304

Composites materials for aerospace industry, 302–310
  challenges and potentials in, 307–310
    lack of 3D printable composite materials, 309–310
    nonstructural applications of composites, 309t
    poor out-of-plane strength, 310
    scaling big area additive manufacturing, 309
  composite industry, 303–304
  composites for three-dimensional printing, 304–307
  conventional fabrication technique, 302f
Compression testing, 263–266, 267t
Computational fluid dynamics (CFD), 58–59, 82–84, 154, 187, 191–193
  CFD–DEM coupling model, 206–207
  simulation, 73, 83–89
    boundary conditions, 88t
    domain properties and simulation settings, 88t
  software, 84–85
Computer and Automated Systems Association of the Society of Manufacturing Engineer (CASA/SME), 34
Computer numerical control machines, (CNC machines), 16–17, 21, 24, 33, 252, 284
Computer system architecture, 23–26
  hardware, 24–25
  software, 25–26
Computer-aided design (CAD), 26–31, 58–60, 70, 74–76, 91–92, 105, 146, 154, 207, 250–251, 346
  CAD-based design exploration method, 30
  CAD/CAE integration feature, 73
  CAD/CFD integration feature, 77
  design process, 29–31
  system, 27–29
Computer-aided design and manufacturing (CAD/CAM), 33, 35–36
Computer-aided drafting, 60
Computer-aided drawing, 60
Computer-aided engineering (CAE), 28–29, 58
  simulation, 82
  systems, 91–92
Computer-aided geometric modeling, 63–64
Computer-aided manufacturing (CAM), 26, 32–33

Computer-aided process planning (CAPP), 26, 31–32
Computer-aided systems, 9–10
Computer-aided technologies (CAX), 73
Computer-aided transfer processes, 310
Computer-controlled galvanometers, 158
Computer-integrated manufacturing (CIM), 33–36
Computers, 15–18
  CNC, 16–17
  computer-supported collaboration design, 41
  DCS, 18
  DNC, 17–18
  programmable logic controller, 16
Computing systems, 186
COMSOL optimization module software, 215
Concrete, 349
  printing of concrete supports materials, 367–368
  printing technologies, 350–355
    arm-based systems, 352–353
    gantry-based systems, 350–352
    multirobot printing systems, 354
    printing process control, 355
  sustainable raw materials in, 373–380
  three-dimensional printable concrete, 355–368
Concurrent Costing module, 105
Concurrent design evaluation system (CONDENSE), 100
Concurrent engineering, 38–40, 112
Conductive materials for electronic printing, 322–331
  challenges and potential in three-dimensional printing electronics, 328–331
  materials for three-dimensional printing of electronics, 325–328
  three-dimensional printed electronics, 322–325
Conductors, 327
Conformal cooling system design, 78–79
Consorci de la Zona Franca (CZFB), 177
Constraints in TO, 213–214
Construction
  in Singapore, 110
  three-dimensional printing, 347–348
Constructive solid geometry, 64
Constructor, The (Reuleauxs), 97–98
Consumer products, 108–109
  Endress + Hauser, 109
  Hasbro, 108

448 Index

Consumer products (*Continued*)
  Hewlett-Packard, 108
  Motorola Solutions, 109
Contact 3D measurement techniques, 279
Contact measurement methods, 395−396
Contact printing techniques, 322−324
Continuous fiber, 307
Continuous inkjet printing, 324
Contour Crafting, 347, 350
Conventional design process, 30
Conventional industrial systems, 43
Conventional machine controllers, 16
Conventional manufacturing (CM), 100, 151
  design methodologies for, 100−112
    DFMA guidelines, 100, 101*t*
    DFMA procedures, 101−107
Conventional methods, 424
Conventional scanning type SLA process, 158
Convex polygons, 63
Copper NPs, 327
Coronavirus disease (COVID-19), 176
"Correct design decision", 92
Correlation approaches, 71
Cost of assembly, 102−105
Cottage industry, 3
Cotton spinning machine, 4
COVID-19 pandemic, additive manufacturing
    during, 176−180
  agile operations and accelerated
      productions, 178
  mass customizations, 178
  preserving sustainability and continuity,
      178−180
  providing rapid emergency responses, 177
CPU. *See* Central processing unit (CPU)
Cracking process, 273
Cradle-to-cradle design, 44−45, 51
Cradle-to-grave approach, 10
Cradle-to-grave economic model, 43
Craft production, 3
Crank-slider mechanism, 81
Cross-link density, 227−228
Cross-linked polymers, 223, 227−228
Crystal morphology, 204
Crystalline polymers, 223
Crystallization
  behavior of printed materials, 187
  crystallization/microstructure simulation,
      198−206
  Granasy model, 204−206
  time-dependent Ginzburg−Landau
      model, 200−204

kinetics, 203
morphology, 203, 203*f*
CTE. *See* Coefficient of thermal expansion
    (CTE)
Curing, 224−225
  characteristics, 225−229
    cross-link density, 227−228
    curing depth, 228−229
    curing rate, 228
    monomer conversion, 226−227
  gelation, 224
  release of heat, 225
  shrinkage, 225
Current signals, 402−403
Curves, 63−64
Customized orthopedic components, 172
Cybe RC 3Dp (mobile-based robotic arm
    system), 352
Cyber-physical security, 50−51
Cyber-physical system (CPS), 49−50
Czech Institute of Informatics, Robotics and
    Cybernetics with the Czech Technical
    University (CIIRC CTU), 179−180
CZFB. *See* Consorci de la Zona Franca
    (CZFB)

D

Dassault Systèmes SE Tosca Structure
    software, 215
D-shape method, 350−351
Data registration, 394
Database management system, 77−78
Datacenters, 50−51
DCP. *See* Digital Construction Platform
    (DCP)
DCS. *See* Distributed computer system (DCS)
Decentralized computer-aided design/
    computer-aided engineering network,
    93−94
Decision trees, 425
Decision-making
  process, 38
  systems, 81−82
DED. *See* Directed energy deposition (DED)
Deep learning (DL), 425
Deep-keyhole melt pool, 196
Defects
  defect categories, 270−277
    cracking, 273
    distortions, 273−275
    excessive residual stresses, 271

Index **449**

large surface roughness, 275–277
pores, 271–273
defects detection techniques, 277–281
porosity detection, 279
printing accuracy determination, 279–280
residual stress measurement, 277–278
surface topography measurement, 280–281, 282*t*
internal, 398
in metallic additive manufacturing components, 268–281
process
directed energy deposition, 415
material extrusion, 420
powder bed fusion, 411–414, 412*t*
type, 187
Degree of conversion, 226
Degree of crystallinity, 223
DEM. *See* Discrete element method (DEM)
Density-based testing methods, 279
Dental restoration framework, 315–316
Dentistry, 291
Dentures, 163
Depth sensing, 61–63
Design efficiency, 104, 107
Design elements, 70
Design engineering team, 102–105
design freedoms afforded by additive manufacturing, 114–115, 115*t*
difficulties in integration of, 111
Design engineers, 31, 34–35, 42
Design evaluation, 68–69
Design feature database, 118
Design for Additive Manufacturing (DFAM), 112–115
Design for Assembly (DFA), 98, 99*f*
criteria to examine against each component of assembly, 103*f*
DFA Product Simplification program, 105
Design for Manufacturing and Assembly (DFMA), 39, 99–100
applications of, 108–110
aerospace and automotive industries, 109–110
construction in Singapore, 110
consumer products, 108–109
challenges of, 136–140
lack of universal validation for common framework and AM–related standards, 137–138
limited studies on AM process–machine–material selection, 138

need for additive manufacturing–specific software tools, 138–139
need to update design and engineering education, 139–140
design stages of general DFMA framework, 119–136
guidelines, 100, 101*t*
limitations of, 110–112
different objectives and project-specific constraints for each product, 111–112
difficulties in integration of design engineering teams, 111
getting management on board, 110–111
lack of resilience in DFMA software, 111
procedures, 101–107
Boothroyd-Dewhurst Design for Manufacturing and Assembly, 102–106
Hitachi assemblablity evaluation method, 106–107
Lucas-Hull Design for Assembly, 107
Design for Manufacturing methodology (DFM methodology), 98–100, 99*f*
Design for X (DFX), 112–113, 113*f*
Design freedoms afforded by additive manufacturing, 114–115, 115*t*
Design methodologies for additive manufacturing, 97–98, 117–140
challenges of DFMA, 136–140
conceptual design, 125–128
design methodologies for conventional manufacturing, 100–112
detail design, 131–132
DFA, 98
DFM, 98–100
embodiment design, 128–131
manufacturing and postprocessing, 132–136
paradigm shift, 112–117
product planning, 121–125
Design optimization, 59
Design parameters, 70–71
Design pattern selection, 59
Design process, 29–31
Design research, 97–98
Design responses, 213
Design rules, 70
Design space, 212
Design structure matrix (DSM), 71
Destructive methods, 277–278

450 Index

Detail design, 131–132
  dimensional specifications of features, 132, 133t
  specific selection of process–machine–material combination, 132
Detailed computer-aided design part modeling, 72–73
Dielectrics, 328
Differential equations, 82–83
Differential scanning calorimetry (DSC), 226
Diffusers, 393
Diffusion coefficient, 193
Digital 2D drawings, 60
Digital Construction Platform (DCP), 352
Digital factory, 51
Digital form generation, 126–128
Digital light processing (DLP), 156, 303–304
"Digital materials", 167
Digital object, 61
Digital product design, 80–81
Digital product design, 89
  3D digital form creation, 59–69
  current challenges, 89–94
    change propagation and information consistency management, 92
    decentralized CAD/CAE network and cloud processing algorithms, 93–94
    interoperability among different design and engineering analysis stages, 91–92
    standardization of model formats in unified approach, 93
  intent-based systemic design, 69–82
  product design process, 58–59
Digital sculpting methods, 64
Digital single-lens reflex cameras (DSLR), 391
Digital transformation in building and construction, 345–347
"Digital-twins", 49
Digitalization, 9–10
Dimensional accuracy, 321–322
Direct ink writing (DIW), 318
Direct laser deposition (DLD), 252–253
Direct laser sintering on HA, 319
Direct metal deposition (DMD), 175
Direct metal laser sintering. *See* Laser powder bed fusion (LPBF)
Direct modeling methods, 64
Direct numerical control (DNC), 17–18

Direct part replacement, additive manufacturing for, 152
Direct-write techniques, 324–325
Directed energy deposition (DED), 156, 158, 174–176, 186–187, 189, 249, 252–254, 286–287, 393, 402, 404, 414–415
  advantages and disadvantages of, 414t
  applications, 176
  benefits, 175
  DLD, 253
  drawbacks, 176
  EBAM, 253
  material extrusion processes, 415–420
    monitored signatures, 417–419
    monitoring systems, 415–417, 417f
    process defects, 420
  monitored signatures, 415, 416t
  monitoring system, 414–415
  process defects, 415
  range of materials available, 176
  WAAM, 253–254
  working principles, 175
Discrete element method (DEM), 154, 186–187
Discretization methods, 84–85
Displacement
  sensor, 402
  signals, 402
Display monitors, 24
Distortions, 273–275
Distributed computer network, 33
Distributed computer system (DCS), 18, 21–22
Distributed NC, 18
Division of labor approach, 6–7, 37
DIW. *See* Direct ink writing (DIW)
DL. *See* Deep learning (DL)
DLD. *See* Direct laser deposition (DLD)
DLP. *See* Digital light processing (DLP)
DMD. *See* Direct metal deposition (DMD)
DNC. *See* Direct numerical control (DNC)
Double-well function, 205
Drafting process, 19
Drilling operations, 18–19
Drop-on-demand inkjet printing, 324
DSLR. *See* Digital single-lens reflex cameras (DSLR)
DSM. *See* Design structure matrix (DSM)
Dynamic light scattering, 261
Dynamic Mechanical Analysis, 231–232
Dynamic simulation, 28–29

Index   451

E

E-score, 106
Eco-effectiveness approach, 44–46
Eco-effective industrial systems, 46
Eco-efficiency strategy, 43–44
Economic order quantity (EOQ), 8
EDS. *See* Energy dispersive spectroscopy (EDS)
Electric arc–based DED, 414
Electricity, 6
Electron beam, 171, 252
  electron beam based DED, 414
Electron beam freeform fabrication.
    See Electron-beam additive
    manufacturing (EBAM)
Electron spectroscopy, 262
Electron-beam additive manufacturing
    (EBAM), 175, 252–253
Electron-beam melting (EBM), 156, 170, 189,
    250–252
Electronic Numerical Integrator and Computer
    (ENIAC), 15
Electronic printing
  conductive materials for, 322–331
  process, 330
Electronics, 322
Elevated temperature performance, 305–307
Embodiment design, 128–131
  function integration and part consolidation,
    128–130
  structural and mechanical optimization,
    130–131
Enclosure Acts, 4
Enclosure chamber system, 250
Endress + Hauser, 109
Energy
  difference, 205
  interpolating function, 205
  and material bonding, 236–240
    adhesive energy, 239–240
    mechanical and thermal energy,
      237–239
    particle energy, 236–237
Energy dispersive spectroscopy (EDS), 262
Engineering
  operations, 28–29
  simulation, 82
Engineering analysis, 59
  case study, 86–89
    computational fluid dynamics simulation,
      86–89
    parametric computer-aided design model
      build-up, 86

shape optimization, 89
computational fluid dynamics simulation,
    83–85
current challenges, 89–94
systems, 73
techniques, 82–89
Entanglement, 230–231
  polymer chains in melt, 230*f*
Enterprise resource planning process (ERP
    process), 35–36
Environment Protection Encouragement
    Agency (EPEA), 44–45
Environmental and safety issues, concerns
    with, 330–331
Environmental pollution, 3
Epoxy resins, 305
Equilibrium melting temperature, 203
Evolutional algorithms, 69
Evolutionary method, 211
Evolutionary structural optimization (ESO),
    210
Extended assemblability evaluation method,
    106–107
Extrusion technique, 27–28
Extrusion-based 3D printing technique, 310

F

Face shields, 178
Fatigue performance testing, 268
FDM. *See* Fused deposition modeling (FDM)
FEA. *See* Finite element analysis (FEA)
Feature learning, 425
Feature-based computer-aided design
    modeling, 73–79
  easy to represent relationships among parts/
    faces/interfaces, 74–77
  friendly to manufacturing analysis and
    additive manufacturing optimization,
    78–79
  good way to manage information flow in
    concurrent engineering, 77–78
  object-oriented, 74
Feature-based cyclic CAD/CFD interaction
    process, 85
Feature-based modeling, 64–66, 73–76
Feedback control, 422–430
  AM life cycle and associated data, 422*f*
  process parameters, 422, 423*t*
    conventional methods, 424
    ML approach, 424–428
FEM. *See* Finite element method (FEM)

452 Index

FFF. *See* Fused filament fabrication (FFF)
Fiber, 223
  reinforcement, 304
  sheets, 302
Fibrous materials, 303
Filters, lenses and, 393
Finite element analysis (FEA), 28—29,
    58—59, 82, 207
Finite element analysis, 154
Finite element method (FEM), 63, 82,
    186—187, 206—207
Finite volume method (FVM), 84—85
First Industrial Revolution, 4—6
  Laissez-faire capitalism, 4—5
  mechanization, 4
  social and environmental impact, 5—6
5G connectivity, 50—51
Fixed-based printing systems, 352
Flexible manufacturing cell. *See* Multiple
    machine cell
Flexible manufacturing system (FMS), 18
Flexographic printing techniques, 324
Flory—Huggins interaction parameter, 227
Flory—Rehner theory, 227
FLOW-3D software, 194
Flowability of ceramics, 320
Fluid dynamic analysis, 28—29
Fluxion-solidification-shrinkage process,
    76—77
Fly ash, 356, 375—376
  cenosphere, 378—379
Form modeling, 64—66
  case study, 66—69
    design evaluation, 68—69
    generative modeling and result, 69
    initialization and parametric
      modeling, 68
    modeling approaches, 64—66
Form-then-bond method, 169
Formative manufacturing, 146
Formwork, 349
Forth Industrial Revolution, 46—52
  CPS, 49—50
  factory of future, 50—52
  Industrie 4.0, 47—49
4D printing, 167
Fourier-transform infrared spectroscopy
    (FTIR), 226
Fourth Industrial Revolution (IR4), 46—47
Free flow materials, layering of, 235—236
Free-form designs, 154
Free-form fabrication, 146—147

FTIR. *See* Fourier-transform infrared
    spectroscopy (FTIR)
Fuel nozzle on General Electric's GE9X
    engine, 172
Function-and process-dependent design rules,
    117—118
Functional design approach, 71
Functional feature approach, 70—73
  abstract geometry features modeling,
    71—72
  detailed computer-aided design part
    modeling, 72—73
  functional analysis of design, 71
Functional feature design process, 70
Functional feature modeling cube, 70
Functional prototyping, 163
Functional three-dimensional printing
  3D printed parts, 301*f*
  biomaterials for bioprinting, 310—315
  ceramics for biomedical implants, 315—322
  composites materials for aerospace
    industry, 302—310
  conductive materials for electronic printing,
    322—331
Functional-physical modeling approach, 70
Fused deposition modeling (FDM), 149—150,
    164, 237—239, 402
  printers, 165
Fused filament fabrication (FFF), 164, 307,
    402
Fusion 360 software, 154
FVM. *See* Finite volume method (FVM)

G

Gantry-based systems, 350—352
Gas atomization (GA), 258—259
Gas pycnometry, 279
Gelatin, 311, 313
Gelatin methacrylate (GelMA), 312
Gelation phenomenon, 224, 225*f*
General Electric Company (GE), 285—286
Generative design (GD), 59, 66, 80—81,
    126—127, 207
Generative modeling, 69
Generic feature, 74—76
Geometric constraints, 70
Geometric elements, 61
Geometric modeling, 73
Ginzburg—Landau models, 187
Glass fibers, 304—305
Glass systems, 315—316

Index **453**

Global additive manufacturing market, 148–149
Global equations, 82–83
Globular and stretched states, 230
Graded lattice, 208
Gradient-based method, 215
Granasy model, 187, 204–206
Graphene, 327
Graphic user interface (GUI), 27–28
Grasshopper 3D programming system, 126–127
Gravitational force, 189
Gravure printing technique, 324
Green parts, 254
"Greener" manufacturing, 43
Ground-granulated blast-furnace slag (GGBFS), 376
Group technology (GT), 18–23, 34–36
    automated flexible manufacturing system, 20–23
    cellular manufacturing, 19–20

H

Hardness testing, 266–268, 269t
Hardware, 24–25
Harrowell–Oxtoby relation, 201
Hasbro, 108
Hausner ratio, 235
Healthcare, metal additive manufacturing application in, 150–151, 290–293
    dentistry, 291
    orthopedic implants, 291–293
Heat conduction equation, 202–203, 206
Heat of polymerization, 226
Heat transfer
    analysis, 28–29
    dynamics, 191–195
        software, 194–195
Heat treatment, 156, 283
Heritage digitalization, 61–63
Heterogeneity, 242–243
Heuristic principles and guidelines, 125
Hewlett-Packard (HP), 108, 177–178
High energy source system, 249
High-performance components, 287
High-speed cameras, 391
Hitachi assemblability evaluation method, 40, 106–107
Homopolymers, 221–222
Hot cracking, 273
Hot isostatic pressing (HIP), 283
    postprocessing, 171

Housing and Development Board (HDB), 110
HTC Vive tools, 127–128
Hyaluronic acid (HA), 311
Hybrid manufacturing
    production, 116–117
    systems, 78–79
Hybrid polymers, 222
Hydrolysis of collagen, 311
Hydroxyapatite (HA), 319

I

Illumination devices, 393
In-house DFMA project, 109
Inductively coupled plasma optical emission spectroscopy (ICP-OES), 262
Industrial engineers, 50
Industrial revolution, 4
Industrial robotic arm, 21
Industrial robots, 14–15, 33
Industrial symbiosis, 44
Industrial/business machines, 150–151
Industrie 4.0 (I4.0), 47–49
Industry Transformation Map (ITM), 110
Information and Communication Technology (ICT), 9–10, 33
Information flow in concurrent engineering, 77–78
Infrared cameras, 395–396
Infrared light spectra, 389
Infrared spectroscopy (IR), 226
Inheritance/abstraction, 74–76
Initial graphics exchange specification (IGES), 76
Initialization and parametric modeling, 68
Injection casting method, 151
Injection molding method, 151
Ink formulation, 300
Inkjet powder printing. *See* Binder jetting (BJ)
Inkjet printing, 324–325
Inorganic dielectrics, 328
Inorganic polymers, 222
Inorganic semiconductors, 327
Input devices, 24
Input/output module (I/O module), 24
Insertion index, 107
Insertion ratio, 107
Inspirational decision-making system, 80–82
Installation method, 393
Integrated circuit (IC), 16
Intelligent computer systems, 36
Intelligent decision-making systems, 58–59

454 Index

Intelligent design systems, 59
Intelligent sensors, 50–51
Intent-based systematic modeling, 78
Intent-based systemic design, 69–82
    feature-based computer-aided design
        modeling, 73–79
    functional feature approach, 70–73
    two typical decision-making types, 80–82
Interface gradient, 201–202
Interfacing, 24
Interference inspection, 78
Intermolecular forces, 223
Internal defects, 398
International Committee F4, 249
International Data Corporation, 148–149
International Standards Organization (ISO),
    155, 263
    ISO 12111, 268
    ISO/ASTM 52910 standard, 118–119
Internet of Things technology (IoT
    technology), 47–48, 50–51
Intersected lattice, 209
Invisible spectra, 389
Ionic-cross-linked alginate bioinks, 311
IoT technology. *See* Internet of Things
    technology (IoT technology)
Iron, 4
Isostatic polymers, 223
Isostatic polystyrene (iPS), 203
Iterative design process, 80–81
Iterative product design process, 73

J

Job sequencing process, 8–9
"Job shops", 8

K

k-epsilon model, 84–85
k-nearest neighbor algorithm (*k*-NN), 425
K-score, 106
Kevlar fibers, 304–305
Kinematics of Machinery, The (Reuleauxs),
    97–98
Kinetic energy, 229

L

Lack-of-fusion pores, 271–272
Laissez-faire capitalism, 4–5
Laminated object manufacturing (LOM),
    168–169, 307

LAMMPS Improved General Granular-
    Granular Heat Transfer Simulation
    (LIGGGHTS), 190
Laser
    beams, 250–251
    laser-based DED, 414
    power, 194–196
    sensor, 352
    shock peening process, 284
    ultrasonics, 400
Laser 3D scanners, 280
Laser confocal scanning microscope (LCSM),
    280–281
Laser engineered lens shaping (LENS), 175,
    253, 414
Laser metal deposition (LMD), 414
Laser powder bed fusion (LPBF), 250–251
Laser powder deposition (LPD), 414
LaserCUSING system, 285
Laserpowder bed fusion (LPBF), 250
Lattice structures, 209
Lattice-Boltzmann method, 187
Layer-by-layer fabrication approach, 308
Layering process, 234–236
    effect on multimaterial printing, 234–235
    of free flow materials, 235–236
    layering of sheet materials, 236
Lead zirconate titanate materials (PZT
    materials), 399
Learning methods, 424–425
Leitat, 177
Lenses and filters, 393
Level set method, 211
LiDAR, 61–63
Light detector. *See* Photodiode
Light source simulation, 195–198
    melt pool simulation using ray-tracing
        model, 196–198
    melt pool simulation using simplified heat
        source, 196
Light-emitting diodes (LED), 158
Linear discriminant analysis, 425
Linear polymers, 223
Linear regression, 425
Liquid cracking, 273
Liquid-dispensing printing, 324–325
Lithium phenyl-2,4,6-
    trimethylbenzoylphosphinate (LAP),
    312
Loads and boundary condition model,
    212–213
Local energy density, 201

Logical relationships, 70
Logistic regression, 425
Lucas-Hull Design for Assembly, 101–102, 107

M

Machine learning (ML), 47, 50, 186, 424–428
  applications in additive manufacturing process and process monitoring, 428–430, 430t, 431t
  learning algorithms, 425, 426t
  learning methods, 424–425
  learning steps, 425–428
Machining, 156
Macroscale modeling in AM, 186–187
Magnesium oxide (MgO), 379
  magnesium oxide–based cement, 379–380
Magnetic technique, 278
Mainframe computer, 15, 17–18
Manual assembly analysis, 102–105
Manual drafting process, 27–28
Manufacturing, 132–136
  analysis, 78–79
  build parameters, 134–136
  costs, 98
  engineers, 32, 50
  industry, 46
  planning, 23
  postprocessing, 136
  process, 7, 23, 157
Market globalization, 40
Mask fitter, 178
Mass conservation, 193
Mass production, 7–8
Massachusetts Institute of Technology (MIT), 12–13
MAT. *See* Minimum assembly time (MAT)
Material extrusion (MEX), 158, 164–165, 320, 404, 410t, 415–420
  applications, 165
  benefits, 164–165
  drawbacks, 165
  monitored signatures, 417–419
  monitoring systems, 415–417
  process defects, 420
  range of materials available, 165, 166t
  working principles, 164
Material jetting (MJ), 158, 165–167, 404, 405t. *See also* Binder jetting (BJ)
  applications, 167

  benefits, 167
  drawbacks, 167
  range of materials available, 167, 168t
  working principles, 166–167
Materials
  challenges involving, 329–330
  scientists, 50
  selection, 325
MATLAB software, 200
Matrices, 305
Matrix materials, 305
Matrix resin, 305
MBD. *See* Multibodydynamics (MBD)
McDonnell Douglas Corporation, 109
MD. *See* Molecular dynamics (MD)
ME. *See* Material extrusion (MEX)
Mechanical arm, 14
Mechanical energy, 237–239
Mechanical properties standard testing for metal additive manufacturing, 263–268
  compression, 263–266
  fatigue performance, 268
  hardness, 266–268
  tension, 263
Mechanical strength of three-dimensional concrete-printed sample, 363–364
Mechanization, 4
Media blasting process, 283
Melt Flow Index (MFI), 231
Melt pool dynamics, 191–195
Melt pool simulation
  using ray-tracing model, 196–198, 199t
  using simplified heat source, 196
    melt pool depths and widths, 197t
    melt pool geometry of sample, 198f
Melting temperature (T_m), 201
Mesh surface, 63
Meshing model, 212
Mesoscale modeling in AM, 187
Metal, 260
  alloys, 260
  AM, 291
  printing, 198
Metal additive manufacturing, 248
  applications, 284–293, 284f
    aerospace, 285–288
    automotive industry, 288–290
    healthcare, 290–293
  classification of, 249–257
    binder jetting, 254–255
    DED, 252–254

456 Index

Metal additive manufacturing (*Continued*)
PBF, 249–252
sheet lamination, 255–257
defects in, 268–281
mechanical properties standard testing for, 263–268
postprocessing, 281–284
preparation and characterization techniques for, 258–263
powder characterization techniques, 260–263
powder preparation techniques, 258–260
Metallic components, 248
Metallic materials, 263, 303
Metallic orthopedic implants, 291–293
Metal–oxide–semiconductor (MOS), 390
Method of moving asymptotes, 215
MEX. *See* Material extrusion (MEX)
MFI. *See* Melt Flow Index (MFI)
Microcomputers, 23
Microprocessor chipset, 23
Microscale modeling in AM, 187
Microscopic cross-sectional method, 279
Microstructural scale, 154
Mid-infrared spectroscopy, 226
Milling operations, 18–19
Minibuilders, 354
Minicomputers, 16, 32
Minimum assembly time (MAT), 104
MIT. *See* Massachusetts Institute of Technology (MIT)
Mixed approaches, 100
MJ. *See* Material jetting (MJ)
MJF. *See* Multijet fusion (MJF)
MJT. *See* Material jetting (MJ)
ML. *See* Machine learning (ML)
Modeling
approaches, 64–66
properties of common software and, 67t
macroscale, 186–187
mesoscale, 187
microscale, 187
process, 61
of architectural concept design, 66
Modern pyrometer, 395–396
Mohs scale, 266
Molds, 163
Molecular dynamics (MD), 187
Molecular material–related classifications, 221–223
Molecular structure–related classifications, 223

Momentum equation, 193
Monomer conversion, 226–227
MOS. *See* Metal–oxide–semiconductor (MOS)
Motorola Solutions, 109
Motorsports, 290
Multibodydynamics (MBD), 59, 82–83
Multifaceted feature, 77
Multijet fusion (MJF), 170, 189
Multimaterial printing, layering effect on, 234–235
Multiple machine cell, 19
Multipurpose product design process, 78–79
Multirobot printing systems, 354
Multiscale-reinforced composites, 309–310
"My Face Mask" tool, 178

N

Naive Bayes algorithm, 425
Nanomaterials, 325–327, 330–331
Nanoparticles (NPs), 325–326
Nanoribbons, 327
Nanotubes, 327
Nanowires inks, 327
Nanyang Technological University (NTU), 371
3DP technology developed by, 372
Natural bioinks, 311
Natural fibers, 304–305
Natural material, 313
Natural polymers, 221–222
Natural resources, 4
Natural sand replacement, sustainable materials for, 376–378
NC machines. *See* Numerical control machines (NC machines)
Net-zero manufacturing strategies, 51
Neural networks (NN), 425
Neutron diffraction technique, 278
Newton's second law, 189
Noncontact measurement methods, 395–396
Noncontact 3D measurement techniques, 279
Nondesign space, 212
Nondestructive techniques, 277–278
Non-destructive testing (NDT), 398
Nonlocal free energy density, 201–202
Nonoptimal printing parameters, 277
Nonparametric forms, curvy edge, 63–64
Nonuniform Rational B-Spline Curve (NURBS Curve), 63–64
nTopology Element Pro software, 215
Numerical control machines (NC machines), 12

O

Object digitization, 61–63
Object-oriented, feature-based computer-aided design modeling, 74
Objective functions, 213
Oculus Rift tools, 127–128
Oligomers, 224
One-dimension (1D), 327
  time-of-flight distance sensor, 355
Onshape software, 154
Ontology speaking, 73
OpenFOAM open-source software, 154, 200, 203
OpenSCAD software, 126
Operating system (OS), 23, 25–26
Optical microscope (OM), 261
Optical monitoring system, 391–392
Optical signals, 389–394
  advantages and disadvantages, 394
  applications, 394
  practical use, 391–394
    data registration, 394
    illumination devices and diffusers, 393
    installation method, 393
    lenses and filters, 393
    sensor parameters, 392
    system testing and calibration, 393–394
  sensor types, 389–391
Optimization, 154
Organic dielectric materials, 328
Organic polymers, 222
Organic semiconductors, 327–328
Orientation equation, 206
Oscillatory testing, 231–232

P

Pahl and Beitz's design process, 119–121
Palladium NPs, 327
Paper lamination technology, 168–169
Paradigm shift, 112–117
  DFAM, 113–115
  DFX, 112–113
  trend of hybrid manufacturing production, 116–117
Parallel interfacing, 25
Parameters optimization in AM, 188
Parametric computer-aided design model, 86
Parametric dimensional restrictions, 70
Parametric forms, 63–64
"Parametric modeling". See Feature-based modeling

Parametric variable, 63–64
Parametric/NURBS-based CAD systems, 138
Part designs, 19
Part scale, 154
Partial differential equation, 84–85
Particle energy, 236–237
Particle size, 235
Particle size distribution (PSD), 258, 260–261
Parts consolidation, additive manufacturing for, 153
Parts digitalization, 66
Paste role in three-dimensional concrete printing, 359
Patient-specific products, 290–291
PBF. See Powder bed fusion (PBF)
PCB. See Printed circuit board (PCB)
PCL. See Polycaprolactone (PCL)
PEEK. See Polyether ether ketone (PEEK)
PEG. See Polyethylene glycol (PEG)
Penalization method approach, 80–81
Performance
  constraints, 213–214
  economy, 44
"Personal computer", 23
Personal protective equipment (PPE), 176
Phase transition temperatures, 232
Phase-field equation, 201–202
Phase-field model, 200, 204–206
Photo energy, 236–237
Photodetector. See Photodiode
Photodiodes, 389–390
Photogrammetry, 61–63, 280
Photoinitiator (PI), 312
Photopolymerizable hydrogel, 312
Photopolymers, 226
Photosensitive resins, 158–161
Photosensor. See Photodiode
Physical laws, 70
Physical simulation, 63
Physics-based process models, 154, 185
Pinion-rack mechanism, 81
Plasma arc-based DED, 414
Plasma atomization (PA), 258, 260
Plasma rotating electrode process (PREP), 258, 260
Plastic viscosity in three-dimensional printing, 360–361
PLM. See Product lifecycle management (PLM)
Pluronic F127 (PF127), 312
Point cloud, 61–63, 154
Poisson's ratio, 190

458 Index

Poly (ethylene oxide) (PEO), 312
Poly (propylene oxide) (PPO), 312
Poly (vinylpyrrolidone), 328
Polycaprolactone (PCL), 312
Polyester, 328
Polyether ether ketone (PEEK), 303–304
Polyethylene glycol (PEG), 312
Polyimide, 328
Polyjet 3D inkjet printing technology, 166
Polymer, 221, 223, 233, 328
  materials
    characteristics of 3D printed parts,
      240–243
    classifications of polymers, 222*f*
    molecular material–related
      classifications, 221–223
    molecular structure–related
      classifications, 223
    polymer classification for additive
      manufacturing, 223–224
    printability in 3D printing, 234–240
    thermoplastics, 229–234
    thermosets, 224–229
  melt, 229–232
    entanglement, 230–231
    globular and stretched states, 230
    random coil, 229–230
  networks, 227
  pellets, 239
  printing process, 198
  semiconductors, 328
Polymer chains, 223, 229–230, 233
Polymer classification for additive
    manufacturing, 223–224
Polymeric materials, 304
Polymerization
  process, 224, 237
  rate, 228
Polymethylmethacrylate, 328
Polystyrene, 328
Polyvinyl alcohol, 328
Polyvinyl phenol, 328
Poor out-of-plane strength, 310
Porosity
  characteristics of 3D printed parts, 241
  detection, 279
Postcuring process, 162
Postprocessing, 136
  of 3D printed parts, 301
  approach, 281–284
    heat treatment, 283
    removal of adhesive powders, support
      structures, and substrate plates, 281–282

    surface finishing, 283–284
  methods, 155–156
  of optimization results, 214
Powder
  characterization techniques, 260–263
    chemical composition, 262–263
    flowability, 262
    PSD, 260–261
  deposition simulation, 189
  flowability, 235, 262
  interaction, 189–191
    software, 190–191
  scale, 154
Powder bed fusion (PBF), 158, 170–172,
    249–252, 320, 404–414, 405*t*, 410*t*
  applications, 172
  benefits, 171
  drawbacks, 171
  electron-beam melting, 251–252
  LPBF, 250–251
  monitored signatures, 404–411
  monitoring system, 404
  process defects, 411–414, 412*t*
  range of materials available, 172, 172*t*
  working principles, 171
Powder preparation technique, 248, 258–260
  GA, 258–259
  PA, 260
  PREP, 260
  systematic review and comparison, 261*t*
Power grids, 6
Power supply and roller system, 250
Power-law model, 230–231
Pozzolanic materials, 375
Pozzolans, 375
PPE. *See* Personal protective equipment (PPE)
PPO. *See* Poly (propylene oxide) (PPO)
Prefab methods, 110
Preindustrialization, 3–4
  agricultural revolution, 3–4
  craft production, 3
PREP. *See* Plasma rotating electrode process
    (PREP)
Print resolution, limitation in, 314–315
Printability
  in 3D printing, 234–240
    energy and material bonding, 236–240
    layering, 234–236
  studies with slump and slump flow test,
    361–363
Printable file format, conversion to, 154–155
Printable inorganic semiconductors, 327
Printed circuit board (PCB), 322

Index **459**

Printed electronics, 322
Printed parts, 254
Printing
accuracy determination, 279–280
of concrete supports materials, 367–368
parameters effects on concrete filament, 366
process
control, 355
limitation on, 330
Pro/Engineer tools, 105
Procedural modeling, 66
Process monitoring, 157
applications in additive manufacturing
processes, 404–421
quality and feedback control, 422–430
signals, sensors, and techniques for process
monitoring, 388–403
acoustic signals, 399–402
optical signals, 389–394
thermal signals, 394–397
X-ray signals, 397–398
standards, 431–434, 431t
toolkits, 435–439, 436t
Process planning, 8–9, 31–33
Processability, 307
Process–machine–material combination
broad selection of, 123–125
limited studies on AM
process–machine–material selection,
138
specific selection of, 132
Process–structure–property relationship,
241–242, 309–310
Product
assembly, 104–105
of consumption, 45
customization, 10
design
and development, 23
process, 58–59, 93–94
designers, 50, 59
development process, 36–41
collaborative engineering, 40–41
concurrent engineering, 38–40
sequential engineering, 37–38
geometric entities, 76
planning, 121–125
additive manufacturing utilization,
122–123
broad selection of
process–machine–material
combinations, 123–125

design requirements, 121–122
part selection, 122
of service, 46
Product lifecycle management (PLM), 73,
93–94
Production operators, 50
"Programed article transfer" prototype, 14
Programmable automation, 11–12
Programmable logic controller, 16
Prototypes, 163
Prototyping, 78
PSD. *See* Particle size distribution (PSD)
PTC Generative Topology Optimization
extension software, 216
Punched cards, 12
Pyrometers, 395–396
PZT materials. *See* Lead zirconate titanate
materials (PZT materials)

Q

Qualitative approaches, 100
Quality control, 422–430
AM life cycle and associated data, 421*f*
process parameters, 422, 423t
conventional methods, 424
ML approach, 424–428
Quality function deployment (QFD), 71
Quantitative approaches, 100

R

Random coil, 229–230
Random-access memory (RAM), 24
Rapid manufacturing, 146–147
Rapid melting and cooling of ceramics, 321
Rapid prototyping, 146–147
additive manufacturing for rapid
prototyping and tooling, 152
Rapid solidification process, 320–321
Rating system, 102
Ray-tracing model, 195
melt pool simulation using, 196–198
Reaction rate constant, 228
Read-only memory (ROM), 24
Recycled glass, 356
Reinforcement, 348–349
learning, 80–81, 425
materials, 304–305
Residual stress(es), 271, 272*f*, 320–321
measurement, 277–278
"Resin", 221–222
Resistance temperature detectors (RTDs), 395

460 Index

Resistors, 16
Retrieval decision-making system, 80–82
Reynolds number (Re), 86–89
Rheological properties of thermoplastics, 231–232
Rheometer, 361
Rhinoceros 3D, 126, 154
Robotic arm, 24, 352
Roll-to-roll printing technique (R2R printing technique), 322–324
Rule-based automatic CFD model selection, 85
Rule-based retrieval system, 81
Ruthenium (Ru), 312

S

Sand-blasting postprocessing, 171
Scaled lattice, 209
Scaling big area additive manufacturing, 309
Scanning electron microscope (SEM), 261
Scanning laser sintering, 170
Scanning motion system, 250
Scanning speed, 194–196
Scientific data, 63
Scleroscope, 266
Screen printing technique, 322–324
Screw-nut mechanism, 81
Second Industrial Revolution, 6–9
    batch production, 8–9
    division of labor, 6–7
    mass production, 7–8
Second-generation transistor computers, 15–16
Selective lamination composite object manufacturing 1 (SLCOM1), 303–304
Selective laser melting (SLM), 149–150, 170, 189, 315–316, 393
Selective laser sintering (SLS), 149–150, 170, 189, 250, 307
Self-compacting concrete (SCC), 361
Self-supervised learning, 425
SEM. *See* Scanning electron microscope (SEM)
Semiconducting materials, 15, 327
Semicross-linking bioink, 313
Semi-supervised learning, 425
Semisynthetic polymers, 221–222
Sensor
    parameters, 392
    for process monitoring, 388–403
    technology, 331

types, 389–391, 395–396
    CCD and CMOS, 390–391
    infrared cameras, 396
    photodiodes, 389–390
    pyrometers, 395–396
    spectrometers, 391
    thermometers, 395
Sequential engineering approach, 37–38
Service-level agreement (SLA), 51, 158, 162
Shape optimization, 89
    result comparison changing design parameters, 90t
Sharing economy, 44
Sheet lamination (SL), 158, 168–170, 236, 249, 255–257, 404, 405t
    applications, 170
    benefits, 169
    drawbacks, 169
    metallic additive manufacturing techniques, 256t
    range of materials available, 170
    working principles, 169
Sheet materials, layering of, 236
Shot peening, 156
Shrinkage, 225, 233, 274–275
    anisotropy, 321
Siemens NX software, 86, 154
Siemens solid edge generative design software, 216
Signals for process monitoring, 388–403
Signatures, monitored
    additive manufacturing processes, 420–421
    DED processes, 414–415
    directed energy deposition processes, 415, 416t
    material extrusion process, 417–419
    other additive manufacturing processes, 420–421
        monitored signatures and process defects, 421, 421t
        monitoring system, 420–421
    powder bed fusion process, 404–411
Silhouette template, 68
Silica (SiO_2), 316–318, 328
    fume, 376
    silica-based powder mixture, 315–316
Silica–alumina (SiO_2/Al_2O_3), 318–319
Silicon dioxide. *See* Silica (SiO_2)
Silicon nitride, 328
Similarity learning, 425
SIMP. *See* Solid isotropic material with penalization method (SIMP)

Simplified heat source, 195
  melt pool simulation using, 196
Simulation, 50, 154
  model, 185–186
Singapore, construction in, 110
SketchUp (digital sculpting methods), 64
SL. *See* Sheet lamination (SL);
  Stereolithography (SL)
SLA. *See* Service-level agreement (SLA);
  Stereolithography apparatus (SLA)
SLCOM1. *See* Selective lamination composite
  object manufacturing 1 (SLCOM1)
"Slicing" process, 275
SLM. *See* Selective laser melting (SLM)
SLS. *See* Selective laser sintering (SLS)
Slump and slump flow test, printability studies
  with, 361–363
Social cloud manufacturing, 41
Socialism, 5–6
Sodium persulfate, 312
Software, 64
  control systems, 248
  heat transfer and melt pool dynamics,
    194–195
  powder interaction, 190–191
    powder bed deposition model, 192*f*
    simulation parameters and material
      properties of SS316L, 191*t*
  technology, 154
Software, 25–26
Sol–gel derived-siloxane-based hybrid
  polymers, 328
Solid 3D modeling, 138
Solid isotropic material with penalization
  method (SIMP), 210–211
Solid modeling, 64, 73
Solidification cracking. *See* Hot cracking
Solidification system, 201
SOLIDWORKS software, 154
Solution algorithms, 84–85
SP. *See* Stylus profilometer (SP)
Specific selection of process–machine–
  material combination, 132
Spectrometers, 391
Spectrophotometer. *See* Spectrometer
Spherulite, 198
Spinning jenny, 4, 11
Spinning process, 11
Spray-based three-dimensional printing,
  sustainable materials in, 378–380
  fly ash cenosphere, 378–379
  magnesium oxide–based cement, 379–380

Stainless steel 316 L (SS316L), 190–193
Standard for exchange of product data
  (STEP), 76
"Standard Tessellation Language", 154–155
Standard Triangle Language (STL), 154–155,
  250–251, 275
State-of-art technologies, 52
Steam, 4, 11
  steam-powered machinery, 4
Steel, 6
Stefan–Boltzmann law, 395–396
Step-by-step product digital design process, 86
Stereolithography (SL), 303–304, 315–316
Stereolithography apparatus (SLA), 148
Stratasys Ltd., 178
Stress analysis, 28–29
Structural optimization in AM, 208–210
  graded lattice, 210
  intersected lattice, 209
  topology optimized lattice-based structure,
    209
Structured-light 3D scanners, 280
Stylus profilometer (SP), 280–281
Substrate plates removal in postprocessing
  stage, 281–282
Subtractive manufacturing, 146
"Subtractive" approach, 322
Supervised learning, 424
Supplementary cementitious materials (SCM),
  375
Support structures removal in postprocessing
  stage, 281–282
Support vector machines (SVM), 425
Surface finishing process, 283–284
Surface model, 63–64, 73
Surface roughness, 275–277, 277*t*
  surface topography of metallic AM
    components, 276*f*
Surface Tessellation Language (STL), 154
Surface topography measurement, 280–281,
  282*t*
Sustainability
  of construction industry, 372
  in manufacturing, 42–46
Sustainable raw materials in concrete printing,
  373–380
  cement replacement, sustainable materials
    for, 375–376
  natural sand replacement, sustainable
    materials for, 376–378
  spray-based three-dimensional printing,
    sustainable materials in, 378–380

462 Index

Synchrotron diffraction technique, 278
Syndiotactic polymers, 223
Synthesis process, 72
Synthetic bioinks, 312−313
Synthetic resins, 221−222
System testing, 393−394
Systematic design methods, 98

T

Tacticity, 223
T-Spline Curve, 63−64
"Take, make, dispose" linear economic model.
 See Cradle-to-grave approach
Taylor's approach, 7
Techniques for process monitoring, 388−403
Technologies of additive manufacturing,
 158−176
 binder jetting, 173−174
 directed energy deposition, 174−176
 MEX, 164−165
 MJT, 165−167
 powder bed fusion, 170−172
 sheet lamination, 168−170
 VPP, 158−163
Temperature-dependent local free energy
 density, 201
Tension testing, 263, 264*t*
Tetrafunctional monomers, 224
Textile industry, 4
Thermal and residual stresses, 320−321
Thermal camera. *See* Infrared camera
Thermal diffusivity, 232
 of material, 202−203
Thermal energy, 237−239
Thermal signals, 394−397
 advantages and disadvantages, 396
 applications, 397
 practical use, 396
 sensor types, 395−396
Thermal stress, 233−234
Thermally reversible and irreversible
 shrinkage, 233
Thermographic camera. *See* Infrared camera
Thermometers, 395
Thermoplastics, 229−234, 307
 materials, 309−310
 polymer, 232
 polymer melt, 229−231
 rheological properties, 231−232
 viscoelasticity, 231−232
 viscosity, 231

thermal properties, 232−234
 CTE, 232−233
 phase transition temperatures, 232
 thermal diffusivity, 232
 thermal stress and warpage,
 233−234
 thermally reversible and irreversible
 shrinkage, 233
Thermosets polymers, 224−229
 curing, 224−225
 curing characteristics, 225−229
 dynamic covalent bonds, 229
Thermosetting polymers, 224
Third Industrial Revolution, 9−46
 additive manufacturing, 41−42
 automation, 11−12
 computer system architecture, 23−26
 computer-aided applications, 26−33
 computer-aided design, 26−31
 computer-aided manufacturing, 32−33
 computer-aided process planning,
 31−32
 computer-integrated manufacturing,
 33−36
 early computers, 15−18
 group technology, 18−23
 industrial robots, 14−15
 modern computers, 23
 numerical control, 12−14
 product development process, 36−41
 sustainability in manufacturing, 42−46
 eco-effectiveness, 44−46
 eco-efficiency, 43−44
3D concrete printing (3DCP), 346
3D-perception in autodriving, 61−63
3D-printed emergency respiration device
 (LEITAT 1 · 3), 177
 companies, 178
 digital forms, 61−64
 creation, 59−69
 of model representations, 65*t*
 modeling, 64−66
 properties, 64
 representations, 61−64
 electronics printers, 324−325
 facial scanning technology, 178
 form creation, 94
 geometric modeling, 30, 61
 measurement techniques, 279−280
 model data, 146
 modeling, 146−147, 151, 178
 print customizable face masks, 178

Index   463

printable composite materials, lack of, 309–310
printable half-mask respirators (CIIRC RP95–3D), 179–180
printed ceramic implants, 315–316
  AM technologies, 317t
  bioglass customized bone implant, 316f
printed parts characteristics, 240–243
  anisotropy, 241–242
  heterogeneity, 242–243
  porosity, 241
printed techniques, 307
printer, 2
printer program, 186
scanners, 146
scanning technology, 154, 178
surfaces, 63–64
systems, 148, 154–155
three-dimensional printable concrete, 355–368
  different materials used and effect on, 356–359
  fresh properties of three-dimensional printable concrete materials, 360–363
  harden properties of three-dimensional printable materials, 363–365
  three-dimensional concrete printing parameters, 366–368
Three-dimensional printed electronics, 322–325
  challenges and potential in, 328–331
  challenges involving materials, 329–330
  concerns with environmental and safety issues, 330–331
  limitation on printing process, 330
  different printing techniques, 323t
  materials for, 325–328
  additive manufacturing technologies, 326t
Three-dimensional printing (3DP), 10, 41–42, 146–147, 226, 236, 254, 300–302, 308, 346, 348–350
  of composite structures, 310
  matrix materials, 309–310
  printability in, 234–240
  process, 155, 225
  for biomedical implants, 316–319, 320f
  challenges and potential in ceramics for, 319–322
  composites, 304–307
  materials for aerospace industry, 306t

concrete printing technologies, 350–355
construction three-dimensional printing, 347–348
digital transformation and automation in building and construction, 345–347
sustainable raw materials in concrete printing, 373–380
technology developed by Nanyang Technological University, Singapore, 372
three-dimensional concrete-printed applications and case study, 363–365
applications of three-dimensional printing for building and construction, 370–371
three-dimensional printable concrete, 355–368
  different materials used and effect on, 356–359
  fresh properties of three-dimensional printable concrete materials, 360–363
  harden properties of three-dimensional printable materials, 363–365
Time (TM), 104
Time gap effect on interlay bond strength, 364–365
Time-dependent Ginzburg–Landau model, 200–204
  local free energy density, 202f
"Time-to-market" process, 48
Tissue engineering, 311
Toolpath
  analysis, 78
  optimization, 78
  planning, 78
Top-down approach, 110–111, 301
Top-down configuration mechanisms, 162
Topology optimization (TO), 80–81, 130, 131f, 188, 207–216
  available commercial software for, 214–216
  structural optimization, 208–210
  TO-based design tools, 139
  types of, 210–211
    ESO, 211
    level set method, 211
    SIMP, 210–211
  workflow for additive manufacturing, 211–214
    creating design responses, 213
    creating design space, 212

464 Index

Topology optimization (TO) (*Continued*)
defining constraints, 213–214
defining loads and boundary condition, 212–213
defining objective functions, 213
meshing model, 212
postprocessing of optimization results, 214
Topology optimized lattice-based structure, 209
Traditional business models, 151
Traditional formwork cast technique, 347
Traditional manufacturing processes, 100, 147, 316
Traditional product design methods, 78
Transformation, 147
Transistor computers, 15
Transistors, 16, 327
Trifunctional monomers, 224
Trommsdorff effect, 225
Two dimensional (2D)
camera sensor, 393–394
materials, 327
printers, 24
Two-phase optimization process, 209

U

Ultrasonic additive manufacturing (UAM), 168–169, 255
Ultrasonic technique, 278
Ultrasonic waves, 399–400
Ultraviolet (UV)
curing, 156
spectra, 389
Uniaxial tensile testing, 263
Unified approach, standardization of model formats in, 93
Unimate robots, 14
United Kingdom Acts of Parliament, 4
United Nations Food and Agriculture Organization, 2
Unsupervised learning, 425
Upwind differencing scheme (UDS), 84–85
User interfaces (UIs), 74–76

V

Validation, 157
Vapor deposition techniques, 328
Variable density lattice, 209
Variant type, 32

Vat photopolymerization (VPP), 158–163, 404, 405t
applications, 163
benefits, 162
drawbacks, 162–163
range of materials available, 163, 163t
working principles, 158–162
Venn diagram, 112
Vibration signals, 402
Vibratory/rotary/tumbling finishing methods, 284
Virtual prototyping, 30–31, 50
Virtual reality (VR modeling), 127–128
Viscoelastic fluid, 231–232
Viscoelasticity, 231–232
Viscometers for measuring viscosity, 231
Visible spectra, 389
Volatile organic compounds (VOCs), 330–331
Volume generation methods, 78
Volume of fluid model (VOF model), 186–187
Volumetric 3D mesh grid, 63
Volumetric shrinkage, 227
Voxel-based model, 63
VPP. *See* Vat photopolymerization (VPP)

W

Warpage, 233–234
Water-soluble bioink, 312
Water-soluble polymer, 311
Weber number, 240
Western industrialization, 4
White-light interferometer (WLI), 280–281
Wire arc additive manufacturing (WAAM), 175, 252–254, 414
Wire feedstocks, 258
Wire-fed DED systems, 252–253
Wireframe modeling, 60–61, 73
World's Advanced Saving Project (WASP), 178

X

X-ray, 262
signals, 397–398
advantages and disadvantages, 398
applications, 398
practical use, 398
working principle, 397
X-ray diffraction technique, 278

X-ray fluorescence (XRF), 262
X-ray photoelectron spectroscopy (XPS), 262

Y

Yade software, 190
Yield stress in three-dimensional printing, 360–361
Young's modulus, 190

Young–Dupre equation, 240
Yttrium-stabilized zirconia (YSZ), 316–318

Z

ZBrush (digital sculpting methods), 64
Zirconia (ZrO_2), 316–318
Zirconia–yttria (ZrO_2/Y_2O_3), 319

Printed in the United States
by Baker & Taylor Publisher Services